KU-280-519

Experts in Uncertainty

Environmental Ethics and Science Policy Series
Kristin Shrader-Frechette, *General Editor*

ACCEPTABLE EVIDENCE
Science and Values in Risk Management
edited by Deborah Mayo and Rachelle D. Hollander

EXPERTS IN UNCERTAINTY
Opinion and Subjective Probability in Science
Roger M. Cooke

EXPERTS IN UNCERTAINTY

Opinion and Subjective Probability in Science

ROGER M. COOKE

New York   Oxford
OXFORD UNIVERSITY PRESS
1991

Oxford University Press

Oxford New York Toronto
Delhi Bombay Calcutta Madras Karachi
Petaling Jaya Singapore Hong Kong Tokyo
Nairobi Dar es Salaam Cape Town
Melbourne Auckland

and associated companies in
Berlin Ibadan

Copyright © 1991 by Roger M. Cooke

Published by Oxford University Press, Inc.
200 Madison Avenue, New York, New York 10016

Oxford is a registered trademark of Oxford University Press

All rights reserved. No part of this publication may be reproduced,
stored in a retrieval system, or transmitted, in any form or by any means,
electronic, mechanical, photocopying, recording or otherwise,
without the prior permission of the publisher.

Library of Congress Cataloging-in-Publication Data
Cooke, Roger.
Experts in uncertainty:
opinion and subjective probability in science/
by Roger M. Cooke.
p. cm. — (Environmental ethics and science policy)
Includes bibliographical references (p.   ) and index.
ISBN 0-19-506465-8
1. Science—Philosophy. 2. Science—Methodology.
3. Uncertainty (Information theory)
4. Probabilities. 5. Decision-making.
I. Title. II. Series.
Q175.C699 1991 501—dc20 90-22493

9 8 7 6 5 4 3 2

Printed in the United States of America
on acid-free paper

To Pam, Tessa, and Dorian

Contents

PART II   SUBJECTIVE PROBABILITY

PART III   COMBINING EXPERT OPINIONS

Introduction

Broadly speaking, this book is about the speculations, guesses, and estimates of people who are considered experts, in so far as these serve as "cognitive input" in some decision process. The questions studied include how expert opinion is in fact being used today, how an expert's uncertainty is or should be represented, how people do or should reason with uncertainty, how the quality and usefulness of expert opinion can be assessed, and how the views of several experts might be combined. The subject matter of this book therefore overlaps a large number of disciplines, ranging from philosophy through policy analysis up to some rather technical mathematics.

Most important, we are interested in developing practical models with a transparent mathematical foundation for using expert opinion in science. The models presented have been operationalized and evaluated in practice in the course of research sponsored by the Dutch government, by the European Space Agency, by the European Community, and by several research laboratories, including The Netherlands Organization for Applied Scientific Research, Delft Hydraulics Laboratory, DSM Research, and Shell Research.

This book presupposes a "general science background," some background in college level mathematics, a nodding acquaintance with probability theory that the reader is willing to extend, and an interest in policy analysis, particularly risk analysis. The reader must be willing to assimilate concepts outside his or her specific field.

The material is organized into three parts. Part I, "Experts and Opinions," studies the various ways in which expert opinion is being used today. It is important to have a good feel for the problems and promises of expert opinion before embarking on mathematical modeling, and this is the purpose of Part I. Readers not interested in *using* expert opinion, but simply wishing to see how the scientific and engineering community is dealing with it, will benefit most from Part I, perhaps supplemented with Chapters 8, 10 from Part II, and Part III.

Part II, "Subjective Probability," focuses on the representation of uncertainty for an individual rational subject. There are many discussions of subjective probability, but none of them oriented to the use of expert subjective probabilities. The present discussion is distinguished by an informal exposition of Savage's theory of rational decision and the representation of preference in terms of expected utility. New proofs of Savage's representation theorem suitable for undergraduate students in science and engineering are included in a mathematical supplement to

Chapter 6. De Finetti's theory of exchangeable events and the relation between subjective probabilities and relative frequencies, which is omitted in introductory and intermediate books on decision theory, are discussed informally. This is made possible by focusing on finite sequences of exchangeable events. In a mathematical supplement to Chapter 7 a standard proof of De Finetti's theorem for infinite sequences is given, and the theory of inference based on it is developed. This material is essential for Bayesian models for combining expert opinion. Further the notion of eliciting, scoring, and evaluating probability assessments is given ample treatment. Much of this material is new, but builds on standard literature. It finds application in the "classical" models for combining expert opinion. Again, an effort is made to keep the discussion informal and place details in a mathematical supplement to Chapter 9. Two recent psychometric experiments involving expert probability assessments illustrate the ideas of scoring.

Part III, "Combining Expert Opinion," reviews the literature on this subject and discusses three types of models: classical, Bayesian, and psychological scaling. These models have been operationalized and applied in the course of the above mentioned research projects.

Techniques of psychological scaling have been around for over 50 years. However, their application to the problem of assessing subjective probabilities is quite recent. The most important work was performed under contract with the U.S. Nuclear Regulatory Commission for the assessment of human error probabilities.

The "classical models" owe their name to the fact that they treat experts much as statistical hypotheses are treated in classical statistical hypothesis testing. The theory behind these models is grounded in contemporary developments in subjective probability, in particular in the theory of proper scoring rules. These models were developed at the Delft University of Technology and are currently being implemented in diverse fields such as risk analysis and project management.

The "Bayesian models" are the most familiar from the literature. Some versions of these models have found substantial application in risk analysis, although these versions tend to place strong constraints on the form in which experts express their probabilities and place heavy demands on the decision maker. Particular attention is given to a "nonparametric" Bayesian model developed and currently being implemented at MIT/UC Berkeley. Some experimental results with this model are available at present.

Case study applications are discussed in some detail, to show that these tools can be fruitfully applied to a diversity of real problems in engineering and planning. This area is still under active development, and these models are surely not the last word in this area. However, they serve to demonstrate the value of a more methodological approach to expert opinion.

I
EXPERTS AND OPINIONS

1

Think Tanks and Oracles

To the Greek philosopher Plato, the title "Experts in Uncertainty" would have been a contradiction. With his famous "divided line" he partitioned knowledge into four categories. The lowest category was "eikasia" which is best translated as "conjecture." After this comes "pistis" (belief), followed by "dianoia" (correct reasoning from hypotheses, as in mathematics), and "episteme" (knowledge). A line divides the lower two categories, belonging to the realm of appearances and deception, from the upper two, for which rigorous intellectual training is required. "Uncertainty," whatever it may be, certainly belongs beneath the line, whereas "expert" denotes a result of rigorous intellectual training. What conceivable purpose could be served by studying the uncertainties of experts?

The purpose is really twofold. First, people in general and decision makers in particular do in fact tend to place great weight on the uncertain opinions of experts. This is presently done in a rather unmethodological way. Given that this is taking place, it is relevant to ask if there are "better" or "worse" ways of using expert opinion. Second, and more importantly, there is a growing body of evidence that expert opinion can, under certain circumstances, be a very useful source of data. Indeed, expert opinion is cheap, plentiful, and virtually inexhaustible. However, the proper use of this source may well require new techniques. Expert opinion is not the same as expert knowledge, and the consequences of ignoring this distinction may be grave.

This chapter encircles the subject of expert opinion with a wide compass. Selected topics from the 1950s and 1960s are treated at some length. These have receded far enough into the past to offer us the advantage of hindsight, but they have not yet lost their relevance. The era of think tanks and expert oracles is roughly the period between the Second World War and the Vietnam war. This period produced two principal techniques for using expert opinion in science, scenario analysis and the Delphi method. Before setting off in search of new techniques we must review these. This is done below. A final section philosophizes on the role of expert opinion in science.

BACKGROUND

The phenomenon of experts is not new; however, the notion that the musings, brainstorms, guesses, and speculations of experts can be significant input in a structured decision process is relatively recent. We may effectively date the inception of this phenomenon with the establishment of the RAND Corporation after World War II. A period of unbridled growth and almost unlimited faith in expert opinion came to a close in the United States sometime in the early 1970s, as suggested by Figure 1.1.

The period between the World War II and the Vietnam War witnessed a rapid growth in research and development. A few facts and figures serve to illustrate this development. Whereas the U.S. federal budget between 1940 and 1963 grew by a factor of 11, the federal outlays for R&D (research and development) in this period grew by a factor of 200 (Kevles, 1978, p. 394). Research contracts to universities and research laboratories soared. In 1964 MIT had a banner year and bagged $47,000,000 in research contracts, the highest for any academic institution in that year. The National Science Foundation reported that by the late 1960s there were

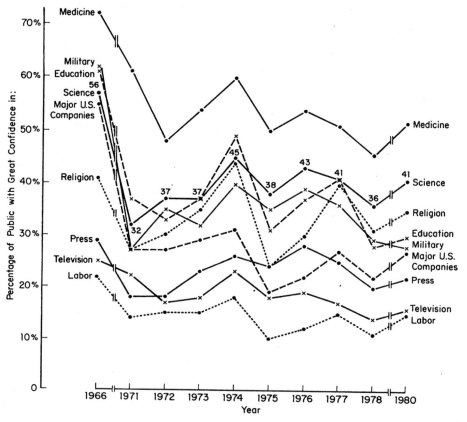

Figure 1.1 Percentage of the American public expressing "great confidence" in the leaders of various institutions. (*Mazur, 1981*).

upwards of 11,000 independent "think tanks" advising the U.S. government. The advice ranged from strategic planning, through the war on poverty, the role of birds in warfare, diarrhea in horses, psychological differences between tatooed and nontatooed sailors, up to and including a classified study of the vulnerabilities of communists with respect to music (*Toekomstonderzoek*, 1974, pp. 6.6.1.02, 6.6.302).

As the war in Vietnam intensified, the honeymoon between science and government in the United States came to an end. In 1966 the R&D growth started to decelerate (Kevles, 1978, p. 411). Most of the President's science advisers opposed the war. Many talked of resigning, but only Los Alamos veteran George Kistiakowsky did. Steven Weinberg later resigned from the elite and secretive Jason Division because of its Vietnam involvement. In 1969, following a debate by the American Physical Society on the proposed Anti-Ballistic Missile system, 250 physicists staged a march on Washington protesting its deployment (Kevles, 1978, p. 405). Scientists opposed to the war were blacklisted from advisory committees (*Science*, p. 171, March 5, 1971). In 1973 Nixon abolished the Science Advisory Committee because of their opposition to the war in Vietnam and to the Supersonic Transport project. Jerome Weisner, president of MIT and member of Kennedy's Science Advisory Committee, was even placed on Nixon's "enemies list" and the Nixon administration considered cutting all research grants to MIT (Kevles, 1978, p. 413).

THE RAND CORPORATION

In the halcyon years between World War II and the Vietnam War two methodologies dictated the form in which structured expert opinion was conveyed to decision makers, namely the Delphi method and scenario analysis. As both were developed at the RAND Corporation, and as RAND (for R-and-D) represents the think tank par excellence, it is appropriate to focus this chapter on the two pink (no pun intended) buildings in Santa Monica, California, which are home to "Mother RAND," as she is called by her employees.

The RAND corporation originated in a joint project, called "Project RAND" of the U.S. Air Force and Douglas Aircraft in 1946. RAND was set up as a 300-man department of Douglas. In its first year, RAND produced a report that predicted that the first space satelite would be launched in the middle of 1957. The first Russian Sputnik was launched on October 4, 1957.

In 1948 RAND split off from Douglas and became an independent corporation, the first think tank. In 20 years its yearly budget grew to $30,000,000, with over a thousand employees. It is reported that in some years as many as 70% of the yearly U.S. crop of mathematics Ph.D.s applied for jobs at RAND. (*Toekomstonderzoek*, p. 6.6.4.06). Throughout the 1950s and into the 1960s RAND worked almost exclusively for the Air Force, but has diversified since then.

The diversification was aimed to demilitarize their public image, attract new clients, and deflect a swelling wave of criticism. As one example of the latter, John Gofman and Arthur Tamplin had revealed that a 1966 RAND study advised post-nuclear-war leaders to forego all measures to help the elderly, the weak and the mentally and physically handicapped, as these groups would only present a further burden and impede the reconstruction of the country. The country would be better off without them (Gofman and Tamplin, 1970). It is pointed out that RAND had never done a study on care for the elderly, except for this study on caring for the elderly after nuclear war.

RAND's research has been concentrated in four broad areas: methodology, strategic and tactical planning, international relations, and new technology. Our interest is primarily in the first. However, as method without matter is empty, it is useful to consider an example of strategic planning at RAND.

HERMAN KAHN

We have chosen for this purpose the extraordinary researches into civil defense and strategic planning, as related by RAND project leader Herman Kahn. During the 1950s, Kahn led several large RAND studies into the effects of thermonuclear war and civil defense for the U.S. Air Force. In 1960 he was the top ranking authority on strategic planning and civil defense. Kahn's research became available to the general public in several books, of which the most important is *On Thermonuclear War* published in 1960. Many of the concepts and themes of this book have entered the folklore through the movie *Dr. Strangelove*. The tragic but distinguishable postwar states, the doomesday machine, the flexible response, these and many more come from *On Thermonuclear War*.

It is not clear to what extent the conclusions and recommendations in *On Thermonuclear War* are representative of thinking at RAND. In any case, Kahn had a falling out with the director over fallout shelters (Kahn wanted them, the director did not), and he became the first RAND alumnus to start his own think tank. In 1961 he founded the Hudson Institute, located in New York at Croton-on-Hudson. People soon began speaking of Herman-on-Hudson. Herman Kahn died in 1983, and *The Futurist* of October 1983 contains several reminiscences by friends and colleagues.

On Thermonuclear War opens with a discussion of possible strategic postures. The official posture in 1960 was called "finite deterrence." It involves the ability to inflict unacceptable damage on an enemy after absorbing a surprise nuclear attack. Finite deterrence can be upgraded by adopting various "counterforce" measures. A counterforce measure is one designed to limit the damage inflicted by an enemy. Building fallout shelters and removing attractive targets from population centers are counterforce measures. So is an antiballistic missile system that destroys enemy missiles in flight. So is a highly accurate attack weapon capable of knocking out enemy missiles before they are launched. When enough counterforce measures are adopted such that one is willing to absorb a retaliatory blow from the other side, one possesses a "credible first strike capability." *On Thermonuclear War* is an extended argument for abandoning finite deterrence and striving for a credible first strike capability:

> Since we wish to be able to limit Soviet behavior to deter them from acts which cannot be met by limited means, *we must have a credible first strike capability* . . . The only time in which the credible first strike capability would be destabilizing would be when we

were being cruelly provoked or in a desperate crisis—a circumstance in which, destabilizing as it would be, we would feel we would need an ability to rescue ourselves from a more dire eventuality either by increasing our bargaining power or *by actual use of the credible first strike capability* (emphasis added).   (Kahn, 1960, p. 158)

Kahn advised that in a "desperate crisis" it is prudent and morally acceptable to "use the credible first strike capability," that is, to initiate a nuclear war.

Kahn's argument for the credible first strike capability is based on the claim that a retaliatory blow can be limited to an acceptable level. This is what the RAND researchers believe to have discovered, and this is what *On Thermonuclear War*, in over 600 pages, seeks to demonstrate.

Kahn and his colleagues have calculated the length of time needed for economic recuperation as a function of the number of deaths under a nuclear attack. Table 1.1 illustrates the results (the captions are taken verbatim from the book).

The crucial question is the one at the bottom of Table 1.1. Kahn reports:

Despite widespread belief to the contrary, objective studies indicate that even though the amount of human tragedy would be greatly increased in the postwar world, the increase would not preclude normal happy lives for the majority of survivors and their descendants.   (Kahn, 1960, p. 21)

Kahn considers two possible nuclear attacks. The "early attack" (representing the nuclear threat through the 1960s) would involve 1500 megatons directed to the 53 largest metropolitan areas and various military targets. The "late attack" (the nuclear threat in the 1970s) involves 20,000 megatons. The early attack is analyzed in greater detail. Kahn concludes

...Our calculations indicate that even without special stockpiles, dispersal, or protection, the restoration of our prewar gross national product should take place in a relatively short time—*if we can hold the damage to the equivalent of something like 53 metropolitan areas destroyed.*   (Kahn, 1960, p. 84)

These calculations are based on seven optimistic assumptions, though Kahn hastens to add that he and his coworkers think the calculations are more likely to be pessimistic than optimistic. The assumptions are shown in Table 1.2. The first

Table 1.1 Tragic But Distinguishable
Postwar States

Dead	Economic Recuperation
2,000,000	1 year
5,000,000	2 years
10,000,000	5 years
20,000,000	10 years
40,000,000	20 years
80,000,000	50 years
160,000,000	100 years
Will the survivors envy the dead?	

Source: (Kahn, 1960, p. 34)

Table 1.2   Seven Optimistic Assumptions

1. Favorable political environment
2. Immediate survival and patch-up
3. Maintenance of economic momentum
4. Specific bottlenecks alleviated
5. "Bourgeois" virtues survive
6. Workable postwar standards adopted
7. Neglected effects unimportant

Source: (Kahn, 1960, p. 84)

assumption means that the United States has not lost the war. The second assumption entails that society starts to function immediately after the attack: "debris has been cleared up, minimum communications restored, the most urgent repairs made, credits and markets reestablished, a basic transportation system provided, minimum utilities either set up or restored, the basic necessities of life made available, and so on" (Kahn, 1960, p. 84).

All these assumptions are uncertain. We shall not go into the details of his argument, but just give an impression how Kahn assesses and deals with the uncertainties inherent in his assumptions. For example, he claims that the destruction of the 53 largest metropolitan areas would not be an economic catastrophe, but "may simply set the nation's productive capacity back a decade or two . . . . This statement strikes most people as being very naive" (Kahn, 1960, p. 77). After all, destroying only a small percentage of the cells of an organism can result in the organism's death and hence the subsequent destruction of all cells. However, the analogy with an organism "seems to be completely wrong . . . . The creating (or re-creating) of a society is an art rather than a science; even though empirical and analytic 'laws' have been worked out, we do not really know how it is done, but almost everybody (Ph.D. or savage) can do it" (Kahn, 1960, p. 77).

The fifth optimistic assumption appears bold to many people. Kahn asks "would not the shock of the catastrophe so derange people that every man's hand would be against every other man's?" Again: "This seems entirely wrong. . . . Nations have taken equivalent shocks even without special preparations and have survived with their prewar virtues intact" (Kahn, 1960, p. 89). The fact that the destruction would happen quickly only strengthens his case: "While many normal personalities would disintegrate under hardships spread over a period of many years, the habits of a lifetime cannot be changed for most people in a few days. . . . It is my belief that if the government has made at least moderate preparations, so that most people whose lives have been saved will give some credit to the government's foresight, then people probably will rally round . . ." (Kahn, 1960, pp. 89–90).

The following passage puts the foregoing in perspective:

In spite of the many uncertainties of our study, we do have a great deal of confidence in some partial conclusions—such as, that a nation like the United States or the Soviet Union could handle each of the problems of radioactivity, physical destruction, or likely levels of casualties, if they occurred by themselves." (Kahn, 1960, p. 91).

Kahn explains that in analyzing each problem, he and his RAND colleagues assumed that this was the *only* problem. The RAND scientists did not consider the eventuality that radiation, destruction, and death might happen *together*.

A review in *Scientific American* conjectured that the whole book was a staff joke in poor taste (Newman, 1961). The book was not meant as a joke, and we must try to understand how an advice to initiate nuclear war could be given, and presumably taken seriously, on the basis of such reasoning. Part of the explanation may lie in the unfamiliarity of reasoning with uncertainty. The following reconstruction of Kahn's reasoning suggests itself:

Kahn wants to establish the conclusion that we will *probably* restore our gross national product (GNP) quickly after a small attack in which 53 metropolitan areas are lost. It is sufficient, he thinks, if it is *likely* that we can handle the problems of radiation destruction and death. Hence, he considers in isolation the three propositions: "It is *likely* that we can handle radiation," "it is *likely* that we can handle death," and "It is *likely* that we can handle destruction." Having convinced himself of each of these, he concludes "It is *likely* that we can handle radiation, destruction, and death." QED

Of course, this is probabilistically invalid; if proposition A is likely, and proposition B is also likely, it does not follow that proposition "A and B" is likely. However, Kahn seems to be reasoning with uncertainty in the same way that most of us do, if we don't stop to think about it. That is, his normal pattern of reasoning is adapted to "cope with uncertainty" by simply adding a number of qualifiers such as probably, likely, seems, or maybe. There is no real attempt to say *how* uncertain a given statement is, and no attempt to propagate uncertainty through a chain of reasoning. The subtleties involved in probabilistic reasoning will become apparent in Chapter 3.

This gives a sufficient impression of Kahn's reasoning under uncertainty. Before turning to methodology, we add one final detail to this picture of expert advice. At the end of the book, Kahn considers a number of possible objections to his proposals. One of these is

7. *Objection:* 'It is pointless to fight a war even if victory is achieved, since the country will be so debilitated that any third- or fourth-rate power could take it over . . .'

Answer: Saving lives is in itself a reasonable objective, but the program is intended to be much more than a reduction in the number of immediate casualties. If the U.S. does win the war promptly, it will have had an infuriating experience that is likely to create a problem quite the opposite of that envisaged by this objection. The country will be very well sheltered, having lost all its soft spots. It will be in a 'mean temper,' having just fought a war to save the world. It will be in no mood to be pushed around by other countries. It will have a large strategic air force because a properly defended air force cannot be thrown away in a single strike or two. Far from being weak or vulnerable, the U.S. might be able to 'take over the world'—even though such a goal is utterly inconsistent with our political institutions and values. The least the U.S. could expect, and insist on, is that the rest of the world help, not hinder, her reconstruction effort and cooperate in organizing the peace." (Kahn, 1960, p. 646)

Since the publication of *On Thermonuclear War*, two problems have been of particular concern to military strategic planners. One is the phenomenon of electromagnetic pulse (see Federation of American Scientists, 1980, and *Science*, vol. 212, no. 12, June 1981). The second concerns the climatological changes brought about by injecting millions of tons of smoke and dust into the upper atmosphere, the so-called nuclear winter. According to the report of 40 prominent atmospheric, biological and physical scientists, who studied the effects from nuclear exchanges ranging from 100 to 10,000 total megatons, "the population size of homo sapiens conceivably could be reduced to prehistoric levels or below, and the extinction of the human species itself cannot be excluded" (*Science*, vol. 222, no. 4630, p. 1293, December 23, 1983). Even a 100-megaton exchange could induce subfreezing land temperatures for months. If the war occurred in spring or summer a year's crops in the northern hemisphere would probably be lost. Neither electromagnetic pulse nor nuclear winter was considered by the RAND scientists.

SCENARIO ANALYSIS

Herman Kahn is regarded as the father of scenario analysis. This method underlies the type of "systems analysis" represented in *On Thermonuclear War*, though it became more explicit as a methodology in Kahn's later futurological research. The most prominent example of the latter is the book he coauthored with Anthony Wiener in 1967, *The Year 2000*. (The maker of the film *Dr. Strangelove*, Stanley Kubric, also made the film *2001*.) In discussing scenario analysis directly after *On Thermonuclear War* we do not mean to suggest that practitioners of this method are implicitly responsible for the views expressed in that book. On the other hand, the choice of methodology must be made in awareness of the consequences to which such choices might lead.

In *On Thermonuclear War* Kahn speaks of a "technological breakthrough in the art of doing Systems Analysis and Military Studies" (Kahn, 1960, p. 119). Previously, analysts would have tried to optimize a single objective, or perhaps attempt to weigh different objectives and assess probabilities of outcomes in order to arrive at a best expected outcome. Kahn, who understood the basis of mathematical decision theory quite well, describes these methodologies as inadequate.

The reason is rooted in the foundations of statistical decision theory. The assessment of probabilities is necessarily subjective. Moreover, programs and proposals must eventually be decided on by committees, whose members will generally have conflicting goals and differing assessments of probabilities. Within classical decision theory there is no way of getting groups to arrive at a rational consensus on subjective probabilities or on objectives. It is a fundamental feature of statistical decision theory, which is usually forgotten in casual discussions, that the concepts of subjective probability and utility are only meaningful for an *individual*. In general, they cannot be meaningfully defined for a group, and it is impossible for a committee to come to a decision on the basis of maximal expected utility.

It follows that if the analyst seeks to develop proposals that will be approved by some committee, he should not attempt to maximize his (or someone else's) expected utility, but should try to develop proposals on which a sufficient number of committee members can agree.

Kahn acknowledges that this is little more than common political sense, but it does have far-reaching consequences. It entails, for example, that the scientist doing systems analysis should think, not as a scientist traditionally thinks, but rather as a politician. Knowing that Kahn's strategic planning studies were contracted by the Air Force, one might charge that he was merely telling the generals what they wanted to hear. If we take his methodological statements seriously, this is exactly what he was *trying* to do. In any event, RAND's "technological breakthrough" entails that probability can play at best a subordinate role in the scenario-analytic approach to policy analysis. Although our focus is on scenario analysis as practiced in the 1950s and 1960s, this latter feature is characteristic of most scenario studies up to the present. The probabilities of the scenarios are not explicitly taken into account.

What is exactly a scenario, and what role can probability play in scenario analysis? For answers to these questions we must turn to *The Year 2000*.

> Scenarios are hypothetical sequences of events constructed for the purpose of focusing attention on causal processes and decision-points. They answer two kinds of questions: (1) Precisely how might some hypothetical situation come about, step by step? and (2) What alternatives exist, for each actor, at each step, for preventing, diverting, or facilitating the process. (Kahn and Wiener, 1967, p. 6)

The method as applied in projecting the year 2000 works basically as follows. The analyst first identifies what he takes to be the set of basic long-term trends. These trends are then extrapolated into the future, taking account of any theoretical or empirical knowledge that might impinge on such extrapolations. The result is termed the *surprise-free scenario*. The surprise-free scenario serves as a foil for defining *alternative futures* or *canonical variations*. Roughly speaking, these are generated by varying key parameters in the surprise-free scenario.

What about the probability of the various scenarios. Kahn is very explicit about this. In doing long-range projections the trouble is, he says, that no particular scenario is much more likely than all the others:

> The subjective curve of probabilities often seems flat.... In order to avoid the dilemma of Buridan's ass, who starved midway between two bales of hay because he could not decide which one he preferred, we must then make arbitrary choices among almost equally interesting, important, or plausible possibilities. That is, if we are to explore any predictions at all, we must to some extent 'make them up.' (Kahn and Wiener, 1967, p. 8)

The surprise-free scenario is *salient* because of its relation to the basic long-term trends, not because it is probable, or more probable than its alternatives. In fact, the surprise-free scenario may be judged to have a very low probability.

One may well ask, what is the point of devoting an extensive analysis to one scenario, plus a few canonical variations, if all the scenarios have a very low inherent probability.

Consider a very simple mathematical analogy. Suppose we have an uncertain quantity, which might take any integer value between zero and one million. Suppose that no particular number is much more probable than any other. Would it make sense to pick one particular number, say 465,985, affirm that it was not more probable than a host of others, and then study extensively the properties of 465,985 plus a few "canonical variations"? Of course not. Is Kahn's method of scenario analysis less absurd? Well, the surprise-free scenario is salient, not because of its probability but because of its relation to basic trends. If we bear in mind that saliency has nothing to do with *prediction*, then focusing on a few scenarios might help study the basic trends. However, it is easy to be misled into believing that scenario analysis yields predictions. Proof of the latter comes from Kahn's own blurb on the book jacket: "Demonstrating the new techniques of the think tanks, this book projects what our own world most probably will be like a generation from now—and gives alternatives."

If the value of a quantity is uncertain, then probability theory presents us with a variety of methods for representing this uncertainty and making predictions. Ideally, we should state the probability distribution of the uncertain quantity. However, we need not describe the distribution completely in order to make meaningful predictions. In the above example we might say that the probability was 90% that the quantity's value would be between 200,000 and 800,000. What sort of prediction is this? Under certain circumstances, which we shall study extensively in Part II, this entails the following: If we give such "90% subjective confidence intervals" for a large number of uncertain quantities, then we expect 90% of the actual values to fall within their respective confidence intervals. We cannot say that a given prediction is right or wrong, but if all the values were to fall outside their respective confidence intervals, then one could rightly conclude that the person giving the predictions was not a very good predictor. In the language of succeeding chapters, he would be called poorly calibrated.

In their book, Kahn and Wiener do give a number of probabilistic predictions. Table 1.3 lists 25 technological innovations that, with characteristic clarity, are described as "even money bets, give or take a factor of five" before the year 2000. If Kahn and Wiener are "well-calibrated assessors" we should expect roughly one-half of these predictions to come true by the year 2000.

THE DELPHI METHOD

The Delphi method was developed at the RAND corporation in the early 1950s as a spin-off of an Air Force–sponsored research project, "Project Delphi." The original project was designed to anticipate an optimal targeting of U.S. industries by a hypothetical Soviet strategic planner. Delphi was first brought before a wider audience in a 1963 RAND study "Report on a Long-Range Forecasting Study," by Olaf Helmer and T. J. Gordon (Rand Paper P-2982, later incorporated in Helmer, 1966). It is undoubtedly the best-known method of eliciting and synthesizing expert opinion.

In the middle 1960s and early 1970s the Delphi method found a wide variety of applications, and by 1974 the number of Delphi studies had exceeded 10,000 (Linstone and Turoff, 1975, p. 3). Although most applications are concerned with technology forecasting, the method has also been applied to many types of policy analysis.

Policy Delphi's differ from technology forecasting Delphi's with respect to both purpose and method. In technology forecasting, the team conducting the Delphi study seeks experts who are most knowledgable on the issues in question,

Table 1.3 25 Technological Innovations

1. "True" artificial intelligence
2. Practical use of sustained fusion to produce neutrons and/or energy
3. Artificial growth of new limbs and organs (either in situ or for later transplantation)
4. Room temperature superconductors
5. Major use of rockets for commercial or private transportation (either terrestrial or extraterrestrial)
6. Effective chemical or biological treatment for most mental illnesses
7. Almost complete control of marginal changes in heredity
8. Suspended animation (for years or centuries)
9. Practical materials with nearly "theoretical limit" strength
10. Conversion of mammals (humans?) to fluid breathers
11. Direct input into human memory banks
12. Direct augmentation of human mental capacity by the mechanical or electrical interconnection of the brain with a computer
13. Major rejuvenation and/or significant extension of vigor and life-span—say 100 to 150 years
14. Chemical or biological control of character or intelligence
15. Automated highways
16. Extensive use of moving sidewalks for local transportation
17. Substantial manned lunar or planetary installations
18. Electric power available for less than 0.3 mill per kilowat-hour
19. Verification of some extrasensory phenomena
20. Planetary engineering
21. Modification of the solar system
22. Practical laboratory conception and nurturing of animal (human?) fetuses
23. Production of a drug equivalent to Huxley's soma
24. A technological equivalent of telepathy
25. Some direct control of individual thought processes

Source: (Kahn and Wiener, 1967, pp. 54, 55)

and seeks to achieve a high degree of consensus regarding predicted developments. Policy Delphi's on the other hand seek to incorporate the views of the entire spectrum of "stakeholders" and seek to communicate the spread of their opinions to the decision maker. We shall be chiefly concerned with the forecasting type of Delphi's in this brief discussion.

The Delphi method has undergone substantial evolution and diversification, but the basic approach may be described as follows (see, e.g., Helmer, 1968). A monitoring team defines a set of issues and selects a set of respondents who are experts on the issues in question. A respondent generally does not know who the other respondents are, and the responses are anonymous. A preliminary questionnaire is sent to the respondents for comments, which are then used to establish a definitive questionnaire. Typical questions take the form "In what year will such-and-such take place?"

This questionnaire is then sent to the respondents and their answers are analyzed by the monitoring team. The set of responses is then sent back to the respondents, together with the median answer and the interquartile range, the range containing all but the lower 25% and the upper 25% of the responses. The respondents are asked if they wish to revise the initial predictions. Those whose answers remain outside the interquartile range for a given item are asked to give arguments for their prediction on this item.

The revised predictions are then processed in the same way as the first responses, and arguments for outliers are summarized. This information is then sent back to the respondents, and the whole process is iterated. A Delphi exercise typically involves three or four rounds.

The responses on the final round generally show a smaller spread than the responses on the first round, and this is taken to indicate that the experts have reached a degree of consensus. The median values on the final round are taken as the best predictions.

Table 1.4　A Delphi Questionnaire

Questionnaire #1

This is the first in a series of four questionnaires intended to demonstrate the use of the Delphi Technique in obtaining reasoned opinions from a group of respondents.

Each of the following six questions is concerned with developments in the United States within the next few decades.

In addition to giving your answer to each question, you are also being asked to rank the questions from 1 to 7. Here "1" means that in comparing your own ability to answer this question with what you expect the ability of the other participants to be, you feel that you have the relatively best chance of coming closer to the truth than most of the others, while a "7" means that you regard that chance as relatively least.

Rank	Question	Answer*
☐	1. In your opinion, in what year will the median family income (in 1967 dollars) reach twice its present amount?	☐
☐	2. In what year will the percentage of electric automobiles among all automobile in use reach 50 percent?	☐
☐	3. In what year will the percentage of households that are equipped with computer consoles tied to a central computer and data bank reach 50 percent?	☐
☐	4. By what year will the per-capita amount of personal cash transactions (in 1967 dollars) be reduced to one-tenth of what it is now?	☐
☐	5. In what year will power generation by thermonuclear fusion become commercially competitive with hydroelectric power?	☐
☐	6. By what year will it be possible by commercial carriers to get from New York's Times Square to San Francisco's Union Square in half the time that is now required to make that trip?	☐
☐	7. In what year will a man for the first time travel to the Moon, stay for at least 1 month, and return to Earth?	☐

*"Never" is also an acceptable answer.

Please also answer the following question, and give your name (this is for identification purposes during the exercise only; no opinions will be attributed to a particular person).

Check one:

　　　　　☐ I would like　　　　　⎫
　　　　　☐ I am willing but not anxious ⎬ to participate in the three remaining questionnaires
　　　　　☐ I would prefer not　　　⎭

Name (block letters please):

. . . .

Source: Helmer (1968).

Table 1.5 Comparison of Delphi Results

	Comparison with Other Studies (Medians)		
	Present Conference Forecast	RAND* Pretest	1963† LRF Study
1. Family income doubled	1985	1985	—
2. Electric autos 50%	1997	1988	—
3. Home computer consoles	2010	2002	2005 Facsimile magazines in home
4. Credit card economy	1987	1995	1974 Credit link bank to stores
5. Economical fusion power	1988	1990	1986 Controlled fusion achieved
6. N.Y. → S.F. in ½ time	1975	1980	—
7. Man on moon 1 month	1976	1980	1975 Two men on moon 1 month

* A 1966 Delphi pretest, using 23 RAND employees as participants.
† A 1963 RAND Long Range Forecasting Study. A report of this study appeared as an appendix to *Social Technology*, by O. Helmer (Basic Books, 1966).
Source: Helmer (1968).

It may be interesting to have a look at a real Delphi questionnaire and real results. Table 1.4 shows a Delphi questionnaire presented to delegates of the First Annual Technology and Management Conference (Helmer, 1968). Note the guarantee of anonymity, characteristic of Delphi exercises, and the use of self-rating. Table 1.5 shows the forecasts of the conference members and compares these, where possible, with the results of two earlier RAND Delphi's. The reader will draw his own conclusions from the Delphi predictions.

The Delphi method has undergone many variations. One of the most important variations involved letting the experts indicate their own expertise for each question. In the above questionnaire, experts rated their expertise on a scale of 1 to 7. Only the opinions of the experts claiming the most expertise for a given item are used to determine the distribution of opinion for that item. This method is claimed to improve accuracy (Dalkey, Brown, and Cochran, 1970). It has been found that women consistently rate themselves lower than men (Linstone and Turoff, 1975, p. 234).

The claim of improved accuracy resulting from self-ratings was challenged in an extensive study by Brockhoff (1975). It was found that self-ratings of participants did not coincide with "objective expertise" as measured by relative deviation from the true value on fact-finding and forecasting tasks.[1] Brockhoff also compared the results of Delphi groups with groups using "face-to-face" confrontation, but could not draw firm conclusions, owing partly to a large number of dropouts.

Furthermore, the geometric mean of the individual errors in Delphi groups of different sizes did not show any convincing overall tendency to decline through successive Delphi rounds. Table 1.6 shows the geometric means of individual errors for fact-finding and forecasting questions through five Delphi rounds in Brockhoff's study.

The Delphi method was developed by mathematicians and engineers, and enjoyed considerable popularity among research managers, policy analysts, and

[1] If T is the true value and T' is the estimate of T, then the relative deviation is defined as $|T' - T|/T$.

Table 1.6   Geometric mean of the individual errors

Group Size	Round (Fact-Finding Questions)					Round (Forecasting Questions)				
	1	2	3	4	5	1	2	3	4	5
5	0.20	0.19	0.18	0.23	0.27	0.28	0.27	0.28	0.35	0.35
7	0.13	0.09	0.07	0.10	0.11	0.44	0.42	0.43	0.36	0.36
9	0.22	0.14	0.09	0.17	0.16	0.25	0.20	0.16	0.16	0.16
11	0.24	0.29	0.23	0.28	0.22	0.12	0.09	0.10	0.10	0.10

Source: Brockhoff, 1975, p. 314.

corporate planners in the late 1960s and early 1970s. By the middle 1970s psychometricians, people trained in conducting controlled experiments with humans, began taking a serious look at the Delphi methods and results.

Perhaps the most significant study in this regard is Sackman's *Delphi Critique* (1975). Sackman approaches the Delphi exercises as psychometric experiments (which, strictly speaking, they are not) and concludes that the Delphi method violates several essential methodological rules of sound experimental science. For one thing, he notes that the questionnaire items are often so vague that it would be impossible to determine when, if ever, they occurred. Furthermore, the respondents are not treated equally. People whose predictions fall inside the interquartile band are "rewarded" with a reduced workload in returning the questionnaires, whereas those whose predictions fall outside this band are "punished" and must produce arguments. Moreover, in many Delphi exercises there seem to be a significant number of dropouts—people who simply don't return their forms. Delphi exercises do not publish the number of dropouts, they make no attempt to discover the reason for dropping out, and do not assess the influence of this negative selection on the results. This all raises the possibility that the Delphi convergence may owe more to boredom than to consensus. Finally, Sackman argues that experts and nonexperts generally produce comparable results in Delphi exercises.

Of course, the proof of the pudding is in the eating. Methodological critiques like that of Sackman would be of limited impact if the predictions from Delphi exercises were good. This latter question has been examined, so far as possible in a laboratory setting, in a number of psychometric studies. Delphi is compared with a "nominal group technique" (Delbecq, Van de Ven, and Gusstafson, 1975) in which participants confront each other directly in a controlled environment after giving initial assessments, and with a "no interaction" model in which initial assessments are simply aggregated mathematically. Gustafson et al. (1973) found the nominal group technique superior to the others, and found Delphi to be the worst of the three. Gough (1975) found a similar pattern of results. Fischer (1975) found no significant differences between the various techniques. Finally Seaver (1977) compared different group techniques as well as different techniques of mathematical aggregation. He found a general lack of difference in scores, regardless of group technique and regardless of method of mathematical aggregation. One significant conclusion was the following. In estimating probabilities, group interaction (either Delphi or nominal group technique) tended to result in more extreme probability estimates (i.e., estimates closer to zero or one). In this sense, the group interaction tended to make the participants more confident. However, this increased confidence did not correspond to an increased relative frequency of correctness. In other words, the group interaction tended to make the participants overconfident. We shall return to the question of overconfidence in discussing measures of quality in probabilistic assessments.

The results cited above have put the oracles of Delphi on the defensive. The blank check extended to experts in the 1960s has generally been rescinded and the burden of proof has been placed on the experts and their monitors. As a result, the whole question of evaluating expert opinion and developing methodological guidelines for its use has moved into the foreground. The Delphi exercises seem to have disappeared, and play almost no role in contemporary discussions of expert opinion (for a recent survey see Parenté and Parenté, 1987). However, one important legacy of the Delphi age does survive in the "benchmark exercises" used to gage the "reproducibility" of risk analyses. Examples are discussed in Chapter two.

CONCLUSIONS: TOWARD A METHODOLOGY FOR EXPERT OPINION

Plato's theory of knowledge, expressed in the divided line, provided the basis for his ideal state, *The Republic*. *The Republic* was governed by "guardians" who had undergone a lifetime of rigorous training in science, statecraft, and ethics. The most important conclusion emerging from this brief review is this: Experts are not the guardians Plato had in mind. They are ordinary mortals with ordinary foibles. They may have interests, they may have biases and predilictions, and they may be unable adequately to assess their own uncertainty. An expert who knows more about a particular subject than any one else may be able to talk an opponent under the table. This does not mean that his opinions are more likely to be true.

The "received view" of science, as articulated by philosophers shortly before and after World War II, paints a picture of science in which the scientific method plays a central role. Science aims at rational consent, and the method of science must serve to protect and further this aim. Hence, the scientific method shields science from idiosyncracies of individual scientists. Concomitantly, any individual who scrupulously follows this method can arrive at results that have a legitimate claim to rational consent, regardless of the person actually performing the research. In this sense science is conceived as impersonal.

Within the received view of science, something like expert opinion can play at most a marginal role. It is useful to recall in this regard a distinction first drawn by the philosopher Hans Reichenbach (1951). Reichenbach distinguished a "context of discovery" from the "context of justification" in science. Discovery in science is often nonmethodological. It is frequently driven by factors that are highly subjective and nonrational; hunches, predilictions, biases, and flights of imagination. Everyone has seen the moon move across the night sky and seen apples fall to earth. However, it took a Newton to realize the the same physical laws underlay both phenomena. Once this idea was proposed, it was subjected to tests that could be verified by everyone. Having the idea is not the same as justifying the idea. The former belongs to the context of discovery, the latter to the context of justification. According to this view, the scientific method operates primarily within the context of justification. Expert opinion, if it plays any role at all, can only figure in the context of discovery, according to the received view.

This received view came under attack in the 1960s as people realized that the very features mentioned above, interest, biases, prediliction, uncertainty, play a larger role in science than many philosophers cared to acknowledge. Against this background, the introduction of expert opinion into the scientific method may have seemed like an imminently logical step. Since this time, expert opinion has become increasingly visible within the context of justification in science. An expert's *opinion*, whether sprinkled with "uncertainty modifiers" or cast in the form of quasi-structured input, is increasingly advanced as an *argument*. In the next chapter we shall see to what lengths this has been taken in certain areas of "hard science."

It is appropriate to pause by the question whether this development is really consistent with the fundamental aim of science, rational consensus. Within the received view, rational consent is pursued via the twin principles of *reproducibility* and *empirical control*. Scientific results must in principle be susceptible to empirical control, *by anyone*, hence they must be reproducible by anyone. A result gained by expert opinion cannot in general be reproduced by other experts, and cannot be directly submitted to empirical control. This is surely why opinion plays no explicit role in the methodology of the received view.

This is not to say that expert opinion cannot be used in science outside the context of discovery. In fact this has been going on to some degree since the inception of science. However, it is fair to say that methodological guidelines for the use of expert opinion outside the context of discovery are unknown within the received canons of science. Much recent research in the field of expert opinion has been directed, at least implicitly, toward articulating such guidelines, and this research is the principal focus of the succeeding chapters.

The central theme of this book is that expert opinion *can* find a place within the context of justification. The "emerging technologies" for expert opinion provide us with the tools for rationally analyzing and evaluating heretofore nonrational portions of scientific activity, and bringing them within the purview of methodology.

The most important tool in rationally incorporating expert opinon in science is the representation of uncertainty. Opinion is by its very nature uncertain. Hence, when expert opinion is used as input in a scientific inquiry or report, the question to be addressed is simply this: Is the uncertainty adequately represented? But what does it mean to "adequately represent uncertainty"? When is uncertainty adequately represented and when is it not? These are the questions which a methodology for expert opinion must answer.

2

Expert Opinion in Practice

Expert opinion has been used in a more or less systematic way in many fields. This chapter discusses uses in the aerospace program, in military intelligence, in nuclear energy, and in policy analysis. The latter field is extremely broad and contains many methodologies that are only marginally related to the central theme of this book, namely the representation and use of expert uncertainty. Techniques such as time series analysis and regression analysis have been extensively used in economic forecasting, for example, but cannot be called methodologies for using expert opinion. Hence, the items chosen from the general field of policy analysis will focus on expert opinion and will not constitute a representative sample from this very wide field.

THE AEROSPACE SECTOR

As in the nuclear sector, expert opinion entered the aerospace sector because of the desire to assess safety. In particular, managers and politicians needed to assess the risks associated with rare or unobserved catastrophic events. The likelihood of such events could obviously not be assessed via the traditional scientific method of repeated independent experiments.

The problems of assessing such likelihoods were dramatically brought out on January 28, 1986, with the tragic accident with the Challenger space shuttle. In 1983, E. W. Colglazier and R. K. Weatherwax (Colglazier and Weatherwax, 1986) brought out a report sponsored by the U.S. Air Force, which reviewed an earlier National Aeronautics and Space Administration (NASA)–sponsored estimate of shuttle failure modes and failure probabilities. [In this earlier report, certain critical failure probabilities were simply dictated by the sponsor, in disregard of available data (Bell and Esch, 1989).] Their estimate of the solid rocket booster failure probability per launch, based on subjective probabilities and operating experience, was roughly 1 in 35. The NASA management rejected this estimate and elected to rely on their own engineering judgment, which led to a figure of 1 in 100,000. As the report of Colglazier and Weatherwax is not available to the public at this writing, the only published documentation is to be found in the abstracts for a conference on risk analysis in 1986 at which their results were presented. We quote their abstract in full:

> We estimated in 1983 that the probability of a solid rocket booster (SRB) failure destroying the shuttle was roughly 1 in 35 based on prior experience with this technology. Sponsored by Teledyne Energy Systems for the Air Force, our report was a review of earlier NASA-sponsored estimates of shuttle failure modes and their probabilities. These estimates were to be used in assessing the risk of carrying radioisotope thermoelectric generators (RTGs) aboard the shuttle for the Galileo and Ulysses missions scheduled for 1986. Our estimates of SRB failure were based on a Bayesian analysis utilizing the prior experience of 32 confirmed failures from 1902 launches of various solid rocket motors. We also found that failure probabilities for other accident modes were likely to have been underestimated by as much as a factor of 1000. A congressional hearing on March 4, 1986, reviewed our report and critiques by Sandia National Laboratory and NASA. NASA had decided to rely upon its engineering judgment and to use 1 in 100,000 as the SRB failure probability estimate for nuclear risk assessments. We have recently reviewed the critiques and stand by our original conclusions. We believe that in formulating space policy, as well as in assessing the risk of carrying RTGs on the shuttle, the prudent approach is to rely upon conservative failure estimates based upon prior experience and probabilistic analysis.

The Challenger accident occurred on the twenty-fifth launch. A presidential commission under former Secretary of State Rogers investigated the accident. The personal experiences of one commission member, Richard Feynman, colorfully depict some of the problems the commissioners found at NASA. One passage deserves to be quoted at length.

> Suddenly I got an idea. I said "All right, I'll tell you what. In order to save time, the main question I want to know is this: is there the same misunderstanding, or difference of understanding, between the engineers and the management associated with the engines, as we have discovered associated with the solid rocket boosters?"
>
> Mr. Lovingood says, "No, of course not. Although I'm now a manager, I was trained as an engineer."
>
> I gave each person a piece of paper. I said, "Now, each of you please write down what you think the probability of failure for a flight is, due to a failure in the engines."
>
> I got four answers—three from the engineers and one from Mr. Lovingood, the manager. The answers from the engineers all said, ... almost exactly the same thing: 1 in 200. Mr. Lovingood's answer said, "Cannot quantify. Reliability is determined by studies of this, checks on that, experience here—blah, blah, blah, blah, blah."
>
> "Well," I said, "I've got four answers. One of them weaseled." I turned to Mr. Lovingood and said, "I think you weaseled." He says, "I don't think I weaseled." "Well, look," I said, "you didn't tell me *what* your confidence was; you told me *how* you determined it. What I want to know is: After you determined it, what *was* it?"
>
> He says, "100 percent." The engineers' jaws drop. My jaw drops. I look at him, everybody looks at him—and he says "Uh...uh, minus epsilon!"
>
> "OK. Now the only problem left is, what is epsilon?"
>
> He says, "1 in 100,000." So I showed Mr. Lovingood the other answers and said, "I see there is a difference between engineers and management in their information and knowledge here...". (Feynman, 1987)

A systematic concern with risk assessment methodology began after the fire on the Apollo flight AS-204 on January 27, 1967, in which three astronauts were killed.

This one event set the NASA planning back 18 months, involved considerable loss of public support, cost NASA salaries and expenses for 1500 people involved in the subsequent investigation, and ran up $410 million in additional costs (Wiggins, 1985). This was reason enough to subject the erstwhile safety policy to an extensive review.

Prior to the Apollo accident NASA relied on its contractors to apply "good engineering practices" to provide quality assurance and quality control. Although there was some systematic review of these policies undertaken at the Johnson Space Center in 1965, there was no formal attempt to define "safety" or "reliability."

The problems of a "contractor-driven" approach to safety were brought out in a letter to Senator Gravel from the Comptroller General of the United States (U.S.N.R.C., 1975 pp. XI 3-15, 3-21), which quoted an Air Force report: "... where a manufacturer is interested in having his equipment look good he can, and will, select some of the more optimistic data he can find or generate to use in his reliability predictions. Thus reliability predictions, for several reasons, tend to be generally optimistic by a factor of two to six, but sometimes for substantially greater factors."

Data on mean times between failures (MTBF) for aircraft radar subsystems supporting this contention are given in Table 2.1.

On April 5, 1969, the Space Shuttle Task Group was formed in the Office of Manned Space Flight of NASA. The task group developed "suggested criteria" for evaluating the safety policy of the shuttle program, which contained quantitative safety goals. The probability of mission completion was to be at least 95% and the probability of death or injury per mission was not to exceed 1%. These numerical safety goals were not adopted in the subsequent shuttle program (Wiggins, 1985, p. 9).

In an attempt to structure their safety policy, and protect themselves from self-serving safety assessments, NASA, following a lead from the military, adopted so-called risk assessment matrix tables to "quantify and prioritize" risks. An example of such tables is given as Table 2.2. Of course, these matrix tables are "quantitative"

Table 2.1 Mean Times Between Failure, Specified and Achieved for Aircraft Radar Subsystems

Aircraft	Specified MTBF	Achieved MTBF
F-4B	10	4
A-6A	75	8
F-4C	10	9
F-111 A/E	140	35
F-4D	10	10
A-7 A/B	90	30
A-7 D/E	250	12
F-4E	18	10
F-111D	193	Less than one
F-4J	20	5

Source: (U.S.N.R.C., 1975 p. XI 3-19)

Table 2.2 NASA Risk Assessment Matrix Tables

Hazard Severity Categories

Description	Category	Mishap Definition
Catastrophic	I	Death or system loss
Critical	II	Severe injury, severe occupational illness, or major system damage
Marginal	III	Minor injury, minor occupational illness, or minor system damage
Negligible	IV	Less than minor injury, occupational illness, or system damage

Hazard Probability Ranking

Description*	Level	Specific Individual Item	Fleet or Inventory†
Frequent	A	Likely to occur frequently	Continuously experienced
Probable	B	Will occur several times in life of an item	Will occur frequently
Occasional	C	Likely to occur at sometime in life of an item	Will occur several times
Remote	D	Unlikely but possible to occur in life of an atem	Unlikely but can reasonably be expected to occur
Improbable	E	So unlikely it can be assumed occurrence may not be experienced	Unlikely to occur, but possible

Hazard Risk Management Matrix

	Hazard Categories			
	I	II	III	IV
Frequency of Occurrence	Catastrophic	Critical	Marginal	Negligible
(A) Frequent	1	3	7	13
(B) Probable	2	5	9	16
(C) Occasional	4	6	11	18
(D) Remote	8	10	14	19
(E) Improbable	12	15	17	20

Hazard Risk Index	Suggested Criteria
1–5	Unacceptable
6–9	Undesirable (project management decision required)
10–17	Acceptable with review by project management
18–20	Acceptable without review

Source: (Wiggins, 1985, pp. 86–87)

*Definitions of descriptive words may have to be modified based on quantity involved.

†The size of the fleet or inventory should be defined.

only in the sense that they use the roman numerals I, II, etc. No attempt is made to quantify probabilities. This reflects NASA's pervasive distrust of subjective numerical representations of uncertainty. Although numerical risk assessment including quantification of accident probabilities was in fact pioneered in the aerospace industry, NASA dropped it early on.

The published reason for abandoning quantitative techniques was that low numerical assessments of accident probability do not guarantee safety. Accordingly, NASA has always been suspicious of "absolute reliability numbers." A recent report describing the NASA safety program, contracted by the European Space Agency and prepared by American aerospace experts, puts the matter this way: "... the problem with quantifying risk assessment is that when managers are given numbers, the numbers are treated as absolute judgments, regardless of warnings against doing so. These numbers are then taken as fact, instead of what they really are: subjective evaluations of hazard level and probability" (Wiggins, 1985, p. 85).

This type of statement recurs frequently in the public expositions of NASA safety policy and are sharply at odds with experiences reported by Colglazier and Weatherwax cited at the beginning of this chapter. Here we do not see managers treating risk assessments as absolute numbers, alas. Instead, they relied on their own judgment, which, in hindsight, must be regarded as tragically self-serving.

In the corridors of the aerospace world there are loud and persistent rumors to the effect that the primary motive for abandoning quantitative risk assessment in the U.S. aerospace program was not distrust in overoptimistic reliability assessments. Rather initial estimates of catastrophic failure probabilities were so high that their publication would have threatened the political viability of the entire space program. For example, a General Electric "full numerical probabilistic risk assessment" on the likelihood of successfully landing a man on the moon indicated that the chance of success was "less than 5%." When the NASA administrator was presented with the results, he "felt that the numbers could do irreparable harm, and disbanded the effort" (Bell and Esch, 1989).

As a result of the extensive investigation following the shuttle accident, there are strong signs that NASA will make increasing use of quantitative risk assessment in the future.

MILITARY INTELLIGENCE

On the basis of articles in journals it is impossible to form a picture of the way in which the intelligence community deals with uncertainty. However, at least one member of the intelligence community, the Defense Intelligence Agency/Directorate of Estimates (DIA/DE), has contracted extensive research on this problem and has made the resulting report available to the public (Morris and D'Amore, 1980). This report is rarely referenced in the literature and seems to be largely unknown. This is unfortunate. It gives us a glimpse of an intelligence service actively concerned with problems of dealing with uncertainty at a high level of mathematical sophistication. We shall discuss highlights from this report here.

The background for this research was an "increasing interest in evaluating the quality of DE's work, with particular emphasis on projections of future force levels for the USSR" (Morris and D'Amore, 1980, pp. 1–4). Soviet force levels had been underestimated in the late 1970s. Reading between the lines, one might infer that failure to predict the fall of the Shah of Iran also figured among the motives for this research.

Strategic planning depends critically on the purported intentions and capabilities of a perceived enemy. Needless to say, assessments of another countries' intelligence work speaks of "the enemy," as such work presupposes a malevolent opponent. Science does not recognize enemies, hence "enemy" will be replaced with "another country." This sometimes makes for awkward constructions.

Intentions and future force levels are highly uncertain. The data on which these assessments are based are often of dubious reliability (e.g., testimony of defectors, reports from informants, high-altitude photographs). How are "consumers" of intelligence reports (as they are called) to be made aware of the estimator's uncertainty, so as to take this into account in making decisions?

In the 1960s, national intelligence estimates and defense intelligence estimates contained phrases like "it is likely that," "it is unlikely that," it is probable that," etc. In 1976 more precise forms for expressing uncertainty were introduced by the DE, to wit: "there is a 60% probability that." In some cases these estimates were accompanied by a colored sheet of paper on which was printed:

> Numeric forms are used to convey to the reader this degree of probability more precisely than is possible in the traditional verbal form. Our confidence in the supporting evidence is taken into account in making these quantifications.... All efforts at quantifying estimates are highly subjective, however, and should be treated with reserve." (Morris and D'Amore, 1980, pp. 2–3)

Even after the numeric forms had been introduced, qualitative expressions were given alongside numeric estimates. Some years later, the practice of using a "high," "low," and "best" estimate was introduced. The true value was said to fall between the high and low values with 75% probability.

Estimators were unsatisfied with this system for two principal reasons:

1. The numbers were indeed "highly subjective," and there was no clearly defined method for obtaining them.
2. The numbers were generally ignored by the intelligence consumers. Consumers tend to take the best estimates (or sometimes the highest values) as if it were certain, and disregard the uncertainty attached to them.

The DE's response has been to contract research for improving the quality (and hence usefulness) of numerical assessments of uncertainty. The resulting program surpasses any other described in the public literature in both scale and sophistication. We shall look briefly at the communication and evaluation of uncertainty assessments, as envisioned in this program.

Communication

The problem of communicating uncertainty to consumers in such a way that they make proper use of it, is formidable, and several systems have been implemented and subsequently discarded. One widely used system, combining so-called reliability-accuracy ratings, is reproduced as Table 2.3. Research (Samet, 1975) has indicated that this system was not adequate. In evaluating reports, intelligence officers were much more heavily influenced by the assessed accuracy of reports and did not take account of the reliability of the source. Moreover, there was a wide disparity in the absolute interpretation of the qualitative ratings.

Table 2.3 Reliability and Accuracy Ratings

Source Reliability		Information Accuracy	
A	Completely reliable	1	Confirmed
B	Usually reliable	2	Probably true
C	Fairly reliable	3	Possibly true
D	Not usually reliable	4	Doubtfully true
E	Unreliable	5	Improbable
F	Reliability cannot be judged	6	Accuracy cannot be judged

Source: (Morris, J. M. and D'Amore, R. J., 1980, pp. 5–23)

The DE had made use of so-called Kent charts, which provide a quantitative interpretation of natural language expressions of uncertainty. A Kent chart is shown in Table 2.4.

Evaluation

Kent charts have been abandoned in favor of direct numeric estimations of probability. Such estimations are not without problems, however. The most important question regarding numerical probabilities is that of "validity" or "calibration" (as it will be called here). A subjective probability assessor is said to be

Table 2.4 A Kent Chart for Estimating Terms and Degrees of Probability

This table explains the terms most frequently used to describe the range of likelihood in the key judgment of DIA estimates.

Order of Likelihood	Synonyms	Chances in 10	Percent
Near certainty	Virtually (almost) certain, we are convinced,	9	99
	highly probable, highly likely		90
Probable	Likely	8	60
	We believe	7	
	We estimate	6	
	Chances are good		
	It is probable that		
Even chance	Chances are slightly better than even	5	40
	Chances are about even	4	
	Chances are slightly less than even		
Improbable	Probably not	3	10
	Unlikely	2	
	We believe . . . not		
Near impossibility	Almost impossible	1	1
	Only a slight chance		
	Highly doubtful		

Note: Words such as "perhaps," "may," and "might" will be used to describe situations in the lower ranges of likelihood. The word "possible," when used without further modification, will generally be used only when a judgment is important but cannot be given an order of likelihood with any degree of precision.
Source: (Morris, J. M. and D'Amore, R. J., 1980, p. 5–21)

"well calibrated" if statements with an assessed probability of $X\%$ turn out to be true $X\%$ of the time, for all X. We shall study the notion of calibration extensively in later chapters, from both empirical and theoretical standpoints. Here, we are only interested in how the DE has coped with this question.

A three-pronged effort to improve calibration has been undertaken, involving

- Debiasing
- Use of proper scoring rules
- Use of feedback and systematic evaluation

Drawing on well-known psychometric literature, which will be discussed in later chapters, a number of biases are identified that may tend to impair the calibration of subjective probability assessments. A training manual is provided to estimators advising them of the most common biases, and how they may best be avoided. Procedures for eliciting subjective probabilities are also described. These procedures are designed to make the estimator aware of the meaning of subjective probability during the estimation process. Again, these procedures are all quite standard and will be described later. Computer-supported interactive elicitation incorporating the above techniques are also implemented.

Whereas debiasing efforts are directed toward the process of eliciting subjective probabilities, it is equally important to attend to the way in which probability assessments are rewarded or scored. A rule for rewarding a probability assessor on the basis of a later observed outcome is called a scoring rule. A scoring rule is called "proper" if it rewards assessors for giving their true opinion (a mathematically precise definition is given later).

Most intuitively appealing scoring rules are highly improper and encourage bad probability assessments. Two such rules are discussed in some detail. The "hit or miss" rule applies to projected events and simply counts the number of projections that have come out, and divides this by the total number of projections. Applying such a rule would lead to evaluations like "70% of the projections from estimator A have come out." It is fairly obvious that such a scoring rule encourages estimators to "hedge," that is, to give projections that are more cautious than they really think appropriate.

A second improper scoring rule is called the "direct rule." It rewards a probability assessment like "A will happen with probability 60%" by giving the score 60 if A happens and the score 40 otherwise. This rule is discussed in Part II, and will be seen to encourage overconfident assessments.

Scoring rules are difficult to apply in intelligence estimates due to the fact that many estimates concern events 10 or 20 years hence. However, the training manual warns estimators that they should not let themselves be judged (either by themselves or others) via improper scoring rules.

The last initiative toward improving the quality of probability assessments concerns an elaborate computerized data bank of intelligence estimates called the "institutional memory" (IM). The most important functions of the IM are to support the elicitation process and to provide feedback on past performance. This feedback is not only essential for helping estimators give better estimates, it also is said to be useful in informing the consumer of the quality of the DE's estimates.

On the basis of the elaborate description in Morris and D'Amore (1980), the

IM appears to be a very sophisticated system for handling uncertainty, probably the most advanced system of its kind. It also contains features for aiding a decision maker in combining intelligence reports (with attendant uncertainties) from different sources.

To sum up this discussion, we affirm that the intelligence community is confronted with uncertainty on a large scale and has been forced to confront virtually all the problems known from the literature. The response of the DIA/DE has been to try to improve the numerical estimations of uncertainty and to support the consumers in making the best possible use of these numerical assessments. It would be very interesting to have more recent information on the sucesss of these initiatives.

PROBABILISTIC RISK ANALYSIS

Probabilistic risk analysis was the first "hard science" to introduce subjective probabilities on a large scale, and it has been at the forefront of developments in using expert opinion. For this reason, our treatment of this area will be somewhat more extended. After examining the historical background we focus on four problems that have emerged in connection with "subjective data," namely, the spread or divergence of expert opinion, the dependencies between the opinions of different experts, the reproducibility of the results of risk studies, and the calibration of expert probability assessments.

Historical Background

Probabilistic risk assessment as such is not new. Covello and Mumpower (1985) identify explicit risk assessment in the writings of Arnobius the Elder in the fourth century. More recently, the American Atomic Energy Commission [AEC, the forerunner of the Nuclear Regulatory Commission (NRC)] pursued a philosophy of risk assessment through the 1950s based on the "maximum credible accident." Because credible accidents were covered by plant design, residual risk was estimated by studying the hypothetical consequences of "incredible accidents." An example of such a study is WASH-740 (U.S. AEC, 1957), released in 1957. It focused on three scenarios of radioactive releases from a 200 Megawatt-electric-power nuclear power plant operating 30 miles from a large population center. Regarding the probability of such releases the study concluded that "no one knows now or will ever know the exact magnitude of this low probability."

Design improvements introduced as a result of WASH-740 were intended to reduce the probability of a catastrophic release of the reactor core inventory. Such improvements could have no visible impact on the risk as studied by the WASH-740 methodology. On the other hand, plans were being drawn for reactors in the 1000-MWe range located close to population centers, and these developments would certainly have a negative impact on the consequences of the incredible accident.

The desire to quantify and evaluate the effects of these improvements led to the introduction of *probabilistic* risk analysis (PRA). As mentioned previously, the basic methods of probabilistic risk assessment originated in the aerospace program in the 1960s. The first full-scale application of these methods, including an extensive analysis of the accident consequences, was undertaken in the Reactor Safety Study WASH-1400 published by the American Nuclear Regulatory Commission (U.S. NRC, 1975). This study is rightly considered to be the first modern PRA.

The reception of the Reactor Safety Study in the scientific community may best be described as "turbulent." The American Physical Society (1975) conducted an extensive review of the first draft of the Reactor Safety Study. In the letter of transmittal attached to their report, physicists Wolfgang Panofsky, Victor Weisskopf, and Hans Bethe concluded, among other things, that the calculation methods were "fairly unsatisfactory," that the emergency core cooling system is unpredictable and that relevant physical processes "which could interfere with its functioning have not been adequately analyzed," and that "the consequences of an accident involving major radioactive release have been underestimated as to casualties by an order of magnitude." The final draft of the Reactor Safety Study was extensively reviewed by, among others, the Environmental Protection Agency (1976) and the Union of Concerned Scientists (1977).

In 1977 the U.S. Congress passed a bill creating a special "review panel" of external reactor safety experts to review the "achievements and limitations" of the Reactor Safety Study. The panel was led by Prof. Harold Lewis and their report is known as the Lewis report (Lewis et al., 1979). While the Lewis report recognized the basic validity of the PRA methodology and expressed appreciation for the pioneering effort put into the Reactor Safety Study, they also uncovered many deficiencies in the treatment of probabilities. They were led to conclude that the uncertainty bands claimed for the conclusions in the Reactor Safety Study were "greatly understated."

Significantly, the Lewis report explicitly supported the use of subjective probabilities in the Reactor Safety Study:

> The Reactor Safety Study (RSS) had to use subjective probabilities in many places. Without these, RSS could draw no quantitative conclusions regarding failure probabilities at all. The question is raised whether, since subjective probabilities are just someone's opinion, this has a substantial impact on the validity of the RSS conclusions.
>
> It is our view that the use of subjective probabilities is necessary and appropriate, and provides a reasonable input to the RSS probability calculations. But their use must be clearly identified, and their limits of validity must be defined. (Lewis et al., 1979, p. 8)

In January 1979 the NRC distanced itself from the results of the Reactor Safety Study:

> In particular, in light of the Review Group conclusions on accident probabilities, the Commission does not regard as reliable the Reactor Safety Study's numerical estimate of the overall risk of reactor accident." (U.S. NRC, 1979)

The future of PRA after the NRC's announcement of 1979 did not look bright. However, the dramatic events of March 1979 served to change that. In March 1979 the Three Mile Island-2 nuclear generating unit suffered a severe core damage accident. Subsequent study of the accident revealed that the accident sequence had been predicted by the Reactor Safety Study, although the Reactor Safety Study had conservatively assumed that a "degraded" core would melt entirely. The probabilities associated with that sequence, particularly those concerning human error, do not appear realistic in hindsight.

Two influential independent analyses of the Three Mile Island accident, the Report of the President's Commission on the Accident at Three Mile Island" (Kemeny et al., 1979) and the Rogovin Report (Rogovin and Frampton, 1980) recommended that greater use be made of probabilistic analyses in assessing nuclear plant risks.

Shortly thereafter a new generation of PRAs appeared in which some of the methodological defects of the Reactor Safety Study were avoided. The Zion Probabilistic Safety Study in particular has served as a model for subsequent PRAs, and its methodology has been canonized in a series of articles in the Journal of *Risk Analysis* (see, e.g., Kaplan and Garrick, 1981). In 1983 the U.S. NRC released the *PRA Procedures Guide*, which shored up and standardized much of the risk assessment methodology. The Zion Probabilistic Safety Study uses subjective probabilities. It is interesting to note that in the first article in the first issue of the first journal devoted to risk analysis, we find the following definition of "probability":

> "...'probability' as we shall use it is a numerical measure of a state of knowledge, a degree of belief, a state of confidence."   (Kaplan and Garrick, 1981, p. 17)

Since the Lewis report, expert opinion has been used in a structured form as a source of data in several large studies. Among these are the risk study of the fast breeder reactor at Kalkar, Germany (Hofer, Javeri, and Loffler, 1985); the studies of seismic risk by Okrent (1975), and by Bernreuter et al. (1984) and a study of fire hazards in nuclear power plants (Sui and Apostolakis, 1985). Finally, the Draft Reactor Risk Reference Document (NUREG-1150, 1987) makes massive (and poorly documented) use of expert opinion to assess the risks of five nuclear power plants. The final draft is still under review at this writing, but a report describing a more concerted effort with regard to expert judgment is now available (Wheeler et al., 1989). The improved documentation reveals a laudable, though time consuming and costly, attempt to counter the various problems with expert opinion via an intensive elicitation process. This approach is extended in Bonano, Hora, Keeney, and von Winterfeldt (1990), with particular attention to potential applications in the field of radioactive waste management. In the United Kingdom, Her Majesty's Inspectorate of Pollution has sponsored similar exploratory research in applying expert opinion to the risk analysis of radioactive waste repositories (Dalrymple and Willows, 1990). Characteristic of these approaches is a heavy investment on the elicitation side, with no evaluation of performance and no proposal for combination other than simple arithmetic averaging. The possible cost ineffectiveness of this approach is underscored in Woo (1990). Perhaps the most prominent area for applying expert opinion in risk analysis is the assessment of human error probabilities. The main reference in this area is the *Handbook of Human Reliability* (Swain and Guttmann, 1983). The *PRA Procedures Guide* (U.S.N.R.C., 1983, chap. 4) and Humphreys (1988) give a good review of the literature and methods.

There is no comprehensive review of expert judgment applications. However, an entire issue of *Nuclear Engineering and Design* (vol. 93, 1986) was devoted to the role of data and judgment in risk analysis. A recent report by Mosleh, Bier, and Apostolakis (1987) gives an overview of some of these studies, and a compressed version of this report appeared in *Reliability Engineering and System Safety* (Vol. 20, no. 1, December 1988).

The best way to appreciate the problems of substituting experts' degrees of belief for data is to look hard at some examples. We shall look at subjective data from four viewpoints, namely the spread of expert opinion, the dependency between experts, the reproducibility of the results, and finally the calibration of the results. We shall not review the above-mentioned literature. This chapter is devoted to examples. Later chapters return to methodological issues raised in this chapter.

Subjective Data: Spread

The probability of a core melt in a nuclear reactor is a typical example of a quantity that cannot be computed without substantial use of subjective probabilities. In calculating this probability the Reactor Safety Study had to estimate the average yearly probability of a core melt due to an earthquake. The value used in that report was 4.7E-7 (we use scientific notation, E-7 means 10^{-7}). A different study (Okrent, 1977) estimated the same probability as 8E-5. Yet another study (Lee, Okrent, and Apostolakis, 1978) gave estimates of 1.77E-4 per reactor year for reactors having no design errors and 2.32E-3 per reactor year for systems having a "maximal number of design errors and reduced original safety factors." We see here sampling of estimates which span four orders of magnitude.

Expert probability assessments in risk analysis typically show extremely wide spreads. A good place to study this phenomenon is the Reactor Safety Study. This

Table 2.5 Estimates of Failure Probability per Section-Hour of High-Quality Steel Pipe of Diameter $\geqslant 7.6$ cm

Source	Value
1. LMEC	5×10^{-6}
2. Holmes	1×10^{-6}
3. G.E.	7×10^{-8}
4. Shopsky	1×10^{-8}
5. IEEE, a	1×10^{-8}
6. IEE, b	1×10^{-8}
7. NRTS Idaho	1×10^{-8}
8. Otway	6×10^{-9}
9. Davies	3×10^{-9}
10. SRS	2×10^{-9}
11. IKWS Germany	2×10^{-10}
12. Collins	1×10^{-10}
13. React. Incd.	1×10^{-10}
RSS estimate	1×10^{-10}
90% confidence bounds	$3 \times 10^{-9} - 3 \times 10^{-12}$

Source: U.S. NRC, 1975, p. III-7.

study needed a large number of component failure probabilities that could not be reliably estimated from available data. Thirty experts (the experts were sometimes consulting firms or data banks) were asked to estimate failure probabilities for 60 components. Not every expert chosen estimated each component, but the total matrix gives a good impression of the divergence of expert opinion.

The results for one component, the failure probability of high-quality steel pipe of diameter at least 7.6 cm per section-hour, are given in Table 2.5 (a section is a piece of pipe about 10 meters long). The thirteen responses range from 5E-6 to E-10. The Reactor Safety Study used the value E-10 in its calculations, with 90% confidence bounds of 3E-9 to 3E-12. Eight of the thirteen responses fall above the upper confidence bound. Calling the spread of expert opinion for a given component the ratio of the largest to the smallest estimate, the average spread over the 60 components was 167,820. If one outlier (with a spread of E7) is omitted, then the average of the remaining components spreads was 1173. By comparison, the average ratio between the upper and lower confidence bounds used by the Reactor Safety study for these 60 components was 126. Hence we see that the confidence bounds given in the study tend to be smaller than the spread of expert opinion. We shall return to this feature in another context shortly.

Subjective Data: Dependence

Another interesting fact emerging from the data matrix of the Reactor Safety Study is that the estimates of the experts were not independent. That is, if an expert was a pessimist with respect to one component, there was a substantial tendency for him to be a pessimist on other components as well. Let us call an expert a pessimist for a given component if his estimate lies strictly above the median estimate for that component. Imagine that we have the whole data matrix laid out before us, with each row corresponding to estimates for a given component, and each column corresponding to estimates of a given expert. We now replace each numeric estimate with a "P" (for pessimist) if that estimate is strictly above the median value for the corresponding components, and with an "O" (for optimist) otherwise.

We shall say that an expert is "rank independent" if his responses show no tendency to cluster either toward optimism nor pessimism. A somewhat crude test for rank independence can be conducted in the following way (see Cooke, 1986b). We first restrict attention to those components that were estimated by at least four experts. The number of rows is hereby reduced from 60 to 39. We then regard the P's and O's in each column as the result of a coin-tossing experiment. Owing to ties at the median value, the frequency of P's over the (reduced) matrix is not 50% but 36.4%. Hence, if an expert showed no tendency to cluster toward pessimism or optimism, his responses could be (crudely) modeled with a coin-tossing experiment with a probability of 36.4% of "heads."

If an expert estimated failure probabilities for at least 14 of the 39 components, then the distribution of outcomes in the corresponding coin-tossing experiment can be approximated with a normal distribution. The results for the nine experts who estimated at least 14 components are shown in Table 2.6. The level at which the hypothesis of rank independence would be rejected for a given expert is the probability of observing at least as many P's (if the expert tends toward pessimism)

Table 2.6 Results of a Rank-Independence Test for Expert Estimates in the Reactor Safety Study

Expert	Number of Estimates	Percent Pessimistic (expected value: 36.4)	Rank Independence Rejected at Level, %
U.S. Nucl. operating experience	33	30	—
AVCO	23	30	—
Farada	34	53	2.3
SRS	27	56	2
Holmes HN-190	21	19	5
Shops U.S. nucl.	21	43	—
Bourne U.K.	16	62	1.5
Underakes German	14	50	—
NRTS Idaho	27	19	0.02

Results of a rough rank-independence test for expert estimates in the Reactor Safety Study for items estimated by at least four experts. Owing to ties at the median values, the relative frequency of pessimism, that is, answers strictly above the median value, was 36.4%. The results concern only those experts who estimated at least 14 of the 39 items.
Source: (Cooke, 1986b)

or at least as many O's (if he tends toward optimism) in the associated coin-tossing experiment. We see that rank independence would be rejected at the 5% level for five of the nine experts.

Nuclear energy is not the only area in which experts disagree dramatically. Shooman and Sinkar (1977) conducted a study into the risks associated with grounded-type electric lawn mowers. They collected opinions from 12 experts regarding the probabilities for events which might lead to an electrical accident, with the intention of combining these probabilities to derive the probability for a shock accident. A description of the events estimated together with the results are reproduced as Table 2.7. One of the matrices shows a cluster analysis similar to that described above. The degree of clustering is quite extreme.

Subjective Data: Reproducibility

In the last few years there have been some studies that attempt to determine whether different experts applying the same risk assessment methodology to the same problem do indeed get similar results. Such studies are called *bench mark studies*. One such study was recently conducted by the Joint Research Centre at Ispra, Italy (Amendola, 1986). In the Ispra bench mark study, the new Paluel Auxiliary feedwater system was chosen as a reference system. Ten teams from different European countries were formed to estimate, independently of each other, the probability that this system would not fulfill its design requirements.

The estimates of the "mission failure probability" were collected in four stages (these illustrate a "nominal group technique").

1. In the first stage, the teams carried out a first probabilistic analysis and sent the results to the project leaders. This is the "blind evaluation." Without discussing the results, the teams were then brought together to compare their analyses qualitatively.
2. After this comparison the teams performed an intermediate "fault tree

analysis." A fault tree analysis decomposes the event of interest (the "top event") into combinations of "component events." The probability of the top event is then computed as a function of the probabilities of the component events (and their interactions). The spread in results after this stage were rather large, and the teams were unable to agree on a common fault tree analysis.

3. In the third stage, the project leaders wanted to separate the effects of different fault tree modeling from the effects of different failure data. The teams agreed to proceed with a common fault tree, although this fault tree did not represent a consensus. This led to the "common fault tree with participants' data" results.

4. In the final stage, the goal was to determine whether different methods of calculation also played a significant role. These are the "common fault tree and common data" results. The results are shown in Table 2.8.

A second important bench mark study concerned human reliability (Brune, Weinstein, and Fitzwater, 1983). As one might guess, human reliability is one of the most important and one of the most recalcitrant areas in risk analysis. Human error is implicated in anywhere from 30% to 80% of serious accidents with sophisticated technical systems (Levine and Rasmussen, 1984). Techniques for quantifying human error probabilities in risk analysis were developed at the Sandia National Laboratories and published in the *Handbook of Human Reliability Analysis* (Swain and Guttmann, 1983, a draft of this report had been circulated earlier). A bench mark study was designed to determine to what extent these techniques would lead to similar results when applied by different human reliability experts. A workbook was written describing six human reliability problems. The first two problems were worked out using *Handbook* methods, by way of example. The remaining four problems (some including a and b parts) were worked out by the experts. The *Handbook* contains estimates for human error probabilities for individual tasks, and participants were encouraged to use these estimates. The results are shown in Table 2.9. Also given are the "normative" answers provided by the authors of the *Handbook*, and the 5% and 95% confidence bounds associated with these answers. It is to be noted that about 38% of the experts' responses fall above the upper confidence bound associated with the normative answers.

From the analysis of the experts' responses it emerged that the extreme spread of opinion was not caused by different values for individual failure probabilities (indeed, these were taken from the *Handbook*). The spread was caused by the fact that the experts identified different tasks that might be performed incorrectly. In other words, they analyzed the human reliability problems differently. Note that the authors of the bench mark study suggest that Monte Carlo simulation would produce wider uncertainty bounds, thus causing more peer responses to fall within these bounds. In simulation studies performed at the Delft University of Technology it emerged that this is not in fact the case (see Cooke and Waij, 1986a).

A recent benchmark exercise by the Joint Research Center involving 15 assessment teams studied the reproducibility of results of human reliability modeling, and assessment. The design of the exercise was roughly similar to that of Amendola (1986), and the results support the conclusions drawn above (Poucet, 1989).

Table 2.7 Probability Estimates for Events Leading to Electric Shock in Electric Lawn Mowers

Definition of Hazard Events

E_g: The event that a person is grounded or has a low resistance while operating the lawn mower. (Low resistance is that resistance that is sufficiently small to cause an electric shock for the specified operating voltage.)

E_0: The event that the wire plug of the mower is in while the mower is being repaired or adjusted.

E_1: The event that a person touches a "live" part of the lawn mower. The live part considered here is a part that is normally live when a mower is connected to the supply, and not a part that has become live due to a fault.

E_2: The event that a person touches cut (E_3) or damaged (E_4) cord when the mower is connected to the supply.

E_3: The event that the cord is cut due to some sharp object either while the mower is in operator or stored.

E_4: The event insulation of the cord does not work (damage due to environment, prolonged and/or abusive use, etc.).

E_5: The event that a person touches the conductive part of the body (includes handle, blade, etc.) when the mower is connected to the supply.

E_6: The event grounding does not work. This event comprises one or more of the following:
   (1) Consumer does not have a grounded outlet box.
   (2) Ground wire is either broken or disconnected at the grounding terminal.
   (3) High-impedance grounding circuit exists.

Estimated Probability of the Event
First Round Grounded Type

| | Expert | | | | | | | | | | | |
Event	1	2	3	4	5	6	7	8	9	10	11	12
E_g	0.5	0.5	0.75	0.5	0.5	0.005 1.0	0.9	0.005	0.005	0.3	0.1	0.878
E_0	0.25	0.1	0.1	0.01	0.0001	0.00028	0.66	0.8	0.0001	0.0045	0.0225	0.001
E_1	0.05	0.01	0.001	0.001	0.5	0	0.1	0.2	0.0001	0.0001	0.00028	0.2
E_2	1.0	1.0	0.9	0.3	0.001	0.007	1.0	1.0	0.005	0.0055	0.01	0.0078
E_3	0.1	0.05	0.01	0.01	0.05	0.0007	0.1	0.7	0.0001	0.0001	0.0125	0.01
E_4	0.05	0.1	0.005	0.05	0.002	0.00007	0.05	0.05	0.0001	0.00015	0.01	0.022
E_5	1.0	0.7	0.8	1.0	0.001	0.007	1.0	1.0	0.005	0.006	1.0	0.6
E_6	0.35	0.2	0.2	0.2	0.65	0.0035	0.6	0.75	0.005	0.0015	0.425	
E_7	0.05	0.005	0.001	0.1	0.00002	0.000003	0.05	0.05	0.0001	0.000002	0.002	0.001
E_8	0.05	0.001	0.01	0.05	0.001	0.000003	0.02	0.05	0.0001	0.000002	0.002	0.00035

Probability Summary for First Round

Event	Lowest Estimate	Highest Estimate	Median Expert
E_g	0.005	1.0	0.5
E_0	0.0001	0.878	0.095
E_1	0	0.5	0.001
E_2	0.001	1.0	0.5
E_3	0.0001	0.7	0.0113
E_4	0.00007	0.1	0.01
E_5	0.0001	1.0	0.75
E_6	0.0015	0.75	0.275
E_7	0.000002	0.1	0.0015
E_8	0.000002	0.05	0.00015

Event	E_g	E_0	E_1	E_2	E_3	E_4	E_5	E_6	E_7	E_8	P/10
1	P	P	P	P	P	P	P	P	P	P	1
2	P	P	P	P	P	P	P	P	P	P	1
3	P	P	O	P	P	P	P	P	P	P	0.9
4	P	P	O	P	P	P	P	P	P	P	0.9
5	P	O	P	O	P	O	O	P	O	O	0.5
6	O	O	O	O	P	O	O	O	O	O	0.0
7	P	P	P	P	P	P	P	P	P	P	1
8	P	P	P	O	O	P	O	O	O	P	1
9	O	O	O	O	O	O	O	O	O	O	0.0
10	O	O	O	O	O	O	O	O	O	O	0.0
11	P	P	P	P	P	P	P	P	P	P	0.8
12	P	P	O	O	P	P	O	P	P	P	0.7

Source: Shooman and Sinkar (1977).

Table 2.8 Results of the Ispra Bench Mark Study

Spread of the probabilistic results of the JRC-RBE

Mission failure probability as evaluated on the basis of the common fault tree and data
$P_F \approx 1.4 \times 10^{-3}$

Point Values of the Different Participants	First "Blind" Evaluation	Intermediate Fault tree after Comparison of Qualitative Analysis	Common Fault tree with Participants Data	Common Fault Tree with Common Data
Dispersion range*	8×10^{-4}	7×10^{-4}	1.4×10^{-3}	$\sim 1.4 \times 10^{-3}$
	2×10^{-2}	2.5×10^{-2}	1.3×10^{-2}	1.4×10^{-3}
Ratio* max/min	25	36	9	~ 1

*By excluding some extreme values according to certain evaluation criteria; such a procedure may give rise to somewhat different tables. All probability values are conditional on the occurrence of the initiating event.
Source: Amendola (1986)

Subjective Data: Calibration

Of course, the question whether experts agree is quite distinct from the question whether the experts' assessments are good. The latter question is related to the notion of calibration, which has surfaced in the preceding discussions. To recall, calibration is concerned with the extent to which the assessed probabilities agree with observed relative frequencies.

It is difficult to check for calibration when the events whose probabilities are assessed are rare. Risk analysis is typically concerned with rare events. However, there are many subjective probabilities that go into the calculation of the probability for a rare event. Although calibration for rare events is ruled out, it is sometimes possible to calibrate assessments for ingredient events.

In fact, data are available that enable us to calibrate some of the subjective probability assessments in the Reactor Safety Study. These data emerge from an ongoing project at the Oak Ridge National Laboratory (Minarick and Kukielka, 1982, Cottrell and Minarick, 1984) directed toward evaluating operating experience at nuclear installations. Although there have been no complete nuclear meltdowns in the United States, there have been a number of incidents involving events whose frequencies were estimated in the Reactor Safety Study. For seven such events the frequencies estimated from operating experience by the Oak Ridge project can be compared with the 90% confidence bounds for these frequencies given in the Reactor Safety Study. The results are presented in Table 2.10. It will be observed that *all* the values from operating experience fall outside the Reactor Safety Study confidence bounds. If the Oak Ridge values are taken as correct, then they reveal two sorts of biases in the Reactor Safety Study estimates. There is a "location" or "first moment" bias causing the estimate to be too low (five of the seven estimates fall below the Oak Ridge values), and a "scale" or "overconfidence" bias, causing the confidence bounds to be too narrow.

Table 2.9 Distribution of Probabilities of Failure (P_f) Calculated by Peers for Human Reliability Problems

	Problem						
	3a	3b$_l$	3b$_u$	4a	5a	5b	6a
	0.594	0.4905	0.930	▶0.543	0.995	0.999	0.990
	0.198	0.1875	0.824	0.131	0.830	0.376	0.500
	0.172	0.0440	0.675	0.130	0.536	0.914	0.497
	0.124	0.0440	0.824	0.114	0.441	0.721	0.126
	0.121	0.0364	0.534	0.100	0.280	0.716	0.081
	0.082	0.0320	0.480	0.082	0.241	0.670	
	0.058		0.317	0.080			0.0785
Third quartile		0.0214			0.200 —	0.570	
	0.028			0.281	0.080		0.070
	▶0.025	0.0140			0.180	0.480	
		▶ ·	0.262	0.080			
	0.024	0.0115	▶ ·		0.110	0.455	0.070
			0.224	0.048			
	0.023	0.0104			0.106	0.430	0.070
			0.200	0.042	▶ ·		
	0.020	0.0100	0.195	0.040	0.085	0.306	0.080
	0.018	0.0100	0.196	0.040	0.080	0.224	0.055
	0.016	0.0078		▷	0.085	0.224	0.054
			0.128	0.038		·	
Second quartile	0.016 —	0.0075			0.080 ▶	0.150 —	0.040 —
			0.090	0.037			
	0.014	0.0070			0.050	0.146	0.026
			0.080	0.035			
	0.010	0.0064			0.049	0.120	0.020
			0.078	0.035			
	0.008	0.0040			0.040	0.120	0.017
			0.078	0.034			
	0.008	0.0040			0.037	0.098	0.0079
			0.073	0.032			
	0.008	0.0040			0.033	0.074	0.008
			0.054	0.032			
	0.008	0.0040			0.031	0.064	
	0.008		▷0.050	0.032			0.005
First quartile		0.0040			0.022 —	0.050	
	0.008	·	0.049	0.031			0.001
	0.008	0.0030	0.040	0.031	▷0.020	0.044	
	0.006	0.0030	0.037	0.030	0.020	0.040	0.0005
	0.006	▷0.0025	0.030	0.025	0.020	▷0.032	0.0004
	▷0.005	0.0019	0.030	0.023	▶0.010	0.026	0.0001
		▶ ·		▶ ·		▶ ·	
	0.004	0.0010	0.010	▶0.020	0.006	0.008	·0.000057
	0.004	0.0000	0.005	·0.012	0.004	0.008	▷0.00001
	▶ ·						▶ ·
	N = 29	N = 27	N = 28	N = 28	N = 27	N = 27	N = 25
Sandia Solution							
P_f(▷)	0.005	0.0025	0.05	0.039	0.02	0.032	0.000011
Lower limit (·)	0.0025	0.00125	0.025	0.0195	0.01	0.016	0.0000055
*Upper limit (·)	0.025	0.0125	0.25	0.195	0.1	0.15	0.000055

*These limits underestimate by a factor of two or more the usual uncertainty bounds that would be calculated by a Monte Carlo procedure. Were the wider bounds used, considerably more of the peers' responses would lie within the usual uncertainty bounds of the Sandia solution.

Source: Brune Weinstein Fitzwater.

Table 2.10 Comparison of Estimates of Failure Frequencies
from Operating Experience with 90% Confidence Bounds from
the Reactor Safety Study

Event	Oak Ridge Value	RSS Confidence Bounds	
PWR			
Small LOCA	8.3E-3	3E-4	3E-3*
AFW (failure/demand)	1.1E-3	7E-6	3E-4
HIP (failure/demand)	1.3E-3	4.4E-3	2.7E-2
LTCC (failure/demand)	1.2E-3	4.4E-3	3.2E-2
BWR			
Small LOCA	2.1E-2	3E-4	3E-3*
ADS (failure/demand)	2.7E-2	3.3E-3	7.5E-3
HPCI (failure/demand)	5.7E-2	3E-3	5.5E-2†

*These bands are derived from the bands for coremelt resulting from a small LOCA. RSS
says that the latter uncertainty is principally a result of uncertainty with respect to the
initiating event (i.e., the small LOCA).

†The Oak Ridge value does not include unavailability due to test and maintenance. The
RSS median value for this contribution is 1.3E-2 per demand. The RSS uncertainty bands
including the test and maintenance contribution are 6.8E-2 1.4E-1. The above bands are
derived by assuming the largest possible "error factor" consistent with the last mentioned
upper bound, under the assumption of independence.

Abbreviations:

PWR	Pressurized water reactor
BWR	Boiling water reactor
LOCA	Loss of coolant accident
AFW	Auxiliary feedwater system
HPI	High-pressure injection
LTCC	Long-term core cooling
ADS	Automatic depressurization system
HPCI	High-pressure coolant injection

The news is not all bad. Snaith (1981) studied the correlation between
observed and predicted values of some 130 reliability parameters. The predicted
values include both expert assessments and results of analysis. The correlation was
generally good. Figure 2.1 plots the ratio R of the observed to predicted values, as a
function of cumulative frequency. We see that in 64% of the cases; $\frac{1}{2} \leqslant R \leqslant 2$, while
in 93% of the cases $\frac{1}{4} \leqslant R \leqslant 4$.

Unfortunately, the above data say nothing about how certain the experts were
of their assessments, hence it is impossible to determine whether, for example, a
value of $R = 2$ would be considered surprising, for a given expert.

A better perspective on the relation of observed and predicted values is given
by Mosleh, Bier, and Apostolakis (1987). The experts gave distributions for
maintenance times used in the risk assessments of the Seabrook and Midland
nuclear power plants, and these were compared with observed mean values. The
results, shown in Table 2.10, indicate a ratio of observed to predicted mean values
generally in agreement with the results of Snaith's study. However, Mosleh, Bier,
and Apostolakis also looked at the degree of confidence in the predicted results and
compared this with the observed spread in the observed values. Degree of
confidence is indicated by "range factors," where the range factor associated with
the probability distribution of a given quantity is the square root of the ratio of the

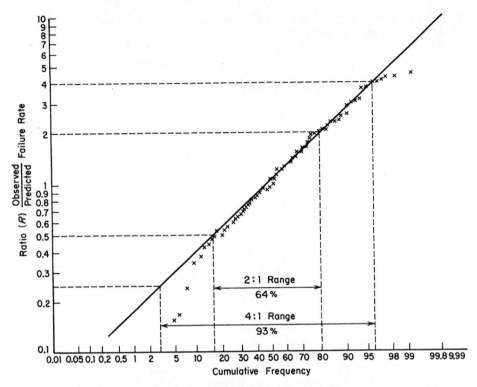

Figure 2.1 Frequency distribution of ratios of observed to predicted reliability parameters. (*Snaith, 1981*)

95th and 5th percentiles of the distribution for this quantity. Hence, if an expert reports of range factor of 3.2, his 95th percentile is a factor 10 larger than his 5th percentile. The range factors associated with the expert distributions are compared with the range factors derived from the observed values. Table 2.11 shows that experts' range factors are consistently too small. If the range factor is increased by a factor of 3, this means the ratio of the 95th to the 5th percentile has been increased by a factor of 10. From Table 2.11 we may conclude that the expert assessments reflect significant overconfidence.

Words of Caution

One must be cautious in drawing general conclusions from examples, as bad news always travels further and faster than good news. In fact, Christensen-Szalanski and Beach (1984) reviewed more than 3500 abstracts on judgment and reasoning. Of the 84 empirical studies, 47 reported poor performance, while 37 reported good performance. However, the Social Sciences Citation Index showed that citations of poor performance outnumbered citations of good performance by a factor of 6.

Regarding the issues of spread, dependence, reproducibility and calibration, the foregoing pretty much capture all that is presently available in the applications literature (new bench mark studies are in progress at Ispra). The conclusion from

Table 2.11 Comparison of Data and Expert Opinion on the Distribution of Component Maintenance Time

Component Type	Technical Specification Time	Data-Based Range Factor	Data-Based Mean	Expert-Estimated Range Factor	Expert-Estimated Mean	Observed/Predicted Range Factor	Observed/Predicted Mean
Pumps	None	22.1	265	6.2	116.0	3.56	2.28
	168 hours	6.2	29	1.8	40.4	3.44	0.72
	72 hours	5.9	11	1.5	20.9	3.93	0.53
	≤24 hours	4.2	7	1.5	10.8	2.80	0.65
Valves	None	26.2	135	6.2	116.0	4.23	1.16
	72 or 168 hours	5.2	19	1.8	40.4	2.89	0.47
	≤24 hours	3.8	4	1.5	20.9	2.53	0.19
Heat exchangers	None	4.6	580	6.2	116.0	0.74	5.03
Other*	None	11.0	39	6.2	116.0	1.77	0.34
	>72 hours	3.0	37	1.8	40.4	1.67	0.92
	48 or 72 hours	7.3	14	1.5	20.9	4.87	0.67
	≤24 hours	5.8	6	1.5	10.8	3.87	0.56

*For example, diesel generators, fans, electrical equipment; also includes heat exchangers with technical specifications.

the above examples is clear and not too encouraging: Expert opinions in probabilistic risk analysis have exhibited extreme spreads, have shown clustering, and have led to results with low reproducibility and poor calibration.

In general, we may affirm that the use of expert opinion to date has been rather ad hoc and bereft of methodological guidance. Indeed, there exists no body of rules for how an analyst should use expert opinion. Different studies have done this in very different ways, and some divergencies in this respect were cited in this chapter. Improvement in this respect is not only urgently needed, but also feasible.

To illustrate the problems created by this methodological vacuum, suppose we are performing a risk analysis and we are confronted with a spread of expert opinion such as that shown in Table 2.5. What are we to do? Are we to take the average value? some weighted average? the median value? the most popular value? Should we try to get the experts to agree more? If so, can we do this without perturbing the result with various more or less overt forms of social pressure? Indeed, such pressures are always present, and there is little reason to believe that they systematically push a group of experts in the direction of the truth.

The presence of clustering in pessimists and optimists makes these problems only more acute. A difference of opinion at the component level in a risk analysis might not be so serious if no one particular component was especially important for the end result. One might hope that the differences of opinion would interfere destructively (i.e., everyone is sometimes a pessimist and sometimes an optimist) so that the uncertainty of the final result would not explode. If the experts cluster, then this hope will be vain.

This leads to a final cautionary note for concluding the discussion of probabilistic risk analysis. After doing a risk analysis, one is expected to perform an "uncertainty analysis," that is, to give some indication of the confidence bounds in which the results are said to lie. In the best case the following procedure is invoked. Probability distributions are introduced reflecting the uncertainty regarding the values of input parameters (i.e., failure frequencies, weather conditions, population densities, etc.). Computer simulations are then used to sample from the distributions of all the input parameters. When a value for each variable is chosen, the whole risk calculation is repeated and the result recorded. This process is then repeated a large number of times to build up a distribution of results, from which confidence bounds can be extracted.

This procedure is perfectly satisfactory, *if the underlying distributions are independent*. If however, these distributions are positively correlated, the above procedure can underestimate the resulting uncertainty. A simple example will make this clear.

Suppose we have three parameters that are uniformly distributed over the range (0, 100). With 90% certainty each variable will be found in the range (5, 95). Suppose now we are interested in the product of these variables. If the variables are *independent*, the value of their product will lie with 90% probability in the range (1600, 440,000). If the variables are completely positively correlated, then this range would be (125, 857,375), and intermediate correlations will result in intermediate ranges. Further, as we shall see later on, the distinctive feature of *subjective* probability distributions (in particular, those of the risk analyst) is exactly that they are positively correlated, even when the physical processes being modeled are not. Hence, uncritical uncertainty analysis (and it is always uncritical) may give the spurious impression of knowing more than one actually knows. At the European Space Agency, a program of uncertainty analysis is under development that accounts for subjective dependence (Preyssl and Cooke, 1989, Cooke and Waij, 1987).

POLICY ANALYSIS

"Policy analysis" is really a catchall for everything that does not fit into the areas discussed above. It will be understood to include macroeconomic modeling, economic forecasting, energy planning, project management, environmental impact studies, etc. Some policy analysis have made explicit use of subjective expert probability assessments. The most conspicuous examples in this regard are Granger Morgan et al. (1984), Granger Morgan and Harrison (1988), and Merkhofer and Keeney (1987). In general this is not the case. For this reason our discussion will be more restricted than in the foregoing. The main service of this section will be to highlight the differences of probabilistic and deterministic methods of using expert opinion.

The typical forecasting problem concerns the future value of some variable. An electric utility may be interested in the peak demand for the coming years, in order to plan their generating capacity. An investor may be interested in the price of some securities, government planners may be interested in the price of oil on the international market, or in the gross national product between now and election time. All these situations involve predicting the values of quantities that become known before all too long and would lend themselves very well to assessment via probabilistic expert opinion. However, in each case there has been very little use of probabilistic assessment.

In economic forecasting various mathematical models have been used, sometimes in conjunction with expert forecasts. Experts' forecasts are combined to produce the subsequent estimate on the basis of past performance and the observed values of various exogenous variables. Such methods are deterministic in the sense that they make no attempt to assess or communicate the uncertainty of the estimate. The decision maker is given no guidance whatsoever as to how the estimate should be factored into his decision problem. Such forecasting methods are therefore more primitive than even the qualitative attempts to represent uncertainty reviewed in the preceding pages. In practice this often means that the decision maker treats the forecast values as certain.

The most important feature of such forecasting models was pointed out by C. W. J. Granger (1980): "There is no really useful procedure for evaluating an individual forecasting technique, as any evaluation cannot be separated from the degree of forecastability of the series being predicted." Any forecasting technique will perform well when estimating some one's yearly age. On the other hand, any method for predicting the outcome of tosses with a fair coin will get about half of the outcomes right. This is perhaps the single most important difference between probabilistic and deterministic methods: as we shall see in succeeding chapters, probabilistic forecasters can most certainly be evaluated independently of the forecastability of the things they are forecasting.

Although it is impossible to evaluate forecasting models or forecasters as such, it is possible to do clever things in given situations. Suppose we have a series $x_1, \ldots$ of realizations of some variable, and suppose f_i, g_i are forecasts of x_i by forecasters f and g. Suppose f and g are "unbiased" in the sense that their expected errors $f_i - x_i$, $g_i - x_i$ are zero (this does not mean that we expect no errors, but rather that the errors are expected to average out to zero). Then Granger (1980, p. 158) shows how to combine f and g in such a way as to make a better forecaster than either f and g alone.

This surprising fact is worth examining. Let r lie in the interval $(0, 1)$, and consider the combined forecaster:

$$c_i = rf_i + (1 - r)g_i$$

Let V_f, V_g, and V_c denote the variance (mean square error) of the respective forecasters. Since the forecasters are assumed to be unbiased, the best forecaster is the one with the smallest variance. A simple calculation shows that

$$V_c = r^2 V_f + (1 - r)^2 V_g + 2r(1 - r) \operatorname{cov}(f, g)$$

where $\operatorname{cov}(f, g)$ denotes the covariance of f and g. Elementary calculus shows that V_c is minimized when

$$r = \frac{V_g - \operatorname{cov}(f, g)}{V_f + V_g - 2 \operatorname{cov}(f, g)}$$

If we substitute this value of r into the definition of c_i, then V_c will generally be smaller than V_f or V_f.[1] For example, suppose that forecasters f and g are independent, so that $\text{cov}(f, g) = 0$. Then

$$V_c = \frac{V_g^2 V_f + V_f^2 V_g}{(V_f + V_g)^2}$$

Dividing this equation by V_g or V_f and recalling that the variance is always nonnegative, we see that the result is always smaller than 1. Hence V_c is always strictly less than V_f or V_g.

Other investigators have discovered similar phenomena in different contexts. For example, Beaver in *Financial Reporting: An Accounting Revolution* (1981) uses this fact as the basis for a theory of compiling investment portfolios. The idea is that the consensus of several experts is generally a better forecaster than any of the experts individually. He illustrates this with an example of forecasting winners of football games. From 1966 to 1968 the *Chicago Daily News* published the predictions of its sports staff as to which team would win the weekend game. From this data it was possible to define the "consensus predictor" by choosing that team which was picked by the majority of the sports staff. Table 2.12 shows that the consensus predictor outperformed all the other predictors, for the entire period. Beaver's idea is then that the advice of several experts should be used in assembling a portfolio of investments, rather than following exclusively the tips of any one expert.

If it were really this easy to produce good forecasters, the problems of using expert opinion would be essentially solved. In practice it is usually not this easy.

Table 2.12 Forecasting Outcomes of Football Games

	1966	1967	1968
Total forecasters (including consensus)	15	15	16
Total forecasts made per forecaster	180	220	219
Rank of consensus*	1 (tie)	2	2
Medium rank of forecasters	8	8	8.5
Rank of best forecasters:			
J. Carmichael (1966)	1 (tie)	8	16
D. Nightingale (1966)	1 (tie)	11	5
A. Biondo (1967)	7	1	6
H. Duck (1968)	8	10	1

Source: Taken from "Here's How Our Staff Picks 'em," *Chicago Daily News*, November 25, 1966, p. 43; November 24, 1967, p. 38; and November 29, 1968, p. 43.

*When all three years are combined, the consensus outperforms every one of the forecasters (i.e., ranks first).

[1] However, to apply this technique the variances and covariances must be estimated from the data. Estimates of the covariance on small data sets tend to be unstable, particularly when the correlation between f and g is high, and this can seriously degrade performance. Examples of this phenomenon can be found in Clemen and Winkler (1986). A similar variance reduction technique, facing a similar problem of assessing an empirical covariance matrix is developed by the Electric Power Research Institute (1986).

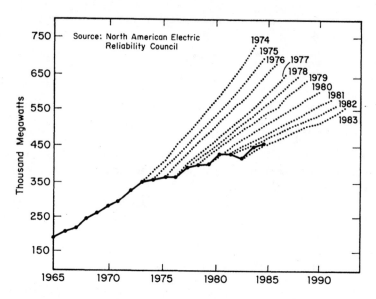

Figure 2.2 Summer peak electric demand 1965–1984. (*Flavin, 1984*)

For example, Figure 2.2 shows yearly projections of summer peak electric demand of the North American Electric Reliability Council. These projections were used by utility planners throughout the United States and Canada. The difference between the summer peak demand in 1983 and that projected for 1983 a decade earlier was equivalent to the output of 300 large nuclear plants, representing an investment of about $750 billion at 1984 prices (Flavin, 1984).

Equally dramatic are the projections of world oil prices, compared with actual prices. Figure 2.3 is an example, showing the oil price projections made by the Dutch Ministry of Economic Affairs, a steering committee of experts advising the ministry, and a political economist specializing in this field.

From these examples, it is easy to see the rub in Granger's combined expert. The argument showing that the combination of expert forecasts is better than each forecaster assumes that the forecasters are unbiased. In the real world, they are not always unbiased. They are not "just as often too high as too low." Moreover, they can be all biased in the same way. In such cases the combined expert will not perform better than the best expert.

Nonetheless, there may be cases in which such biases are absent and in which these techniques may work well. Beaver cites the example of compiling investment portfolios. If the underlying market forces determining the prices are "felt" by each expert in a way which involves a great deal of random errors, then a consensus forecast might well feel the underlying forces better than each expert. Indeed, in the consensus forecast the random errors will cancel out.

The market analogy is less plausible in the areas of science and policy analysis. Science is not a question of majority rule. The voice in the wilderness is sometimes right, and good new ideas usually come from the wilderness. In deciding matters of policy one has to reckon with very powerful interests that can impose a significant

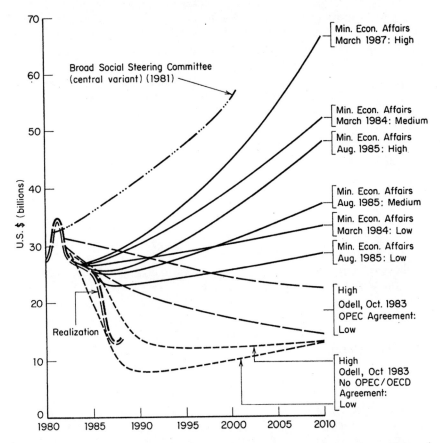

Figure 2.3 Oil prices in 1985 dollars and projected prices from Dutch experts. (Kok, private communication)

"drift" on experts' opinions. In both cases it is essential, at least in principle, to make normative distinctions in expert performance. This is exactly what the forecasting models do not admit.

Granger acknowledges this by noting that performance is relative to "forecastability." Another way to appreciate this point is as follows. When the value of a variable is highly uncertain, simply giving a point estimate conveys little information about the state of uncertainty regarding the variable, too little to perform an evaluation. For example, if the projections in Figures 2.2 and 2.3 represent central tendencies of very broad probability distributions, then the realizations might be compatible with the projections.

In science and policy analysis, expert point estimates are often used to fix the values of exogenous parameters in mathematical models. If the uncertainties regarding the true values of these parameters are large, then reliance on point estimates can render the models very brittle: If the experts are independent and unbiased, then fine, but if the experts are biased, and if their biases are correlated, then performance can be poor.

A good illustration of economic modeling with uncertain exogenous parameters emerged in the "Broad Social Debate" over nuclear energy in the Netherlands (see Vlek and Otter, 1987). Various interested parties (actors) were invited to develop scenarios for energy production in the year 2000. The actors generally used their own models and fed in values for the relevant model parameters to derive predictions for the year 2000. Many of the input parameters are highly uncertain. This uncertainty is not represented in the individual outputs, as these are point estimates. One parameter in particular was highly uncertain and very critical for the model outputs, namely the cost per kilowatt-hour of producing electricity from nuclear power plants in the year 2000. Different actors used very different values for this parameter and came up with very different results.

Kok (1985) obtained subjective probability distributions from various actors for the kilowatt-hour cost of producing electricity from nuclear generating plants in the year 2000, as presented in Table 2.13. These distributions are quite different. A's distribution is diffuse, reflecting substantial uncertainty. C's is rather concentrated on the higher values and E's is quite concentrated on the lower values. E is quite certain that the true value lies below $0.048. Note also how sharply the assessments differ for the lowest and highest cost brackets.

Table 2.13 contains much more relevant information than could be acquired simply by point estimates. Not only can the central tendencies of the actors' distributions be extracted, the probabilities of large deviations can be ascertained as well. This information can be useful in evaluating and combining their assessments.

Table 2.13 Subjective Probabilities of Five Actors for Production Costs (in 1982 Dollars) per Kilowatt-Hour of Electricity from a 1000-MW Nuclear Plant in the Year 2000

Production Costs (1982$), Cents/kWh	Actors' Subjective Probabilities				
	A	B	C	D	E
2.4–3.6	0.10	0.40	0.05	0.25	0.70
3.6–4.8	0.35	0.35	0.15	0.25	0.25
4.8–6	0.35	0.20	0.30	0.20	0.04
6–7.2	0.20	0.05	0.50	0.05	0.01

A Center for Energy Conservation
B Federation of Netherlands Industries
C Trade Union Federation
D Ministry of Housing, Physical Planning and Environment
E Ministry of Economic Affairs
The distributions do not represent offical standpoints of these organizations.
Source: Kok 1985.

3

Probabilistic Thinking

This chapter discusses the general problem of representing reasoning under uncertainty. The introduction of artificial intelligence systems, or expert systems, for modeling scientific reasoning under uncertainty on computers has made this a problem of great practical importance. The first section is devoted to "thinking" in general, and attempts to show why and how probabilistic thinking differs from ordinary logical inference. The second and main section of this chapter discusses developments in artificial intelligence relating to the representation of reasoning under uncertainty. It is shown why developers of expert systems turned away from probability theory in the 1970s, and how they have attempted to model uncertain reasoning. It is argued that such attempts have been unsuccessful. In this connection the theory of fuzzy sets is briefly discussed.

THINKING

In one sense thinking is as easy as breathing, we do them both all the time. We are also inclined to believe that logical thinking is easy. In our daily life we seldom need a paper and pencil to draw logical conclusions from premises. However, logical thinking has been an explicit object of study at least since Aristotle. In spite of the fact that we have a large and well-understood body of rules governing logical thinking, there are a number of situations in which the majority of us can be expected to commit elementary logical errors.

Consider the following argument:

premises   Only the fittest species for their biological niche survive.
             Cockroaches are one of the fittest species for their biological niche.

---

conclusion   Cockroaches survive.

Most people will judge that the conclusion of the above argument follows logically from the premises. Indeed, about 90% of third year mathematics students at the Delft University of Technology, most of whom have had a course in mathematical logic, regard this as valid.

However, the argument is invalid. The first premise does not say that the fittest survive, but that if a species survives, then it is one of the fittest. Maybe nothing survives at all. The first premise is logically equivalent to "For any species x, if x

survives then x is one of the fittest species for its biological niche." If we throw the above argument into the logically equivalent form:

> For all species x, if x survives then x is one of the fittest for its biological niche.
>
> Cockroaches are one of the fittest for their biological niche.
>
> ---
>
> Cockroaches survive.

then most people would recognize that the argument is invalid. It is an example of the classical fallacy of "affirming the consequent." The difference is simply this: In the first premise of the second argument the grammatical order in which the clauses appear corresponds to the direction of inference—from "survive" to "fittest." In the first premise of the first argument this correspondence does not hold and the reader is tricked into mistaking the grammatical for the logical order.

Suppose this same elementary error is committed by experts in evolutionary biology, and suppose we set out to represent expert reasoning among evolutionary biologists. Should we distinguish between the above two, logically equivalent arguments? Or should we "reconstruct" their reasoning in such a way that the two arguments are the same? In the latter case we would allow the representation to deviate sometimes from observed inferential behavior, in order to conform to the rules of logic. Our choice would obviously depend on our purpose. If we were interested in *describing* the inferential behavior of evolutionary biologists we might opt for the first alternative. If we were writing a textbook on evolutionary biology, or designing an expert system to do evolutionary biology, then we should choose the second.

This same choice confronts us when we want to represent probabilistic reasoning. In this case, however, the choice is much more difficult. Probabilistic reasoning is much more subtle than nonprobabilistic reasoning. Whereas nonprobabilistic reasoning has been studied for more than 2000 years, the study of probabilistic reasoning is of very recent origin. The first formal "probability logic" was given in 1963 (Los, 1963; for a simple exposition see Cooke, 1986). Moreover the rules for probabilistic reasoning cannot even be formulated in nonmathematical language, and they are not well understood, either by those whose reasoning is to be represented, or by those doing the representing.

This can be illustrated with some simple examples. In one sense these are just curiosities, but they do illustrate the subtleties involved in probabilistic reasoning. For this purpose we introduce a few abbreviations:

$$x = \text{an arbitrary species}$$

$$F(x) = \text{"}x \text{ is one of the fittest species for its biological niche"}$$

$$S(x) = \text{"}x \text{ survives"}$$

$$c = \text{cockroaches}$$

The following argument is logically valid:

> For all x, if $F(x)$ then $S(x)$
> $F(c)$
>
> ---
>
> $S(c)$

and it is logically equivalent to

> For all x if not-$S(x)$ then not-$F(x)$
> $F(c)$
> _____
> $S(c)$

Now let us "probabilize" the above arguments, that is, we treat the premises as reflecting highly probable, though not certain, knowledge. A rendering might read as follows:

> For all x, if $F(x)$ then probably $S(x)$
> Probably $F(c)$
> _____
> Probably $S(c)$

> For all x, if not-$S(x)$ then probably not-$F(x)$
> Probably $F(c)$
> _____
> Probably $S(c)$

Are these arguments "probabilistically valid," and are they equivalent? These questions cannot be answered without some serious reconstructing and some pencil and paper calculations. A formalism adequate to this task is presented in Cooke (1986), but we can get by for the present with the following notation. Let

$$p(S(x)\,|\,F(x)) \geq r = \text{"The conditional probability of } S(x) \text{ given } F(x) \text{ is greater than}$$
$$\text{or equal to } r\text{"}$$

Then for r and t close to 1, the above arguments could be reconstructed as

> For all x, $p(S(x)\,|\,F(x)) \geq r$
> $p(F(c)) \geq t$
> _____
> $p(S(c)) \geq rt$

> For all x, $p(\text{not-}F(x)\,|\,\text{not-}S(x)) \geq r$
> $p(F(c)) \geq t$
> _____
> $p(S(c)) \geq \dfrac{r+t-1}{r}$

Both arguments are valid in the sense that the conclusions hold whenever their respective premises hold.[1] However, as the lower bounds for the conclusions are not the same, and both bounds can be obtained, the arguments are not equivalent. In fact, the second argument is generally better than the first in the sense that its lower bound on the probability of the conclusion is higher, for r and t close to 1.

_____

[1] For the first argument, observe that $p(S(c)) \geq p(S(c)$ and $F(c)) = p(S(c)\,|\,F(c))p(F(c)) = rt$. For the second, observe that $(1-r)[1-p(S(c)] + p(S(c)) \geq p(F(c)\,|\,\text{not-}S(c))p(\text{not-}S(c)) + p(F(c)\,|\,S(c))p(S(c)) = p(F(c)) \geq t$. Solving for $p(S(c))$ yields the inequality in the conclusion of the second argument. Observe also that equality can hold in both arguments, though not simultaneously.

Things can get much worse. There are valid elementary arguments whose probalizations are not valid. Let $I(x)$ be another predicate, for example, $I(x) =$ "x influences the future ecosystem," and consider:

For all x, if $F(x)$ then $S(x)$

For all x, if $S(x)$ then $I(x)$

For all x, if $F(x)$ then $I(x)$

This argument is logically valid. However, for r and t arbitrarily close to, but not equal to, 1, its probabilization:

For all x, $p(S(x) \mid F(x)) \geqslant r$

For all x, $p(I(x) \mid S(x)) \geqslant t$

For all x, $p(I(x) \mid F(x)) \geqslant v$

is not valid for any v greater than 0. For given values of r and t we can concoct a situation in which for all x, $p(I(x) \mid F(x)) = 0$. Here is a counterexample, expressed in natural language:

If x has 30 letters in his/her last name, then x is probably less than 90 years old.

If x is less than 90 years old, then x probably has less than 25 letters in his/her last name

---

If x has 30 letters in his/her name, then x probably has less than 25 letters in his/her last name ????????

These logical calisthenics make clear why probabilistic reasoning is hard. Put in somewhat metaphorical mathematical language:

Logic is not continuous in certainty.

Arguments that are valid when the premises are known with certainty are not "almost valid" when the premises are "almost certain." Premises that are equivalent when known with certainty are not "almost equivalent" when the premises are "almost certain." Rather, discontinuities arise, and just the slightest bit of uncertainty can change a valid argument into an invalid one or can make equivalent propositions inequivalent. This lies behind the problems with probabilistic reasoning noted in Chapter 1. This fact becomes especially important when one attempts to represent experts' reasoning with uncertainty on a computer.

REPRESENTING UNCERTAINTY IN ARTIFICIAL INTELLIGENCE

Artificial, or machine, intelligence has almost become synonymous with expert systems. Expert systems have been given a very high profile in the Japanese "fifth-generation" computing project, and in the American strategic computing initiative of the Defence Advanced Research Projects Agency (DARPA). Both projects are budgeted at about one billion dollars over 10 years (in comparison, Star Wars has been projected to cost 30 billion in 5 years). The DARPA project in particular envisions many futuristic applications of expert systems, including integrated battle management systems with "speech input," a "pilot's associate" (a sort of R2D2 for fighter pilots), and "autonomous land vehicles" for waging automated warfare. [For a description of this project's goals see Stefik (1985).] In all these projects, reasoning with uncertain information is essential.

Experts systems are being rapidly deployed in a wide variety of fields, and many software houses now offer inexpensive "shells" allowing the user to build his own expert system. The user fills in his own "rules" and supplies his own "certainty factors," without being told what these might mean, and the shell is transformed into an expert system that relieves the user of the task of reasoning with uncertainty.

An entrance into the active areas of research regarding the representation of uncertainty on intelligent machines can be gained from the February 1987 issue of *Statistical Science*, in which some of the principal protagonists lock horns. It is not the purpose of the ensuing discussion to pursue active areas of research in this exciting field, but rather to review some of the standard literature and become acquainted with the problems that researchers in this field are facing. Hence we shall set our focus not on what is being contemplated and discussed, but what has actually been implemented and evaluated. This entails, perhaps regrettably, that we shall not discuss the theory of belief functions, as these are still in the discussion phase. On the other hand, a final section of this chapter will look at the theory of fuzzy sets, as these are being applied. For mathematical details the reader must consult the original sources.

Computer-Aided Diagnosis

The first expert system was developed by E. Feigenbaum and B. Buchanan at the Stanford Heuristic Programming Project, started in 1965. This system is called DENDRAL and is used to identify organic compounds. The reasoning in this system is largely nonprobabilistic. Probabilistic inference systems were initiated in the early 1970s as an aid for diagnosis. The early systems were designed by decision theorists with a background in probability, and the inference mechanisms used the probability calculus. These early attempts were generally unsuccessful, and it is essential to understand the reasons for this lack of success, before we can appreciate the contribution which artificial intelligence techniques have made to the field. The following discussion of these early systems follows Szolovits and Pauker (1978).

The generic decision problem in diagnosis can be described as follows. In a given area we distinguish a number of possible diseases that a patient might have. These are traditionally called "hypotheses," and will be denoted $h_1, \ldots, h_n$. In a typical application there may be 15 such hypotheses. A patient may be given any of m tests $t_1, \ldots, t_m$. We assume that the result of a test is either "yes" or "no." A test may be anything from asking the patient's sex, to determining his/her response to a treatment. In a typical application there may be 10 tests. A patient is given a number of tests, in a particular order, and the results are recorded as a "case history." Let Q denote a generic case history.

The generic decision problem is simply this: On the basis of a case history Q determine which hypothesis is the most likely. For the decision theoretically oriented analyst, this is simply a question of determining the "posterior probability" $p(h_i \mid Q)$. The theorem of Bayes (see Chap. 4) states that

$$p(h_i \mid Q) = \frac{p(Q \mid h_i)p(h_i)}{p(Q)}$$

It is a simple matter to calculate the posterior probability if the quantities on the right-hand side are known. The "prior" probability $p(h_i)$ of hypothesis h_i can be determined from population data on the frequency of disease h_i. The term $p(Q)$ drops out if we compare the ratio of posterior probabilities of competing hypotheses. The real problem is the "likelihood" term $p(Q \mid h_i)$. The computer must "maintain" assessments for each possible likelihood term. These assessments may come from data or from expert judgment, or from some combination of the two. The problem is simply the number of such terms. In the typical application, with 15 hypotheses and 10 tests, there are about 90,000,000,000 such likelihood terms![2]

This is too much for any computer. The problem only becomes tractable when the number of likelihoods to be maintained is reduced. To accomplish this reduction a number of simplifying assumptions may be contemplated.

The first simplifying assumption is that the order of the tests is irrelevant. This assumption is not harmless, as a test is frequently a patient's response to a treatment. Having had a particular treatment may alter the patient's probabilities for responding to other treatments, for certain ailments. Thus, the order in which treatments are tried may not be irrelevant. In any case, introducing this simplification brings the number of likelihoods that must be maintained down to about 900,000.[3] Still too many.

A sufficient reduction is accomplished by assuming that the tests are all *conditionally independent*, that is, for t_j not in Q:

$$p(t_j = \text{"yes"} \mid h_i \text{ and } Q) = p(t_j = \text{"yes"} \mid h_i)$$

for all i and j. Under this assumption, it can be shown that the number of likelihoods in the typical problem is 300.[4]

---

[2] There are $m!/(m-j)!$ ordered sets of tests of length j, and each test can have one of two outcomes. This gives $2^j m!/(m-j)!$ possible case histories of length j, for each hypothesis h_i. Summing over j and multiplying by n, we find that the number N of likelihoods is

$$N = n \sum_{j=1,\ldots,m} \frac{2^j m!}{(m-j)!}$$

[3] The number of likelihoods N' under this assumption is now given by

$$N' = n \sum_{j=1,\ldots,m} \frac{2^j m!}{j!(m-j)!} = n[(2+1)^m - 1]$$

[4] Since

$$p(t_j = \text{yes and } t_k = \text{yes} \mid h_i) = p(t_j = \text{yes} \mid h_i \text{ and } t_k = \text{yes})p(t_k = \text{yes} \mid h_i)$$

$$= p(t_i = \text{yes} \mid h_i)p(t_j = \text{yes} \mid h_i)$$

it suffices to know the likelihoods for single tests. There are 10 tests, 15 diseases, and 2 possible outcomes of each test. This yields 300 likelihoods.

This is a workable number. The expert systems DENDRALL and MYCIN (to be discussed presently) each contain about 400 "rules," and a rule in this context may be roughly compared with a likelihood term. Everyone agrees that the assumption of conditional independence is wrong. In fact, it has recently been shown that if the hypotheses are mutually exclusive and jointly exhaustive, then conditional independence implies that at most one case history can alter the prior probabilities of the hypotheses (Johnson, 1986). The designers of the early systems knew that conditional independence was a very strong assumption, but were unaware of this particular problem. They were forced to adopt it because of hardware constraints.

Artificial Intelligence Enters the Field

The above formal problems are only part of the reason why early efforts were generally unsuccessful. In addition, it is often remarked that doctors had difficulty interpreting the results, and were unused to think in terms of probability. In the mid 1970s artificial intelligence (AI) techniques made their appearance.

Before this time, the paradigm problem for representing intelligent behavior on a computer was felt to be the chess problem—designing computer programs to play chess. In the mid 1970s, urged on by Feigenbaum, AI researchers began attacking real-world problems. The paradigm of intelligent behavior shifted from the chessboard to the laboratory, and representing scientific reasoning became the new problem on which AI programmers would cut their teeth.

The Present Illness Program

A good illustration of an early application of AI techniques is found in the Present Illness Program of Szolovits and Pauker (1978). The program uses hypotheses and findings. Each hypothesis is supplied with a logical frame. Each frame shows (1) relations between the hypothesis and findings, (2) relations between the hypothesis and complementary hypotheses (hypotheses that may occur together with the given hypothesis), and (3) relations between the hypothesis and competing hypotheses. When findings sufficient to guarantee the truth of a hypothesis are found, the program categorically (i.e., nonprobabilistically) decides on the hypothesis. If a "necessarily not" finding occurs, the hypothesis is categorically rejected. A generic frame is shown in Table 3.1.

An hypothesis can be in one of four states:

- *Inactive*: A hypothesis is initially inactive, and returns to inactive if it is rejected via a "necessarily not" finding.
- *Active*: A hypothesis becomes active if one of its "triggered by" findings occurs.
- *Semiactive*: A hypothesis becomes semiactive if it is complementary to or competing with an active hypothesis.
- *Confirmed*: A hypothesis is confirmed when its likelihood (see below) passes a certain threshold.

The likelihood is determined by combining two scores:

1. The *matching score* counts the number of expected findings of an hypothesis

Table 3.1   Logical Frame for the Present Illness Program

Hypothesis

(1) Findings

Triggered by	Expected	Sufficient	Necessary	Necessarily Not
—	—	—	—	—
—	—	—	—	—
—	—	—	—	—

(2) Complementary hypotheses

Caused by	Cause of	Complicated by	Complication of	Associated with
—	—	—	—	—
—	—	—	—	—
—	—	—	—	—

(3) Competing hypotheses

—
—
—

and its complementary hypotheses that are actually observed, and divides this by the total number of possible expected findings.

2. The *binding score* divides the number of observed expected findings of a hypothesis and its complementary hypotheses by the total number of observed findings.

The program calculates the likelihood scores of all hypotheses and inquires about the status of not-yet-observed expected findings of all active hypotheses. After each new finding the process is repeated. When all active hypotheses' expected findings have been queried, it starts with the expected findings of the complementary hypotheses.

Regarding these likelihood scores, Szolovits and Pauker write "we must think of them as an arbitrary numeric mechanism for combining information, somewhat analogous to the static evaluation of a board in a chess-playing program" (Szolovits and Pauker, 1978).

This will give an impression of the inference mechanism. The program has 38 completely developed hypotheses, of which 18 can be confirmed by "sufficient" findings.

We see here a conscious attempt to mimic the reasoning of a doctor in a diagnostic situation. The likelihood scores used to represent uncertainty bear no relation to probabilities, and would not satisfy the axioms of probability theory. It is plausible that this program more closely resembles the actual reasoning of doctors than the decision theoretic programs discussed earlier.

This is not the place to evaluate this program as such. It is surely a very useful tool for the task for which it was designed. However, it is appropriate to cast a critical eye on the inference mechanism, abstracted from this particular application. It is obvious that the scores are strongly influenced by the numbers of expected findings. A hypothesis with few expected findings will have a hard time achieving a high binding score, and a hypothesis with many expected findings may be placed at a disadvantage with regard to the matching score. The number of findings may also be rather ad hoc, reflecting the number of tests that have been devised for the hypothesis in question.

More important, however, is the following. The program takes no account of the prior probability of the hypotheses. Suppose there is a rare disease, say Mongolian tongue-worm, whose initial symptoms are identical with those of the common cold. In every session, the program would necessarily give these two hypotheses the same likelihood. The designer could try to design this anomaly away by adding the expected finding "recently visited Mongolia," but does this solve the problem? Will the slight change in the matching score induced by this addition compensate for neglecting the very low prior probability of Mongolian tongue-worm, even among people who have recently visited Mongolia?

Neglecting the prior probabilities is well known in probabilistic inference and has been given the name "base rate fallacy" (prior probabilities are often termed "base rates"). This is a typical feature of probabilistic thinking, which has no direct analog in nonprobabilistic reasoning. In Chapter 4 we shall see that the base rate fallacy is very common, in particular among doctors.

Production Rule Systems: MYCIN

Production rule systems differ from systems like the Present Illness Program in the following respect. The latter program works with nonprobabilistic input; the user simply feeds in information regarding the presence or absence of findings. From this input, a "probabilistic" output is generated in the form of likelihood scores.

Production rule systems allow for "probabilistic" input, that is, the user, or an expert, feeds in a numerical representation of uncertainty. (This representation will not in general obey the laws of probability.) In a particular consultation, the program propagates this uncertainty through an inference chain.

The best-known production rule system is called MYCIN and was developed by Shortliffe and Buchanan (1975) of the Stanford Heuristic Programming Project to support the diagnosis and treatment of bacteriological infections. The system EMYCIN (for "essential MYCIN") is a skeleton program that allows the user to develop a MYCIN-type expert system for the user's problem field. This illustrates that MYCIN's inference mechanism is regarded by its developers as *generic*. It is thought to present a model for reasoning with uncertainty in a wide class of problems, rather than an ad hoc solution for a particular problem.

The terms "shell" and "tool kit" have been introduced to denote skeletal expert systems, where the user supplies the "flesh and blood." Tool kits are larger and more powerful than shells.

Knowledge Base

MYCIN operates with a "knowledge base" containing a dynamic data base and a "rule base." The dynamic data base contains data of a patient fed in during a consultation. The rule base contains rules of inference, similar to our inferences about cockroaches at the beginning of this chapter. The following is an example of a rule from MYCIN's rule base:

Rule $_i$:   If (i.2) the stain of the organism is gram positive,
         and
         (i.2) the morphology of the organism is coccus,
         and
         (i.3) the growth conformation of the organism is chains;

Then: there is evidence (certainty factor = 0.7) that the identity of the organism is streptococcus.

The value of the certainty factor, 0.7, is obtained by giving an expert the following prompt:

> On a scale of 1 to 10 how much certainty do you affix to the conclusion based on the evidence (i.1), (i.2), and (i.3)?

The answer to this prompt is divided by 10 to obtain the certainty factor.

The evidence (i.1), (i.2), and (i.3) may be present in the dynamic data base if a culture has been taken from the patient. In this case, the conclusion "the organism taken from the patient is streptococcus" would be added to the data base, with certainty factor 0.7. Statements with certainty factors greater or equal to 0.2 may be used as premises in new inferences.

The program will now be able to "fire" other rules, as the data base has been enlarged. However, the certainty factor of the conclusions must be reduced to reflect the uncertainty of the premises of the rules. MYCIN has a general mechanism for compounding uncertainties in "if" clauses, which is derived from the theory of fuzzy sets.

To illustrate this mechanism, let us consider a Rule$_j$ with premise (j.1), (j.2), and (j.3), in the following layout:

Rule$_j$:   If (j.1)
         and
         [(j.2) or (j.3)];

     Then:
         conclusion, with certainty factor CF$_j$.

Let CF(j.1) denote the certainty factor attached to (j.1), etc. The fuzzy set rules applied in MYCIN calculate the certainty factors of conjunctions and disjunctions, as functions of the certainty factors of the conjuncts and disjuncts, as follows:

$$CF(P \text{ and } Q) = \min\{CF(p), CF(Q)\}$$

$$CF(P \text{ or } Q) = \max\{CF(P), CF(Q)\}$$

The above rules can be applied to calculate certainty factor for the "if" clause of Rule$_j$. This latter number would then be multiplied by the number CF$_j$ to give the certainty factor of the conclusion. This new conclusion is then entered in the data base, and the "inference engine" looks for new rules to fire.

One additional feature of MYCIN's inference mechanism deserves mention here. It may arise that the same conclusion can be drawn from firing two or more different rules. Suppose the same conclusion h can be drawn from Rule i and Rule j with certainty factors CF_i and CF_j; and suppose the premises of both rules are known with certainty. MYCIN would then attribute a "combined certainty factor" CF_{i+j} to the conclusion, defined as follows:

$$CF_{i+j} = CF_i + CF_j - CF_iCF_j$$

The above does not constitute an exhaustive description of MYCIN's inference mechanism, but it will give an adequate impression for present purposes.

Probabilistic Interpretation

In their initial publications, Buchanan and Shortliffe proposed a probabilistic interpretation for the certainty factors CF_i attaching to rules. Let E_i denote the "if" clause and h_i the conclusion of Rule$_i$. The CF_i was to be interpreted as

$$CF_i = \frac{p(h_i \mid E_i) - p(h_i)}{1 - p(h_i)} \qquad p(h_i) < 1, \qquad p(h_i \mid E_i) \geqslant p(h_i)$$

The right-hand side of the above expression is derived from the confirmation theory of the philosopher Rudolph Carnap (1950). However, Carnap emphasized that this could be used as a measure of *increased belief* in h_i, on the basis of evidence E_i. MYCIN in effect takes this measure of increased belief as a measure of belief. Buchanan and Shortliffe note, however, that if $p(h_i)$ is small, as one would expect at the beginning of a consultation, then CF_i is approximately equal to the conditional probability $p(h_i \mid E_i)$.

Why not just use $p(h_i \mid E_i)$ instead of CF_i? Buchanan and Shortliffe claim that doctors just don't reason in such a way that "certainty" of an hypothesis and the "certainty" of its negation add to one, as is the case with (conditional) probabilities. Another reason may be found in the following fact. The conditional probability $p(h \mid E_i$ and $E_j)$ cannot be calculated as a function of $p(h \mid E_i)$ and $p(h \mid E_j)$. Even assuming independence and conditional independence (under h) of the evidences, we cannot do better than[5]

$$p(h \mid E_i \text{ and } E_j) = \frac{p(h \mid E_i)p(h \mid E_j)}{p(h)}$$

In order to calculate the conditional probability of h on the combined evidence, we need to know the base rate of h. The certainty factor combination rule ignores the base rate, and hence commits the base rate fallacy.

[5] We have

$$\frac{p(h \mid E_i \text{ and } E_j)}{p(h)} = \frac{p(E_i \text{ and } E_j \mid h)}{p(E_i \text{ and } E_j)}$$

$$= \frac{p(E_i \mid h)p(E_j \mid h)}{p(E_i)p(E_j)}$$

$$= \frac{p(h \mid E_i)p(h \mid E_j)}{p(h)^2}$$

In more recent publications (Shortliffe and Buchanan, 1984), this probabilistic interpretation is abandoned, and other changes in the inference mechanism are announced as well.

The question then arises, if the certainty factors are not to be interpreted probabilistically, how are they to be understood. No clear answer to this question has been forthcoming. Shortliffe and Buchanan (1984) suggest that certainty factors are to regarded as combinations of probability and utility, without further specification. They suggest that the Dempster-Shafer theory of belief functions may provide a more satisfactory representation of uncertainty for expert system applications (Gordon and Shortliffe, 1984).

Critique

It is hard to find anyone presently willing to defend the probabilistic interpretation of certainty factors, or willing to specify any other interpretation. That a formalism that no one is willing to defend is being commercialized on such a massive scale is a fact worthy of contemplation.

The probabilistic interpretation came under sharp criticism rather quickly. Adams (1976) proved that the combination rule for CF_{i+j} given above in combination with the probabilistic interpretation is equivalent to

$$p(E_i \text{ and } E_j) = p(E_i)p(E_j)$$

$$p(E_i \text{ and } E_j \mid h) = p(E_i \mid h)p(E_j \mid h)$$

$$p(E_i \text{ and } E_j \mid \text{not-}h) = p(E_i \mid \text{not-}h)p(E_j \mid \text{not-}h)$$

where h is the common conclusion of Rule_i and Rule_j. These are even stronger than the conditional independence assumed by the decision theoretic approaches discussed earlier. He also showed that the choice of certainty factors in this situation is severely constrained. If the prior probability $p(h)$ is $\frac{1}{11}$, and if $CF_i = 0.7$, then CF_j must be less than 0.035. Expert system shells contain no checks for consistency in this sense, and it is doubtful if the numbers actually used in any expert system satisfy this constraint. This sort of consistency check is not incorporated into the production rule systems. Cendrowska and Bramer (1984) further showed that in some cases the order in which the rules are fired can affect the resulting certainty factors.

MYCIN's advice has been evaluated in a sort of "Turing test" (Yu et al., 1979) ("try to pick out the computer in a blind experiment"). Eight doctors and MYCIN gave advice for treatment of 10 cases. The advices were evaluated blind by eight specialists. Thus, in total there were 80 evaluations per "expert." Thirty-five percent of MYCIN's evaluations were judged "unacceptable," and this was *lower* than the corresponding percentage for the eight human experts.

Simulations described by Shortliffe and Buchanan (1984) indicate that MYCIN's treatment recommendations are very insensitive to the numerical values of the certainty factors. When the values of the certainty factors were coarse grained, so that MYCIN distinguished only three different values, a difference in diagnosis resulted in 5 of 10 cases, but a difference in treatment recommendation in only 1 of 10 cases. Moreover, conclusions are often drawn on the basis of rules whose "if" clauses were all certain, so that the propagation of uncertainty is not really important for the conclusions.

It must be emphasized that this latter conclusion is peculiar to the specific application for which MYCIN was designed. Reading Shortliffe and Buchanan (1984), one gets the impression that the designers fiddled around with the certainty factors and with the combination rules until the recommendations made sense. Of course this is an eminently sensible thing to do. However, it suggests that the MYCIN model is not a generic solution to the problem of reasoning under uncertainty, but another ad hoc solution.

FUZZINESS

Many expert systems use the theory of fuzzy sets to propagate uncertainty, and it is appropriate to make a few remarks on fuzzy sets. The theory of fuzzy sets was introduced by Zadeh (1968). A recent exposition can be found in Zadeh (1986). A well-received and frequently cited introduction to the literature is Dubois and Prade (1980). For a pointed critique, see French (1984, 1987).

There is an immense literature on fuzzy sets, which will not be reviewed here. It is scarcely possible to speak of a "theory" of fuzzy sets, as writers in this field have been unable to agree on a definition of a fuzzy set, and unable to agree on rules for operating with fuzzy sets. Indeed, Zadeh has said that it is not in the spirit of fuzzy sets to ask for precise definitions. We shall confine ourselves to a brief discussion of the philosophy behind this development and a few remarks on the combination rules applied in MYCIN. These are the rules most frequently encountered in applications. The exposition will draw on Zimmermann (1987).

To introduce fuzziness, we may distinguish between *uncertainty* and *ambiguity*. Uncertainty is understood to denote a state of partial knowledge regarding (future) observations; ambiguity denotes a state of partial knowledge regarding the meaning of terms in the language. Put differently, uncertainty is that which is removed by observation, ambiguity is that which is removed by linguistic convention.

Many writers regard subjective probability as a representation of uncertainty in the above sense. Many fuzzy set adherents would regard fuzziness as a representation of ambiguity. Everyone would agree that observations can sometimes be ambiguous, hence the two notions overlap. Three questions will be briefly addressed.

1. Can the probability formalism adequately represent uncertainty for practical purposes?
2. Can the fuzzy formalism adequately represent uncertainty for practical purposes?
3. Can the fuzzy formalism adequately represent ambiguity for practical purposes?

The formulations indicate that we are not interested in philosophical questions like "what *is* uncertainty." The questions are addressed to the needs of practicing scientists and engineers, but we do not go so far as to demand that efficiently computable representations for every conceivable problem actually exist at present.

1. The present work proceeds from the conviction that the probability formalism can adequately represent uncertainty for practical purposes. The articulation of this position occupies the second part of this book, and readers will be able to judge the results for themselves. This position presupposes that ambiguity of observations can be reduced via linguistic conventions to a level sufficient for the practical purposes at hand.

2. The answer to this question depends on what one means by "fuzzy formalism." Many "fuzzicists" regard probability as a subtheory of fuzzy set theory. In this case, the answer consistent with the answer to (1) above is trivially "yes." However, the brief exposure to fuzzy formalisms in the present chapter leads one to anticipate a number of difficulties in representing uncertainty.

The first problem concerns the fuzzy combination rules encountered in the above discussion, or a host of others encountered in the fuzzy literature. Let P and Q denote propositions in the language, and let $U(P)$ denote some real-valued function on propositions representing uncertainty. If U satisfies

$$U(P \text{ and } Q) = F(U(P), U(Q))$$

$$U(P \text{ or } Q) = G(U(P), U(Q))$$

where $F(x, y)$ and $G(x, y)$ are real functions of two variables, then U cannot be an adequate representation of uncertainty. Indeed, the uncertainty of "P and Q" depends not only on the *uncertainty* of P and the *uncertainty* of Q, but also on the interaction of P and Q. Note that we cannot calculate the probability of "P and Q" as a function of the probability of P and the probability of Q. The propositions

The next baby born in the municipal hospital is a boy.

The next baby born in the municipal hospital is a girl.

The next baby born in the municipal hospital has weight greater than the median weight for babies born in the municipal hospital.

each have probability one-half. However, the probabilities of the different conjunctions of pairs of propositions are obviously not the same. The propositions interact, the first is compatible with the third but not with the second. This is an essential feature of uncertainty, and it is not satisfied by the min and max operators encountered in the discussion of MYCIN, nor by the myriad other rules proposed in the fuzzy literature.

There is a second feature of reasoning under uncertainty that seems to defy representation in the theory of fuzzy sets. Uncertainty can be reduced by observations. In fact, if observations cost nothing, then it is generally in our interest to reduce uncertainty before taking decisions (see Theorem 4 of Chap. 6). Any adequate representation of reasoning under uncertainty must explain *why* we bother to perform observations, and *how* our uncertainty changes as a result of observation. The fuzzy set theories provide a mechanism for propagating uncertainty, but provide no mechanism for altering uncertainty on the basis of observations.

There is a third feature of representing uncertainty that is closely related to the second. Since it is in our interest to reduce uncertainty, we often consult experts. Since experts often differ, it is essential that a representation of uncertainty provide tools for evaluating and combining the uncertain opinions of experts. This means that we must be able to give meaning to expressions like "expert A provides a better representation of the uncertainty in this case than expert B." We have repeatedly suggested how meaning can be given to such statements when uncertainty is represented as probability. If uncertainties should be represented as "fuzzy truth values" or as "membership in fuzzy sets" then someone should give criteria for distinguishing "good" from "bad" fuzzy truth values or fuzzy memberships. This has not been done to date.

We summarize by listing three questions which should be addressed to any putative representation of uncertainty:

- Does it account for the interactions between events or propositions?
- Does it explain how uncertainty is affected by observation?
- Does it enable us to distinguish between "good" and "bad" representations of uncertainty?

The above discussion concludes that the fuzzy formalism has difficulty giving positive answers to these questions.

3. Can fuzziness represent ambiguity? This question can be sensibly discussed only if some particular version of fuzziness is put forward. It seems most implausible that a combination rule satisfying the above functional equations could be adequate. Zimmermann (1987) suggests that the choice of combination rules should be decided by future empirical investigation of linguistic behavior. We shall leave the question hanging there. However, the pitfalls of empirical investigations of linguistic behavior should not be underestimated. The example of logic may be useful to recall at this point. It is undoubtedly true that logic has been enriched by the attempt adequately to represent certain *aspects* of linguistic behavior. On the other hand, does anyone believe that empirical investigations of "inference behavior" would lead to an interesting logic? Remember the cockroaches.

CONCLUSION

Logical thinking is not always as simple as it seems. There are many situations in which most of us will make elementary logical mistakes if we do not pay close attention. Probabilistic thinking is much more subtle and tricky than ordinary logical thinking. Logical validity is not continuous in certainty; rather, uncertainty forces us to reason in ways that are not simple transcriptions of logical reasoning. Moreover, validity of probabilistic arguments cannot be assessed without performing calculations.

For deterministic reasoning we have a well-developed theory of logical inference that helps us track down and repair errors in thinking. We also need such a theory for reasoning under uncertainty.

This need is felt most acutely by designers of artificial intelligence systems that model scientific reasoning under uncertainty. Such systems have tried to model uncertain reasoning on an analogy with nonprobabilistic reasoning and have run into problems. Correctly taking account of the base rate and ensuring global consistency are features of reasoning under uncertainty that have no direct analog in nonprobabilistic reasoning.

For any particular problem area, we can design a particular system of rules for reasoning, and if we fiddle with these rules long enough we can "tune" the system such that it commits big errors infrequently. Sooner or later we will thus arrive at a satisfactory ad hoc solution for the particular problem. The expert systems reviewed in this chapter must be seen as ad hoc solutions in this sense. They cannot be regarded as generic solutions to the problem of reasoning under uncertainty.

A generic solution to the problem of reasoning under uncertainty requires a theory that will guide us in tracking down and correcting errors in probabilistic thinking. There are many such theories, or proto-theories presently under discussion. One of these, the theory of fuzzy sets, has been examined briefly. The conclusion was that this theory, at least in its most common form, is not capable of adequately representing reasoning under uncertainty. It is claimed that subjective probability can do the job.

In Chapter 6 it will be shown that the existence of a unique measure of partial belief satisfying the axioms of probability can be derived from principles describing rational preference behavior. If one accepts these principles, then one must necessarily accept that uncertainty is represented by subjective probability. This does not mean that all practical problems involving uncertain inference are solved, and it certainly does not entitle us to downgrade other attempts at solving such practical problems. It must be remembered that probability theory has not yet provided computationally feasible models for the practicing designer of expert systems. This will surely occupy researchers for some time to come.

4

Heuristics and Biases

When called upon to estimate probabilities or determine degrees of belief, people do not usually perform mental calculations, but rely instead on various rules of thumb. Such rules are called *heuristics*. Heuristics may be adequate for everyday life, but they can lead to predictable errors. The word "error" must be used gingerly in this context, since a degree of belief is subjective. By error we shall mean either a violation of the axioms of probability, or an estimate that is not really in accord with the subject's beliefs and that he/she would want to correct if the matter were brought to his/her attention.

When heuristics lead to errors in this sense, we speak of *biases*. There is another sense of the word "bias" that might be freely rendered as "distortions of judgment through ideology." Psychologists also speak of a "motivational bias" when judgment is distorted willfully (as in lying). These are not the senses used in this chapter. The biases studied here have to do with "misperceptions" of probabilities. It is important to be aware of them when designing techniques for eliciting subjective probabilities.

Much of the psychometric research into heuristics and biases has been performed by Kahneman, Tversky, Slovic, and their colleagues, and has been gathered in Kahneman, Slovic, and Tversky (1982). Most of the experiments described therein did not involve experts. Some of these experiments have recently been performed with experts (Thys, 1987). The design of these latter experiments allows us to compare the performance of experienced and inexperienced experts on technical and on general knowledge items.

Four heuristics are discussed in this chapter; *availability, anchoring, representativeness,* and *control.* The *base rate fallacy* can be seen as a bias deriving from the representativeness heuristic, but because of its importance it will be discussed under a separate heading. Perhaps the most important type of bias is *overconfidence.* Overconfidence is not directly associated with an estimation heuristic, though in some cases it may be aggravated by an anchoring. When assessing probabilities, overconfidence expresses itself as poor *calibration.* In this chapter some of the standard literature on calibration of expert subjective probabilities is discussed.

Overconfidence as poor calibration has been studied extensively. This is undoubtedly due to the dramatically poor results often obtained when calibrating expert probabilities. Dramatic results can have the effect of obscuring important aspects of the problem, and this seems to be the case with regard to calibration.

Many important issues remain unclarified. Most measures used to score calibration are mathematically dubious, at best. Failure to appreciate this fact has led to inappropriate experimental designs whose results must remain ambiguous. This in turn has clouded the relation between calibration and expertise or knowledge. Finally, the emphasis on calibration has led to neglect of other important aspects of probability assessments, in particular, the information contained in such assessments.

The above issues are studied in detail in Part II, but it is important to mention them before examining the experimental literature. The results reported in this chapter do not suffer from the problems mentioned above, as far as one can tell from their descriptions in the literature.

AVAILABILITY

When asked to estimate the size of a class (e.g., the number of automobile deaths in a year) subjects tend to base their estimates on the ease with which members of the class can be retrieved from memory. Hence, the frequency with which a given event occurs is usually estimated by the ease with which instances can be recalled. In one study, for example, subjects were told that a group of 10 people are forming committees of different sizes. When asked to estimate the number of committees with two members, the median estimate was 70, for eight members the median estimate was 20 (Tversky and Kahneman 1982c). Of course, there are just as many committees with two members as with eight members, as the eight people not included in the committee of two, can form their own committee. The responses indicate a bias. The correct number for two and eight is 45. The cause of this type of estimation is presumably the following. It is much easier to imagine committees of two and to imagine grouping the ten people into different groups of two, than it is to imagine groups of eight. Committees of size two are "more available.' to the subject.

Another example along the same lines is given by the permutation experiment (Tversky and Kahneman 1982a). Subjects are given the following text:

Consider the two structures A and B which are displayed below:

(A)	(B)
XXXXXXXX	XX
XXXXXXXX	XX
XXXXXXXX	XX
	XX
	XX
	XX
	XX
	XX
	XX

A path in a structure is a line that connects an element in the top row with an element in the bottom row, and passes through one and only one element in each row. In which of the two structures are there more paths? How many paths do you think there are in each structure?

In a typical experiment, 46 of the 54 subjects thought there were more paths in (A). The median estimate for the number of paths in (A) was 40, and in (B), 18. In fact, there are 8^3 paths in (A) and 2^9 paths in (B); and $8^3 = 2^9 = 512$.

Why do people see more paths in (A) than (B)? Kahneman and Tversky speculate that it is much easier to "imagine" paths through three points than paths through nine points, hence the paths in (A) are more available. Another question is why the subjects underestimate the number of paths so severely.

Perhaps the best-known instance of the availability heuristic involves the perception of risks. When asked to estimate the probabilities of death from various causes, subjects typically overestimate the risks of "glamorous" and well-publicized causes (botulism, snake bite) and typically underestimate "unglamorous" causes (stomach cancer, heart disease). Figure 4.1 is typical of the subjects' responses to these types of questions.

ANCHORING

When asked to estimate a probability, subjects sometimes will fix on an initial value and then "adjust" or "correct" this value. Frequently the adjustment is insufficient. In one experiment subjects were asked what percent of the member nations of the UN were African. A number between 1 and 100 was then generated

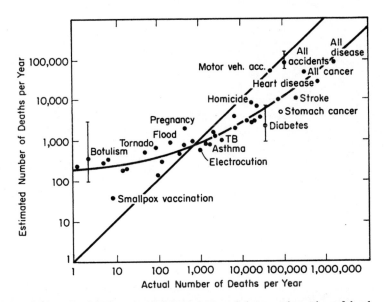

Figure 4.1 Relationship between judged frequency and the actual number of deaths per year for 41 causes of death. If judged and actual frequencies were equal, the data would fall on the straight line. The points, and the curved line fitted to them, represent the average responses of a large number of lay people. As an index of the variability across individuals, vertical bars are drawn to depict the 25th and 75th percentiles of the judgments for botulism, diabetes, and all accidents. The range of responses for the other 37 causes of death was similar. (*Slovic, Fischoff, and Lichtenstein, 1982*)

by spinning a wheel in the subject's presence. The median estimates of the percentage of African countries was 25 and 45 for groups with "anchor" numbers 10 and 65, respectively, from the wheel (Tversky and Kahneman, 1982a).

In another experiment, two groups of high school students were given 5 seconds to estimate one of the following two products:

$$8 \times 7 \times 6 \times 5 \times 4 \times 3 \times 2 \quad \text{and} \quad 2 \times 3 \times 4 \times 5 \times 6 \times 7 \times 8$$

The median estimate for the first sequence was 2250, and for the second was 512. The correct answer is 40,320.

To explain these results, Kahneman and Tversky note that given the short time available for the estimation, students probably perform the first few multiplications and then extrapolate. The extrapolations are insufficient, but the differences reflect differing anchor values gotten from the initial calculation.

Another instance of anchoring, which affects expert estimation, occurs when subjects are asked to estimate the fractiles, or quantiles, of a continuously distributed quantity. The $k\%$ quantile of a continuously distributed quantity X is the smallest value x_k such that

$$p(X \leqslant x_k) \geqslant k/100$$

When subjects are asked to estimate the 5% and 95% quantiles of a distribution, they apparently fix on a central value for X and then adjust. The 5% and 95% quantiles obtained in this way are frequently too close to the central value; that is, the true values will too often fall outside the given quantiles. We meet this phenomenon again under the heading "calibration."

REPRESENTATIVENESS

When asked to judge the conditional probability $p(A \mid B)$ that event A occurs given that B has occurred, subjects seem to rely on an assessment of the degree of similarity between events A and B. It is easy to see that this heuristic can lead to very serious biases. Indeed, similarity is symmetrical. The degree to which A resembles B is also the degree to which B resembles A. However, conditional probabilities are not symmetrical. Applying the definition of conditional probability (the first equality):

$$p(A \mid B) = \frac{p(A \text{ and } B)}{P(B)} = \frac{p(B \mid A)p(A)}{p(B)}$$

This simple result is *Bayes' theorem*. It is legitimate to equate $p(A \mid B)$ and $p(B \mid A)$ only if the *base rates* $p(A)$ and $p(B)$ are equal. The representativeness heuristic leads people to neglect the base rates and behave as if $p(A) = p(B)$. We shall examine the base rate fallacy in more detail shortly. The following example serves to illustrate the representativeness heuristic.

Subjects are presented with description X of Bill and asked to rank the probabilities of various events based on X.

X: Bill is 34 years old. He is intelligent, but unimaginative, compulsive and generally lifeless. In school, he was strong in mathematics but weak in social studies and humanities.

Please rank in order the following statements by their probability, using 1 for the most probable and 8 for the least probable:

A   Bill is a physician who plays poker for a hobby.
B   Bill is an architect.
C   Bill is an accountant.
D   Bill plays jazz for a hobby.
E   Bill surfs for a hobby.
F   Bill is a reporter.
G   Bill is an accountant who plays jazz for a hobby.
H   Bill climbs mountains for a hobby.

(Tversky and Kahneman 1982b)

The text is constructed such that C, being an account, is most representative for the description X. If the subject ranks C higher than E, this means that he regards the conditional probability $p(C|X)$ as larger than $p(E|X)$. In other words, he judges that

$$\frac{p(C|X)}{p(E|X)} > 1$$

He has probably reached this conclusion by the same psychological procedure that he would follow were he asked to rank the conditional probabilities $p(X|C)$ and $p(X|E)$. However,

$$\frac{p(C|X)}{p(E|X)} = \frac{p(X|C)p(C)}{p(X|E)p(E)}$$

To perform this task correctly, he must also weigh the base rates $p(C)$ and $p(E)$. From experiments similar to the above, it is known that subjects generally neglect the base rates. In fact, there are many more people in class E than in class C. Even if the subjects are given this information in the experiment, they fail to use it. Estimates of $p(C|X)$ and $p(E|X)$ are not influenced by information regarding the base rates. Clearly, if $p(E)$ is much larger than $p(C)$, and if $p(X|C)$ is of the same order as $p(X|E)$, then the above ratio will be less than one.

Another interesting aspect of this experiment reveals the strength of the representativeness heuristic. Event G is the conjunction of events C and D. Event C is highly representative for X, and event D is highly unrepresentative. Event G is of intermediate representativeness. Subjects using the representativeness heuristic will therefore rank G between events C and D. However $p(D|X) \geqslant p(G|X)$.

In the experiment cited above 87% of the "statistically naive" subjects ranked G higher than D. The experiment was repeated with subjects taken from graduate students in the Stanford Business School who had had several advanced courses in probability and statistics. Eighty percent of these "statistically sophisticated" subjects ranked G higher than D.

Similar results were found in the experiment of Thys (1987). Experienced operators of sophisticated technical systems did not differ significantly from their inexperienced colleagues in performance on this type of test. About 80% ranked the subset (G) above the superset (D). Moreover, this same pattern was observed on items relating to their technical expertise.

The representativeness heuristic also leads people to ignore effects due to sample size. For example, if a subject is told that an urn contains 500 white and 500 red balls, and is asked to assess the probability of drawing at least six white and four red balls on ten draws, then he will judge this probability by comparing the ratios 4/10 and 500/1000. The same ratios will be compared if he is asked to assess the probability of drawing at least 60 white and at most 40 red balls. For him the 10-ball sample and the 100-ball sample are equally representative of the population being sampled, and he will tend to equate their probabilities. This, of course, is wrong. The 60/40 sample is much less likely than the "6/4 sample."

The following experiment illustrates this effect (Tversky and Kahneman, 1982c):

> A certain town is served by two hospitals. In the larger hospital about 45 babies are born each day, and in the smaller hospital about 15 babies are born each day. As you know, about 50% of all babies are boys. However, the exact percentage varies from day to day. Sometimes it may be higher than 50%, sometimes lower.
>
> For a period of 1 year, each hospital recorded the days on which more than 60% of the babies born were boys. Which hospital do you think recorded more such days?

Of the 95 subjects responding to this question, 21 opted for the larger hospital, 21 for the smaller hospital, and 53 thought that both hospitals recorded about the same number of such days. Of course, the smaller hospital is much more likely to see more than 60% boys on any given day.[1]

CONTROL

Subjects tend to act as if they can influence situations over which, objectively speaking, they have no control whatsoever. This may lead to distorted probability assessments.

In one experiment (Langer, 1975), 36 undergraduates at Yale University were divided into two groups and each group was placed in the following betting situation. Each participant was asked to play a simple betting game with one of the two opponents. The game consisted of cutting a deck of cards. The subject determined the stakes that would be wagered (even money) and whoever cut the highest card would win. One opponent was instructed to be shy and insecure, and the other opponent was instructed to be confident and self-possessed. It was conjectured that the subjects would think they had a better chance of winning against the insecure opponent, and that this would be reflected in the amount of money they were willing to bet. The maximum bet was set at $.25, and each subject played the game four times. The median bet against the insecure opponent was $.16, whereas against the confident opponent this was $.11.

---

[1] Regarding the births as independent tosses of a coin with probability p of "heads" (for "boy"), the variance of the percentage S_n/n of heads in n tosses is $p(1 - p)/n$. Since this variance decreases with n, the probability of a given deviation from the mean value of S_n/n decreases in n.

A second experiment illustrating the effects of apparent control was conducted among office workers in New York. One group of 26 subjects was given the opportunity to buy a ticket for an office lottery at $1 each. The prize of $50 would go to the winning ticket, to be drawn from an urn. The subjects were allowed to choose their own ticket. A second group of 27 subjects was given the opportunity to buy tickets for the same lottery but were simply given tickets and were not allowed to choose. Each subject as then approached by the experimenter and asked:

> Someone in the other office wanted to get into the lottery, but since I'm not selling tickets any more, he asked me if I'd find how much you'd sell your ticket for. It makes no difference to me, but how much should I tell him?

In the group that chose their own ticket, the median resale price was $8.67, whereas in the group that were given their tickets, the median resale price was $1.96 (Langer, 1975).

THE BASE RATE FALLACY

The base rate fallacy is one of the most common and most pernicious biases involved in subjective probability assessments. It arises in a wide variety of contexts involving expert opinion. The best way of learning to recognize this fallacy is to study several examples.

Medicine

In the field of medicine, the base rate fallacy may be described as a veritable epidemic. A study by Eddy (1982) discusses several instances of this fallacy in medical textbooks. One example involves the diagnosis of breast cancer.

A definitive diagnosis for breast cancer is accomplished by means of a biopsy, a surgical operation (usually under complete anesthesia) in which a portion of suspicious breast mass is removed. This is not a trivial operation and is usually performed on the basis of a mammogram, that is, an x-ray photograph of the breast mass.

The symptoms that may lead a physician to order a mammogram may be caused by any number of disorders. It is stated that the frequency of malignant cancer of the breast among women complaining of a painful hardening in the breast is about 1/100. Numerous medical sources estimate the accuracy of mammography at about 90%. This means that the probability of a positive x-ray result given a malignant cancer is about 90% and the probability of a negative x-ray result given no malignant cancer is about 90%.

In deciding whether to order a biopsy given a positive x-ray result in a patient complaining of a painful hardening of the breast, the physician must estimate the conditional probability of cancer given a positive result, $p(C \mid +)$. Using Bayes' theorem:

$$p(C \mid +) = \frac{p(+ \mid C)p(C)}{p(+)}$$

The base rate $p(C)$ is about 1/100. The base rate $p(+)$ can be calculated as

$$p(+) = p(+ \mid C)p(C) + p(+ \mid \text{not-}C)p(\text{not-}C)$$
$$= (.9)(.01) + (.1)(.99) = 0.108$$

This yields

$$p(C \mid +) \approx 0.0833$$

Observing a positive x-ray result raises the probability of cancer from 1% to about 8%. In an informal experiment conducted by Eddy, about 95% of the physicians interviewed stated that the probability of cancer in the above situation would be about 75%.

It is hardly surprising that physicians misestimate this probability, given the advice found in textbooks and medical journals. Eddy cites the following passage from an article entitled "Mammography in its Proper Perspective," from *Surgery, Gynecology and Obstetrics* (vol. 134, p. 98, 1972):

> In women with proved carcinoma of the breast, in whom mammograms are performed, there is no x-ray evidence of malignant disease in approximately one out of five patients examined. If then on the basis of a negative mammogram, we are to defer biopsy of a solid lesion of the breast, then there is a one in five chance that we are deferring biopsy of a malignant lesion.

Thys (1987) using experienced and inexperienced operators of technical systems, provided results similar to Eddy's informal experiment. Moreover, there was no significant difference between the two groups, either on general knowledge items or on technical items. Christensen-Szalanski and Bushyhead (1981) report that doctors are able to take the base rate into account, at least partially, when the base-rate information is obtained by experience in a clinical setting.

Simulator Training

The following example resulted from an analysis of a simulator training program in The Netherlands performed by members of the Department of Mechanical Engineering at the Delft University of Technology (Thys 1987).

A simulator training program was developed at a research center to train operators of large technical installations. A pilot group of trainees was selected from operators currently employed at the type of installation for which the training was designed. Part of the training program worked as follows. The simulator would choose a particular malfunction from its library of malfunctions, and the appropriate instrument readings would appear on the simulator's control panel. The trainee had to diagnose the malfunction from the instrument readings. For example, the malfunction might be an oil leak and the symptoms on the instrument panel would be an alarm indicating oil underpressurization and high temperature readings for the coolant of a particular component. The same symptoms might also be caused by failure in the oil pump. The engineering knowledge that the trainees were being taught enabled them to assess the probabilities that various malfunctions would cause various meter readings.

The frequencies of various malfunctions are determined by installation-specific factors such as maintenance regime, load, weather, supplier, and design. Each installation has its own particular signature of base rate frequencies. The oil pump may frequently have given trouble in the past, or the oil lines may be of inferior quality. The designers of the simulator program could not know this signature of the trainee's particular installation, and they therefore chose to confront trainees with malfunctions chosen *at random*.

An initial evaluation of the results of the training program indicated that the trainees performed worse on particular points after completing the course, than before. The problem was analyzed as follows. Let M be a particular malfunction, and R a particular reading on the control panel. The operator's knowledge enabled him to assess $p(R \mid M)$, whereas his diagnostic task required him to assess $p(M \mid R)$. Let M' be another malfunction which might cause R. In choosing between M and M', the operator has to compare $p(M \mid R)$ and $p(M' \mid R)$. It is not difficult to see that

$$\frac{p(M \mid R)}{p(M' \mid R)} = \frac{p(R \mid M)p(M)}{p(R \mid M')p(M')}$$

Since malfunctions are chosen randomly on the simulator, the bases rates $p(M)$ and $p(M')$ were equal *on the simulator*. Hence

$$\frac{p(M \mid R)}{p(M' \mid R)} = \frac{p(R \mid M)}{p(R \mid M')} \qquad \text{on the simulator}$$

The simulator was teaching the trainees to estimate $p(M \mid R)$ as proportional to $p(R \mid M)$, effectively wiping out the implicit knowledge they had of the base rates at their particular installations. When they return to their installations, this is reflected in degraded performance. The simulator was teaching them to commit the base rate fallacy.

Empirical Bayesians

The final example of the base rate fallacy is a bit embarrassing, as it is committed by Bayesians and seems to have been going on in risk analysis for several years (Martz and Bryson, 1983). It has been used to estimate the spread of failure probabilities in a given population on the basis of the spread in expert estimates of these probabilities (Martz, 1984).

In risk analysis, experts are called upon to estimate failure rates of components. Such failure rates are commonly assumed to be lognormally distributed, that is, the log of the failure frequency per unit time is assumed to be normally distributed over the population of similar components. For simplicity, we assume that the experts are asked to estimate log failure frequencies.

Suppose an expert estimates a log failure frequency as y. As he is not certain this estimate is correct, he gives some indication of his subjective uncertainty. The usual way of doing this is to give the standard deviation σ of his distribution over the possible values of the log failure frequency. The usual interpretation is this: The expert's subjective probability distribution for the log failure frequency in question is a normal distribution with mean y and standard deviation σ, notated $N(y, \sigma)$.

The "empirical Bayes" methodology involves treating the expert's responses themselves as random variables. In other words, we assume there is a probability density $p(x, y, \sigma)$ giving the probability density that the true value is x, that the expert estimates this value as y with standard deviation σ. The expert may be subjected to various biases but in the simplest case he is unbiased. In this case the empirical Bayes methodology assumes that if the true value is x and the expert gives standard deviation σ, then the probability that he estimates x as y, $p(y \mid x, \sigma)$, is normal with mean x and standard deviation σ:

$$p(y \mid x, \sigma) = (2\pi\sigma^2)^{-1/2} e^{-(1/2)[(y-x)/\sigma]^2}$$

Now what does it mean in this context to say that the expert is unbiased? The empirical Bayesians have not given a definition; but it seems reasonable to say this: If an unbiased expert's subjective distribution is normal with mean y and standard deviation σ, with y and σ fixed, for a large number of log failure frequencies, then if we examine the true values for these frequencies, they will indeed be normally distributed with mean y and standard deviation σ.[2] For such an expert,

$$p(x \mid y, \sigma) = (2\pi\sigma^2)^{-1/2} e^{-(1/2)[(x-y)/\sigma]^2}$$

The normal density depends only on the square distance to the mean and the standard deviation, hence

$$p(y \mid x, \sigma) = p(x \mid y, \sigma)$$

However, Bayes' theorem says

$$p(y \mid x, \sigma) = \frac{p(x \mid y, \sigma)p(y \mid \sigma)}{p(x \mid \sigma)}$$

For this unbiased expert, then, we must have

$$p(y \mid \sigma) = p(x \mid \sigma) \qquad \text{for all } x, y$$

which in turn entails that both $p(y \mid \sigma)$ and $p(x \mid \sigma)$ are the improper uniform density[3]. As this holds for all values of σ, it follows that $p(y)$ is also the improper uniform density. In addition to being 'improper,' this is inconsistent with other modeling assumptions made by the empirical Bayesians and highly implausible.

OVERCONFIDENCE AND CALIBRATION

Calibration will be treated systematically in Part II. We content ourselves here with a rough provisional definition: A subjective assessor is *well-calibrated* if for every probability value r, in the class of all events to which the assessor assigns subjective probability r, the relative frequency of occurrence is equal to r. Calibration represents a form of empirical control on subjective probability assessments.

This concept was introduced in Chapters 1 and 2. In Chapter 2, for example,

---

[2] Martz (1986) seems not to accept this interpretation of unbiasedness, but does not propose another (see also Apostolakis 1985, and Cooke, 1986).

[3] Improper probability densities and not 'real' probability densities, as the total probability is infinite.

we saw that the precursor risk studies of nuclear plants could be used to calibrate the probabilistic assessments in the *Reactor Safety Study*.

Calibration can be measured in two types of test, *discrete tests* and *fractile* or *quantile tests*. In a discrete test the subject is presented with a number of events. For each event, he is asked to state his probability that the event will occur. His probabilities are discretized, either by himself or by the experimenter, such that only a limited number of probability values are used. For example, his probabilities may be rounded off to the nearest 10%. It is helpful to think of the subject as throwing events into "probability bins," where, for example, the 20% probability bin contains all those events to which the subject attributes (after discretization) a probability of 20%. The subject is well calibrated if the relative frequency of occurrence in each probability bin is equal to the corresponding bin probability.

Quantile tests are used when the subject is required to assess variables with a continuous range (continuous for all practical purposes). For example, we may be interested in the maximal efficiency of a new type of engine under certain operating conditions. No one would predict this value with certainty, but an expert will typically be able to give a subjective distribution over the real line, reflecting his uncertainty. We can learn something about this distribution by asking the following type of question: "For which x is your probability 25% that the engine's efficiency is less than or equal to x?" The expert's answer is called his 25% quantile. In a risk analysis we are typically interested in 5%, 50% and 95% quantiles, so these are often elicited from experts. Another popular choice is 1%, 25%, 50%, 75%, 99%.

Suppose we ask an expert for his 1%, 25%, 50%, 75%, and 99% quantiles for a large number of variables, for which the actual values later become known. If the expert is well calibrated, then we should expect that approximately 1% of the true values fall beneath the 1% quantiles of their respective distributions, roughly 24% should fall between the 1% and the 25% quantiles, etc. The *interquartile range* is the interval between the 25% and the 75% quantiles. We should expect 50% of the true values to fall within the interquartile ranges. The *surprise index* is the percentage of true values that fall below the lowest or above the highest quantile. For the quantiles given above, we should expect this to occur in 2% of the cases.

Interest in calibrating subjective probabilities seems to have arisen in meteorology. National Weather Service forecasters have been expressing their predictions of rain in probabilistic terms since 1965. An enormous amount of data has been collected and analyzed for calibration. For example, Murphy and Winkler (1977) analyzed 24,859 precipitation forecasts in the Chicago area over 4 years. The results shown in Figure 4.2 indicate excellent calibration. It is reported in Lichtenstein, Fischhoff, and Phillips (1982) that the more recent calibration data is even better.

The data from weather forecasters are highly significant for two reasons. First, it shows that excellent expert calibration is possible, and second, closer analysis reveals that expert calibration improves as the results of calibration measurements are made known. Murphy and Daan (1984) studies the learning effect in a 2-year experiment in The Netherlands. Figure 4.3 compares the overall calibration of the forecasters in the first and second years of the experiment.

The experiment on which Murphy and Daan report was continued up to 1987.

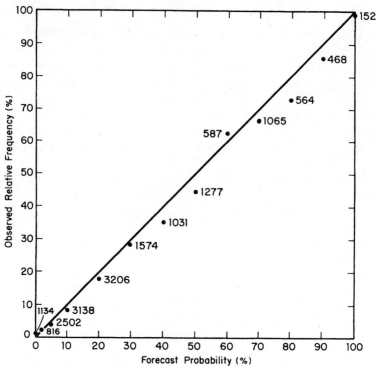

Figure 4.2 Calibration data for precipitation forecasts. The number of forecasts is shown for each point. (*Murphy and Winkler, 1977*)

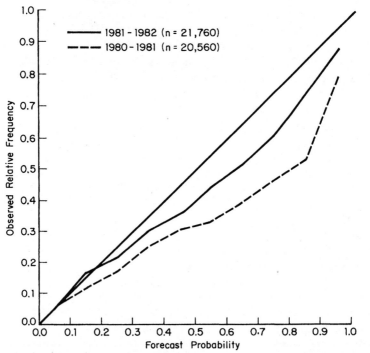

Figure 4.3 Comparison of calibration of weather forecasters before (1980–1981) and after (1981–1982) receiving feedback on their own performance. The number of forecasts in each year is shown as *n*. (*Murphy and Daan, 1984*)

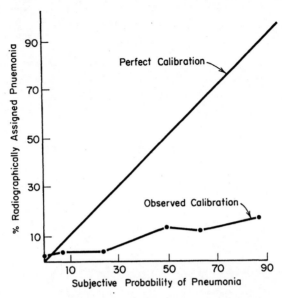

Figure 4.4 Relation between physician's subjective probability of pneumonia and the actual probability of pneumonia. (*Christensen-Szalanski and Bushyhead, 1981*)

The raw data have recently been obtained and analyzed using the models for combining expert opinion developed in Part III. In Chapter 15 we shall see how this combination leads to superior probabilistic forecasts.

Expert calibration is not always so good. Figure 4.4 shows calibration for Army physicians for diagnoses of pneumonia (Christensen-Szalanski and Bushyhead, 1981). The calibration is described as "abysmal." For subjective probability of 90% the corresponding relative frequency is about 20%. It is natural to suppose that the doctors are "erring on the side of safety" in their predictions in these cases. The experimenters were aware of this possibility and instructed the doctors to try to avoid this. Moreover, the doctors were asked to evaluate the effect of making true and false diagnoses given that the patient does or does not have the disease. The mean results, shown in Figure 4.5, indicate that failing to diagnose pneumonia when the patient in fact had the disease was slightly *preferable* to diagnosing the disease when the patient in fact was healthy.

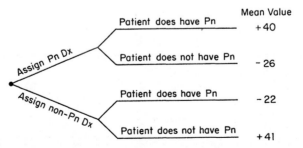

Figure 4.5 Physicians' mean values for outcomes of pneumonia diagnosis (PnDx) decision. (*Christensen-Szalanski and Bushyhead, 1981*)

Table 4.1 Calibration Summary for Quantile Tests

	N	Observed Interquartile Index*	Surprise Index Observed	Surprise Index Ideal
Alpert & Raiffa (1969)				
Group 1-A (.01, .99)	880		46	2
Group 1-B (.001, .999)	500		40	.2
Group 1-C ("min" & "max")	700	33	47	?
Group 1-D ("astonishingly high/low")	700		38	?
Groups 2, 3, & 4				
Before training	2,270	34	34	2
After training	2,270	44	19	2
Hession & McCarthy (1974)	2,035	25	47	2
Selvidge (1975)				
Five fractiles	400	56	10	2
Seven fractiles (incl., 1 & .9)	520	50	7	2
Moskowitz & Bullers (1978)				
Proportions				
Three fractiles	120	—	27	2
Five fractiles	145	32	42	2
Dow-Jones				
Three fractiles	210	—	38	2
Five fractiles	210	20	64	2
Pickhardt & Wallace (1974)				
Group 1				
First round	?	39	32	2
Fifth round	?	49	20	2
Group 2				
First round	?	30	.46	2
Sixth round	?	45	24	2
T. A. Brown (1973)	414	29	42	2
Lichtenstem & Fischhoff (1980b)				
Pretest	924	32	41	2
Post-test	924	37	40	2
Seaver, von Winterfeldt, & Edwards (1978)				
Fractiles	160	42	34	2
Odds-fractiles	160	53	24	2
Probabilities	180	57	5	2
Odds	180	47	5	2
Log odds	140	31	20	2
Schaefer & Borcherding (1973)				
First day, fractiles	396	23	39	2
Fourth day, fractiles	396	38	12	2
First day, hypothetical sample	396	16	50	2
Fourth day, hypothetical sample	396	48	6	2
Larson & Reenan (1979)				
"Reasonably certain"	450	—	42	?
Pratt (1975)				
"Astonishingly high/low"	175	37	5	?
Murphy & Winkler (1974)				
Extremes were .125 & .875	132	45	27	25
Murphy & Winkler (1977b)				
Extremes were .125 & .875	432	54	21	25
Staël von Holstein (1971a)	1,269	27	30	2

Note: N = total number of assessed distributions.
*The ideal percentage of events falling within the interquartile range is 50, for all experiments except Brown (1973). He elicited the .30 and .70 fractiles, so the ideal is 40%.
Source: Lichtenstein, Fischhoff, and Phillips, 1982

An interesting study involving probabilistic assessments of civil engineers was performed by Hynes and Van Marcke (1976). An embankment on a clay foundation, originally part of a projected highway, was overloaded with extra fill until the foundation failed. A number of internationally known geotechnical engineers were invited to predict the height of fill at which failure would occur. Failure was defined as a deformation at the crest of the embankment of at least 0.5 meter. The experts received extensive documentation prior to making their predictions. Seven experts completed the assessment task, and the results are shown in Figure 4.6. The elicitation of experts' opinions took the form of the quantile tests discussed above. The continuous variable whose quantiles were assessed was "added height to failure." The distributions were interpolated on the basis of various quantiles. Figure 4.6 shows the best (median) estimate of each expert, together with the interquartile range, that is, the interval between the 25% and the 75% quantiles of the corresponding distribution. None of the experts managed to catch the true value in their interquartile range. Note also the substantial spread in the experts' assessments. Curiously, the true value is approximately equal to the average of the experts' median estimates.

Table 4.1, taken from an extensive review of (predominantly nonexpert) calibration data (Lichtenstein, Fischhoff, and Phillips, 1982, the original references may be retrieved from this source) shows the results of a large number of

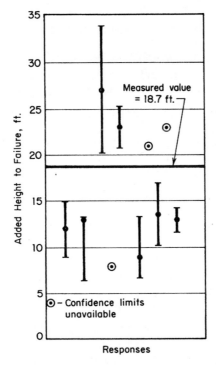

Figure 4.6 Predictors' interquartile ranges and best estimates of added height to failure. (*Hynes and Van Marcke, 1976*)

calibration experiments using continuous variables. Many of these experiments attempted to measure the effect of training on calibration performance.

The "observed interquartile index" is the frequency with which the true values fell in the subjects' interquartile ranges, and the "observed surprise index" is the frequency with which the true values fell outside the interval between the lowest and highest quantiles. The "ideal surprise index" is the value we should expect for the observed surprise index if the subjects were well calibrated. This value is typically 2%. It is remarkable that the surprise index is generally larger than the interquartile index, especially before training. There is some evidence that training is effective in improving calibration.

A rather different type of calibration involved 32 clinical psychologists, and students in training to become clinical psychologists. The subjects were presented with actual case descriptions of psychiatric patients in four successive stages. In each stage each subject was given a sort of multiple choice test. Each question had five possible answers, for example:

> During college, when Joseph Kidd (the patient) was in a familiar and congenial situation he often
>
> a. tried to direct the group and impose his wishes on it
> b. stayed aloof and withdrawn from the group
> c. was quite unconcerned about how people reacted to him
> d. took an active part in the group but in a quiet and modest way
> e. acted the clown and showed off.

The correct answer was known from the case history, but not revealed to the subject. After each stage, the patient chose the answer to each question that he considered most likely and gave his confidence (subjective probability) that the answer was correct. Twenty percent confidence means that the subject is guessing. The results for the group of 32 subjects are shown in Table 4.2.

From the analysis of the results it emerged that there was no significant difference in the performance of the experienced clinicians and the students. Furthermore, there was no significant improvement trend in the accuracy. Giving more information about the patient did not significantly increase the percentage of correct answers. However, increased information did affect the confidence in the answers. Whereas the level of confidence was more or less appropriate at stage 1, the subjects had become quite overconfident in stage 4.

CONCLUSION

There is good news and bad news. The good news is that the probabilistic representation of uncertainty provides clear criteria for evaluating subjective probability assessments. It is scarcely conceivable that a subject would wish to persist in any of the biases discussed in this chapter, when these would be brought to his/her attention. This means that subjective probabilities are not beyond the pale of rational discourse.

The bad news is that people, even experts, do not handle subjective probabilities with any great aplomb. There is ample room for improvement, and

Table 4.2 Performance of 32 Judges on the 25-Item Case-Study Test

Measure	Stage 1	Stage 2	Stage 3	Stage 4
Accuracy (%)	26.0	23.0	28.4	27.8
Confidence (%)	33.2	39.2	46.0	52.8
Number of changed answers	—	13.2	11.4	8.1

Source: Oskamp, 1982.

there is indication that training in "reasoning with uncertainty" can be worthwhile. This having been conceded, experiments discussed in Chapter 10, using suitable measures for calibration and information, how that experienced experts, as a group, significantly outperform inexperienced experts.

5

Building Rational Consensus

The foregoing four chapters have explored a very wide range of subjects, all connected with the use of expert opinion in science. Part II is devoted to assembling the mathematical modeling tools that will be needed in Part III. Before embarking on mathematical modeling, it is useful to gather some conclusions from the previous chapters and to form a preliminary picture of what we would like to accomplish in Part III. Chapter 1 ended with a question, what does it mean to represent uncertainty adequately for purposes of science? We shall not answer this question here. Rather, we return to the fundamental aim of science and derive some general principles that a methodology for using expert opinion in science should satisfy. At the same time, we review the existing practice, as it has evolved, and determine whether science's fundamental aims are presently being served.

RATIONAL CONSENSUS

Science aims at rational consensus, and the methodology of science must serve to further this aim. Were science to abandon its commitment to rational consensus, then its potential contribution to rational decision making would be compromised.

Since the authoritative Lewis report (Lewis, 1979), the use of experts' subjective probabilities in risk assessment has been officially sanctioned and universally recognized. Nonetheless, a subjective probability is "just someone's opinion." Traditional scientific methodology does not explicitly accommodate the use of opinions as scientific data.

The Lewis report did not address the question *how* subjective probabilities should be used in science. Experience with expert opinion to date has demonstrated both the potential value and the potential dangers of expert subjective probability assessments. In short, the use of subjective probabilities could further the aim of rational consensus, but could also obstruct this aim.

One conclusion from the foregoing chapters is overwhelmingly evident. Expert opinion may, in certain circumstances, be a useful source of data, but it is not a source of rational consensus. Given the extreme differences of opinion encountered in virtually every example, it is clear that an unmethodological use of expert opinion will not contribute to rational consensus building.

The following principles represent an attempt to formulate guidelines for using expert opinion in science. These principles have been generated in a research project into models for expert opinion, under the auspices of the Dutch government, and they underly the models presented in Part III.

PRINCIPLES

Reproducibility

It must be possible for scientific peers to review and if necessary reproduce all calculations. This entails that the calculational models must be fully specified and the ingredient data must be made available.

It goes without saying that reproducibility is an essential element of the scientific method. Nevertheless, there is no existing study that fully respects this principle. The Reactor Safety Study (1975) faithfully reproduced the data from experts, including the names of the individuals and or institutions from which the data originated. However, these data were synthesized in a way that was entirely inscrutable. Hence, it was impossible for reviewers to reproduce the resulting assessments. The SNR-300 risk study (Hofer, Javeri, and Loffler, 1985) use weighted combinations of expert opinions. However, the weights are not reported. Again, it is impossible to reproduce the calculations.

Accountability

The source of expert subjective probabilities must be identified.

The notion of accountability is essential to science. Articles for scientific journals are not considered for publication unless the author is identified. In the present context, accountability entails that the decision maker can trace every subjective probability to the name of the person or institution from which it comes. In the cases of public decision making, this information must be made public.

The controversial Reactor Risk Reference Document (Office of Nuclear Regulator Research, 1987) is cited as an example where expert anonymity resulted in controversial assessments (the December 1988 issue of Reliability Engineering and System Safety, devoted to expert opinion in risk analysis, was prompted by this controversy). The Seismic Hazard Characterization of the Eastern United States (Bernreuter et al., 1984) publishes the names of the experts consulted. However, the individual assessments are associated with the *number* of an expert, and the reader is not told which name corresponds to which number. On the other hand, the bench mark studies from the European Joint Research Center at Ispra (Amendola, 1986, Poucet, Amendola, and Cacciabue, 1987) identify the experts by name and give their individual assessments.

A potentially very serious breach of accountability occurs in the various expert system methodologies reviewed in Chapter 3. The expert opinions output by such systems belong to no one. Who is accountable? Can an expert diagnostic system be sued for malpractice?

Accountability to the decision maker enhances quality and credibility and must be insisted upon. Outside the sphere of public decision making, the additional enhancement achieved by publishing names and assessments must be weighed against possible disadvantages. For example, the following situation may easily arise. Experts for judging the reliability of new technical systems will often be found in the employ of contracting firms designing the systems in question. In this case, the decision maker is the contractor. An expert's assessment of reliability parameters may well be in conflict with the firm's financial interests, as perceived by upper management. If the expert works for the firm, then publishing his name and assessment would place him in an unacceptable conflict of interest.

Mathematically speaking, the models presented in the succeeding chapters require that assessments originating from the same source be identified as such. Of course, they do not require that the source be identified by name.

Empirical Control

Expert probability assessments must in principle be susceptible to empirical control.

The requirement of empirical control through observations is universally recognized as a cornerstone of scientific methodology. In contemporary methodology this requirement is given the following expression: Scientific statements and scientific theories should be *falsifiable in principle*. It is recognized that theories can never be conclusively verified, but at least it should be possible *in principle* to discover a reproducible conflict with observations, if the theory is in fact false. The necessary experiments need not be feasible, but they must be physically possible.

In the same spirit, a methodology for using expert opinion should incorporate some form of empirical control, at least in principle. In other words, it must be possible in principle to evaluate expert probabilistic opinion on the basis of possible observations. Moreover, this evaluation should be reflected in the degree to which an expert's opinion influences the end results.

Empirical control ensures that the use of subjective probabilities cannot be construed as a license for the expert to say anything whatever. Without empirical control it is easy to argue that "one subjective probability is as good as another," and subjective probabilities would be of very limited value in reaching rational consensus. Further, the expert would be deprived of a weapon for resisting the institutional and psychological pressures that may well attend the assessments of critical quantities in risk analysis.

There has been no application of expert opinion in risk analysis that utilizes the possibilities of empirical control. Most notable in this respect is the use of expert opinion in Wheeler et al. (1989). After extensive elicitation procedures, expert opinions are combined by simple arithmatical averaging. However, the published literature (Morris and D'Amore, 1981) indicates that such possibilities are utilized in the field of military intelligence. Some of the Bayesian models used in risk analysis (Apostolakis, 1985) incorporate empirical control via Bayesian updating methods.

Neutrality

The method for combining/evaluating expert opinion should encourage experts to state their true opinions.

A poorly chosen method of combining/evaluating expert opinion will encourage experts to state an opinion at variance with their true opinion. Perhaps the best-known examples of this are found in the Delphi techniques. As Sackman (1975) demonstrates, these techniques "punish" experts who deviate strongly from a median value and reward changes of opinion in the direction of the median.

Most methods for forming weighted combinations of expert probability assessments must be criticized from the viewpoint of neutrality. For example, the Seismic Hazard Characterization of the Eastern United States (Bernreuter et al., 1984) and the SNR-300 risk study (Hofer, Javeri, and Loffler, 1985) both use weighted combinations of expert opinion, as mentioned above. The former determines these weights by asking the experts to weight themselves according to how good an expert they think they are for the question at hand (these are termed "self-weights"). The latter employs so called "De Groot weights," whereby each expert weighs the expertise of each other expert (including himself).

The idea of using self-weights was first introduced by practitioners of the Delphi method. Although an initial study (Dalkey, Brown, and Cochran, 1970) indicated that self weights resulted in improved accuracy, a later and more extended study challenged this conclusion (Brockhoff, 1975). Curiously, women consistently rate themselves lower than men (Linstone and Turoff, 1975, p. 234). There is no psychometric measurement underlying the notion of "good expert," hence the scale for self-weights cannot be given any operational meaning (does a rating of 6 mean "twice as good an expert" as a rating of 3?).

A high rating is obviously a form of reward. The system of self-weights asks an expert to punish and reward himself; De Groot weights empower the experts to punish and reward each other. Without wishing to suggest that experts would misuse such rating systems, it must be affirmed that these systems offer no incentive for performing these tasks honestly. Indeed it is not clear what this latter notion would mean, as the scales used in the ratings are not susceptible of operational interpretation.

We note in addition that the use of self-weights or De Groot weights makes it very difficult to satisfy the principles of reproducibility and accountability.

Fairness

All experts are treated equally, prior to processing the results of observations.

Most Bayesian models for combining expert probability assessments require the analyst to "assess the reliability of a given expert." No guidance is given as to how this should be done, the analyst is simply expected to quantify his trust in some inscrutable way. This type of model is clearly ruled out by the above principle. It would indeed be curious to subject the expert assessments to empirical control, but to allow the analyst to nullify this by his own assessments of reliability of individual experts.

Since empirical control is acknowledged as the means for evaluating expert opinions, in the absence of any empirical data there is no reason for preferring one expert to another.

Of course, the analyst must "prefer" one expert to another when he decides which experts to consult. However, it is judged that these decisions must be made initially on the basis of factors that cannot be meaningfully translated into numerical input in the combination models.

CONCLUSION

There is no uniformity in the use of expert opinion at present. There is no method at present which satisfies all of the above principles. There is every reason at present to develop these methods.

II
SUBJECTIVE PROBABILITY

6

Savage's Normative Decision Theory

L. J. Savage's *Foundations of Statistics* appeared in 1954 and provided a powerful synthesis of many ideas surrounding uncertainty and the interpretation of probability. In particular, he linked the problem of representing uncertainty with the problem of representing rational preference and modeling rational decision. He showed that *rational* preference can be uniquely decomposed into subjective probability and (affine unique) utility. That is, a decision maker prefers act *f* to act *g* if and only if the expected utility of *f* is greater than that of *g*. The subjective probability used in this representation is uniquely determined by the decision maker's preferences, and the utility is determined up to the choice of a zero and a choice of units on the utility scale. Utility is therefore determined up to a positive affine transformation.

Savage's theory is an intellectual milestone for at least two reasons. First, it fully clarifies the nature of subjective probability and provides an interpretation of subjective probability in terms of observable preference behavior. Second, it gives us a fully developed quantitative model of what rational decision is. It tells us when we should expect that all rational people will agree, and when we must admit that rational people may hold differing opinions.

The main virtue in knowing what rational decision is, is that we also know what rational decision is *not*. There are many decision contexts in which Savage's model cannot be applied. The most notable example involves group decisions in which the members of the group do not share the same utility functions or the same probability assessments. Rational decision in Savage's sense is just not meaningful in such contexts; it makes no sense to look for a "rational group decision." Put differently, there is no rational way for choosing a particular probability measure or a particular utility function. If this simple fact were sufficiently appreciated it would save much time and much skull bashing. If groups cannot decide *rationally*, then they must decide in some other way, for example, *fairly*.

Savage's theory of rational decision has given rise to a rich body of literature. His axioms of rational preference have inspired much empirical research. Some of this work (Kahneman and Tversky, 1979; Tversky, 1969; Ellsberg, 1961; Allais, 1953, 1979; MacCrimmon, 1968) has shown that people do not in general behave in conformity with Savage's axioms. Of course, this does not threaten the normative status of the theory, it merely shows that rational choice is not so simple as we might like to believe. Perhaps the most interesting work in this direction is that of MacCrimmon (1968). MacCrimmon studied preference behavior in middle- and upper-level executives enrolled in a management training course. He found significant departures from the Savage axioms. However, when these departures were pointed out to the persons involved, they usually admitted that they had "made mistakes" and revised their preferences to conform with Savage's requirements. Moreover, the more experienced executives made fewer "mistakes" than the less experienced executives.

Savage's ideas have been critized on normative grounds (Allais, 1953, 1979; Machina, 1981; Shafer, 1986) and on theoretical grounds (Luce and Krantz, 1971; Blach and Fishburn, 1974; Cooke, 1983, 1986) as well. There have also been attempts to refine and improve the axiomatic basis of rational decision (Jeffrey, 1966; Blach, McFadden, and Wu, 1974; Pfanzagl, 1968; Luce and Krantz, 1971; Krantz, Luce, Suppes, and Tversky, 1971; Blach and Fishburn, 1974). Much current research looks for generalizations and/or alternatives of the "expected utility model" that capture more of observed empirical behavior. Although this work has led to some generalizations and qualifications, it has not fundamentally altered our thinking about rational decision. All theories of rational decision being discussed today have their points of departure in Savage's work. A good elementary discussion of some of these issues is found in Hogarth (1987).

Of course Savage's theory is well known among decision theorists and econometricians, and there would be little point in describing it in a book addressed to them. However, the problems surrounding expert opinion are presently attracting attention from researchers from very diverse backgrounds, and a contemporary review of Savage's model seems eminently appropriate for them. As Savage's original proofs are not tailored for this public, it seems appropriate to include proofs of the main results suitable for undergraduates having taken a course in probability. The first section of this chapter discusses Savage's decision model. The second section sketches the important representation theorem. Proofs are given in the supplement to this chapter. A third and final section outlines the role of observation within the theory of rational preference.

THE DECISION MODEL

Regarded as a theory of rational preference, Savage's most profound insight might well be his answer to the very first question any such theory must answer: What is preference about? Naively, we would probably say that preference concerns *things*, and if we reflected a little on the matter, we might conclude that preference is really about *events*. However, if a theory of rational preference is to be based on observable behavior we immediately become entangled in the question what it means to prefer *thing* A to *thing* B, or *event* A to *event* B. What observable behavior would such preferences entail? When we reflect a little more on the matter, it becomes evident that we can only really *express* preference in our choice behavior. What sorts of "things" do we choose? Do we choose *things*, *events*? Not really, not directly. We don't choose *things*, we choose to *do things*. We choose acts, and acts are the proper objects of our preferences. Accordingly, in Savage's theory preference is regarded as a relation between acts.

Now what is an act? It is a piece of voluntary behavior undertaken for some purpose. In general, however, we are not certain what the outcomes of our acts will be. We choose acts on the basis of our expectations regarding their outcomes, but we are never certain exactly which outcomes our acts will have. This uncertainty is really uncertainty regarding the true state of the world. If we knew all the details of the world in which we act, then we would also be able to predict with certainty the outcome of every action. When we think of acts in this way, we must also ask what we mean by "outcome." The outcome of an act is surely determined by the true state of the world, but it is not *part* of that state, since we can consider performing different and mutually exclusive acts in one and the same world.[1] Knowing the state of the world does not involve knowing which act we are going to perform, since otherwise there would be nothing to deliberate about. Decision making is not a branch of natural science. Although it sounds paradoxical at first, we do acts for the sake of their outcomes, but their outcomes are not part of the real world.

We have seen that a behaviorally well founded theory of rational decision involves three fundamental concepts: acts, states of the world, and outcomes. Let us illustrate these with an example. Suppose that one needs some form of transportation. Among the acts which one can consider are buying a car and buying a motorcycle. Now what is the outcome of the act buying-a-car? Naively, one would answer, a car. But "a car" is not really a single outcome. One may get an attractive new job further away from home thus increasing the "base load" demand on the car. On the other hand, the job is not certain. One may wreck the car driving out of the lot, or one might get years of good service. The price of gasoline might go up to $10 a gallon, or it might stay the same. These are all different states of affairs that would affect the outcome of the act under consideration.

We construct the reduced set of states generated by the events of interest to us in choosing a form of transportation. To keep things simple, we shall represent these events as

A: Base load demand goes up
B: Have an accident
C: Price of gas goes up

It is obvious that these events will have different impacts depending on which act we choose. If the price of gas goes up, this would favor the choice of a motorcycle, if the new job comes through this would favor the choice of a car. Having an accident with a motorcycle is worse than having an accident with a car.

Each of these events could occur in many different ways, and the differences might be very important. For example, one could have an accident in a parking lot or driving off a cliff. However, in each decision problem we have to cut the analysis

---

[1] Strictly speaking, we must consider this as a modeling requirement and not as a theorem of metaphysics. In one legitimate sense of the word, our acts *are* parts of the state of the world. In this sense we can be uncertain which act we shall perform, just as we are uncertain which state we live in. However, this type of uncertainty cannot be represented in Savage's theory, and his "logic of partial belief" is less general in this sense than logic itself. For a discussion of this and other philosophical aspects see Cooke (1986).

off at some point. Let us agree to cut this analysis off with the above events and their complements (denoted as A' etc.). This just means that we are not going to distinguish types of accidents, or prices of gasoline, etc.

A state or possible world for this decision problem is a complete specification of which events occur. Alternatively, it is an atom in the field generated by the events of interest for this problem. There are eight states for this problem:

A and B and C
A and B and C'
A and B' and C
A and B' and C'
A' and B and C
A' and B and C'
A' and B' and C
A' and B' and C'

Now, in deciding whether to buy a car or a motorcycle, we have to consider the outcomes of each act in all these eight states, evaluate the outcomes of each act in each state, and evaluate the probability of each state. What is then the outcome of an act? It is simply the degree of satisfaction associated with a given act in a given state. An outcome understood in this sense is what Savage calls a *consequence*. A consequence is simply a *state of the subject*. Regarded mathematically, actions are functions taking states of the world into states of the subject.

Supposing we have done all this, which act do we choose? According to Savage, we should choose the act with the best expectation. Where S denotes the set of all states, the expectation of an act is

$$\text{Expectation of act} = \sum_{s \in S} p(s) \times \text{consequence of act in } s$$

There is much yet to be explained before we can really understand the above formula. For example, how do we determine the probability of state s and how do we multiply a probability with a "state of the subject"? This will all be explained in the next section; however, there is one feature that must be addressed before going further.

There is something wrong with the above analysis of our decision problem. In the formula for expectation of act, the probability of each state is independent of the act under consideration. The $p(A$ and B and C$)$ used for determining the expectation of the act "buy a car" will also be used to determine the expectation of the act "buy a motorcycle." Now it is reasonable that the price of gasoline and getting a new job do not depend on my choice of mode of transportation. But what about B; "have an accident"? It seems plausible that the probability of this event (and hence of the event A and B and C) *does* depend on my choice of transportation.

The above analysis of the decision problem is not suitable for the decision model which Savage puts forward. However, there is a simple "technical fix" that will convert the above analysis into an analysis for which Savage's model *is* suitable. We simply throw out the event "have an accident" and replace it with two other events:

B$_1$:   Have an accident if driving a car
B$_2$:   Have an accident if driving a motorcycle

We must remember that B$_1$ stands for all possible states in which we would have an accident if we were driving a car and so on. There are some possible worlds in which we would have an accident if we were driving a motorcycle, but not if driving a car, some worlds in which we would have an accident driving a car, but not a motorcycle, some worlds in which we would have an accident regardless what we were driving and some worlds in which we would have no accident regardless.

We now have 16 states instead of 8, and the additional states enable us to express our belief that driving a motorcycle is more dangerous than driving a car. For example, letting B$_2'$ denote "not-B$_2$," etc., we might then have

$$p(\text{A and B}_1 \text{ and B}_2' \text{ and C}) < p(\text{A and B}_1' \text{ and B}_2 \text{ and C})$$

If we buy a car, then states involving "B$_1'$ and B$_2$" are those in which we say "Whew, had I bought a motorcycle instead of a car, I would have had an accident." On the other hand, in those states involving "B$_1$ and B$_2'$" we would say "why didn't I buy a motorcycle." The above inequality means that we are more likely to be in the first situation than in the second, if we buy a car.

Hopefully, this all seems quite elementary—the example is chosen to make things perspicuous. Analyzing a decision problem into states whose probabilities do not depend on our actions is always possible, but seldom really carried through. As a thought experiment try rerunning the above example replacing "buy a car" with "adopt an energy policy based exclusively on fossile fuels," and "buy a motorcycle" with "adopt an energy policy relying on nuclear reactors for supplying the base load."

Supposing now that we have analyzed our decision problem into act-independent states, how do we know which probabilities to use for the states, and how do we arrive at a numerical scale for consequences? Indeed, are there such probabilities, and can we represent consequences with numbers?

THE REPRESENTATION THEOREM

Savage answers both questions in the affirmative. In the "big decision problem of life" Savage proves that a rational agent always behaves in such a way that he prefers act f to act g if and only if the expected utility of f is greater than that of g. The expected utility is calculated with a (subjective) probability that is uniquely determined, and utility is calculated with a utility function on consequences that is unique up to a positive affine transformation. The subjective probabilities and utilities are peculiar to the agent, different agents will in general have different probabilities and utilities. Savage's theory does not prescribe probabilities and utilities; any probability in combination with any utility can lead to rational behavior in his sense, provided the subject chooses according to the principle of maximum expected utility. According to Savage, it it not possible to specify rational behavior further.

The proof of this fact is rather technical, and Savage's original proof is quite delicate. A simplified proof is presented in the supplement to this chapter. In this section we will "walk through" the ideas behind the proof. In the course of doing so we shall indicate what a "rational agent" is. The axioms mentioned in prose are formulated precisely in the supplement, where notation is also summarized (see also the mathematical appendix).

Step 1: Qualitative Probability

The first step in the argument involves showing how a preference relation between acts generates a qualitative probability relation between events. In order to define this notion and guarantee some essential properties, a number of assumptions on the subject's preference order must be made. These assumptions will partially characterize the notion of "rational preference." Let $f \geqslant g$ denote "act f is at least as preferable as act g;" we can 'lift' the preference relation to consequences, by considering preferences between constant acts, that is, acts taking a single consequence in all states. We assume in the first place:

Preference generates a weak order on acts; that is to say:

For any acts f and g, either $f \geqslant g$ or $g \geqslant f$ (or both).
For any acts f, g, and h, if $f \geqslant g$, $g \geqslant h$, then $f \geqslant h$.

The idea behind qualitative probability is quite simple. Suppose we have two consequences, "good" and "bad," and two events A and B. We construct two acts, f and g, as given below:

$$f(s) = \text{good if } s \in A \qquad f(s) = \text{bad if } s \in A'$$

$$g(s) = \text{good if } s \in B \qquad g(s) = \text{bad if } s \in B'$$

We now offer the subject a choice between f and g. Which one will he choose? Well, assuming he would rather have "good" than "bad," he will choose f if he thinks A is "more probable" than B, otherwise he will choose g. One problem arises here. Suppose the subject regards A and B as equally probable? In this case his choice between f and g would be arbitrary. From his choice we could not determine whether his preference was strict, we can only conclude that he regards A as *qualitatively at least as probable* as B, and notate this:

A qualitatively at least as probable as B: $A \geqslant . B$

In order to conclude that A is qualitatively strictly more probable than B, notated $A > . B$, we should have to verify that his preference of f over g was strict, that is, whether $f > g$. We could do this by making the good consequence g a little bit better. If for sufficiently small improvements of this kind, f remained preferable over g, then we conclude $A > . B$.

For the concept of qualitative probability to be well-defined, we must assume the

Principle of definition. The subject's preference is independent of our particular choice of "good" and "bad." That is, if we had used other consequences, say "really good" and "terrible," we would have observed preference behavior leading to the same qualitative probability relation.

We can say more about rational preference. For example, if we define the act BAD:

$$BAD(s) = bad \qquad \text{for all } s$$

then we should expect that for the act f above

$$f \geqslant BAD$$

Since BAD takes the consequence "good" on $\emptyset$, it follows that for any event C, $C \geqslant. \emptyset$. If also $\emptyset \geqslant. C$, then we say that C is *null*. Hence, we guarantee that for all C, $C \geqslant. \emptyset$ by assuming that preference satisfies the

> *Principle of dominance.* If for all s, the consequence $h(s)$ is at least as preferable as the consequence $k(s)$, then $h \geqslant k$, for any acts h, k. If in addition $h(s) > k(s)$ for all $s \in C$, with C nonnull, then $h > k$.

We shall want to assume more. Suppose there is a third event C disjunct from both the events A and B above: $C \cap A = C \cap B = \emptyset$. Then we might expect that the qualitative probability relation is *additive*, that is, that

$$A \cup C \geqslant. B \cup C$$

Why should we expect this? Well, what does this mean in terms of preference? Consider again the acts f and g above. Since C is disjunct from both A and B, we have

$$f(s) = g(s) = bad \qquad \text{for all } s \in C$$

Since f and g agree on C, a preference between f and g should not be determined by the consequences of these acts on C. Hence, if we change the values of f and g on C, still keeping these values equal, the preference between f and g should not be affected. In fact, let us put

$$f'(s) = f(s) \quad \text{for } s \in C' \qquad\qquad g'(s) = g(s) \quad \text{for } s \in C'$$
$$f'(s) = good \quad \text{for } s \in C \qquad\qquad g'(s) = good \quad \text{for } s \in C$$

From what we have just said, it follows that $f' \geqslant g'$. From this it follows that $A \cup C \geqslant. B \cup C$. In other words, the additivity of the qualitative probability relation is implied by what Savage calls the

> *Sure thing principle.* The preference between two acts is independent of the values the acts take on any set on which the two acts agree.

Other authors refer to the sure thing principle as the *principle of independence*, for obvious reasons. Letting S denote the set of all states, it is easy to show that for any A, $S \geqslant. A$ (start with $A' \geqslant. \emptyset$, and "add" A to both sides). It is the additivity property, together with the fact that $\emptyset \leqslant. A \leqslant. S$, that makes the qualitative probability relation "look like" a probability. Any representation of uncertainty that does not have these properties will not be representable as a quantitative probability measure. However, it is not the case that every qualitative probability can be represented by a genuine probability measure (Kraft, Pratt, and Seidenberg, 1959). For this we need more assumptions introduced in the following step of the argument.

The critique of Savage's axioms has been largely directed at the sure thing principle. To give one example, we describe what is known as the "Ellsberg paradox" (Ellsberg, 1961). If subjects are given a choice between an even money wager on "heads" with a coin of unknown composition and an even money wager on "heads" with a fair coin (i.e., for which the probability of "heads" is known to be $\frac{1}{2}$), most subjects wil prefer the latter wager. Hence, they behave as if the probability of heads on the first wager were less than $\frac{1}{2}$. However, changing the first wager so that the subject wins on "tails," they *still* prefer the second wager. This preference behavior is not consistent with Savage's axioms (in particular, it contradicts (e) of Lemma 6.1 in the supplement).

Step 2: Quantitative Probability

In step 2 additional assumptions on the preference relation are introduced, which guarantee that the qualitative probability relation has a unique representation in terms of a probability measure. In other words, for any preference relation satisfying these assumptions, there will be a unique probability measure p over S such that the induced qualitative probability relation satisfies:

For all A, B, A $\geqslant$. B if and only if $p(A) \geqslant p(B)$.

This is the most technical part of the argument, and we will only indicate the general drift here. Suppose we could partition any set A into two subsets A_1 and A_2 such that $A_1 \geqslant . A_2$ and $A_1 \leqslant . A_2$; that is, $A_1 =. A_2$. Starting with S, for each natural number n, we could then construct a partition of S into 2^n sets of equal qualitative probability. Intuitively, each set in the partition would have probability 2^{-n}. Given any set B, we could compare it with successively greater unions of elements of this partition and find the number k, such that, intuitively speaking, the probability of B should lie between $(k-1)2^{-n}$ and $k2^{-n}$. Calling this number $k(B, n)$, we could then define

$$p(B) = \lim_{n \to \infty} \frac{k(B, n)}{2^n}$$

The assumption which enables us to do this is the

Principle of refinement. For any B, C with B $<.$ C, there is a partition $\{A_i\}_{i=1,\ldots,n}$ such that B $\cup A_i <.$ C, $i = 1, \ldots, n$.

The proof that this principle does the job is given in the supplement to this chapter. We record for convenience the result here as

Theorem 6.1: If $\geqslant$. satisfies the above principles, then there exists a unique quantitative probability that represents $\geqslant$ ..

Step 3: Utility

Let C denote the set of consequences and S the set of states. In the simplest version of utility theory, which is adequate for all practical applications, it is assumed that

C is finite. Let **F** denote the set of all possible acts, that is, **F** is the set of functions from **S** to **C**. The simple version of utility requires the slightly

> *Strengthened principle of refinement.* If $f > g$, then for all α sufficiently small, $\alpha > 0$, the preference between f and g is unaffected by altering f and/or g on a set of probability less than α.

The foundation of utility theory is given in the following:

Theorem 6.2: Let **C**, **F**, and **S**, be as above, let **R** denote the real numbers, let $\geqslant$. satisfy the principles set forth above, and let p be the unique quantitative probability representing $\geqslant$.. Finally, let $U:\mathbf{C} \to \mathbf{R}$ be a "utility" function, and let Uf denote the expected utility of f with respect to p:

$$Uf = \sum_{c \in C} p(f^{-1}(c))U(c)$$

Then

1. There exists a utility function $U:\mathbf{C} \to \mathbf{R}$ such that for all $f, g \in \mathbf{F}$,

$$f \geqslant g \quad \text{if and only} \quad \text{if } Uf \geqslant Ug$$

2. If U' is another real-valued function on **C** satisfying (1), then there exist real constants a, b with $a > 0$, such that

$$U' = aU + b$$

The second statement in this theorem says that the utility functions satisfying (1) are related by positive affine transformations, or equivalently, that utility is positive affine unique.

The mathematics behind this theorem are elementary, though Savage's original proof was quite technical. A simplified proof is given in the supplement to this chapter.

OBSERVATION

It was observed in Chapter 3 that uncertainty is that which is removed by observation. Accordingly, it is appropriate to broach the subject of observation from the viewpoint of Savage's decision model. The treatment given here represents the generic mathematical solution to the problem of observation: Why should we observe, and how should the results of observation affect our beliefs. A discussion of an important special case, namely repeated observations, will be postponed until Chapter 7.

Why observe? The question sounds silly; but it is not. The preceding pages describe rational preference, but nowhere in this theory can one discover why a rational person should prefer to make observations, as opposed to acting without the benefit of observations.

Indeed, one might imagine that the representation of uncertainty as subjective probability involves the following paradox. Since one subjective probability "is as good as another," why should one bother to change one's beliefs via observation? For example, suppose I have some money to invest and I am interested in the probability that the price of gold will go up next year. According to Savage's theory, I already have some subjective probability that the price of gold will go up next year. Why shouldn't I just act on my preferences? Most people in this situation would try to get more data, but why should they? Isn't one subjective probability as good as another?

The decision model presented in this chapter gives a satisfactory answer to this question. We shall first examine how observations affect subjective probabilities, and then examine how observations can influence a given decision problem. Again, we shall report on a theorem whose proof is given in the supplement.

What is an observation? In the case of the price of gold next year, I could imagine doing any number of things to get a better idea whether the price of gold will rise. I could consult expert investors, I could look up the gold prices for preceding years, I could examine the political situation in South Africa where much of the world's gold is mined, etc. Whatever I do, I can always express the result as a number. I do not know beforehand what this number will be; that's why I have to look or ask. The value of the number that I find is obviously determined by the actual state of the world (of which I am uncertain). In other words, an observation can be mathematically represented as a real-valued function on the set S of possible worlds.

Mathematically, observations are similar to actions, but as they have a different interpretation we shall distinguish them in the notation. An observation will be called a random variable and will be denoted with capital letters from the end of the alphabet: X, Y, etc. The possible values of observation X will be assumed to be finite in number, and will be denoted $x_1, \ldots, x_n$.

We may consider a partition $B_1, \ldots, B_m$ of S. For simplicity, we shall take B_1 = "the price of gold goes up next year," and B_2 = "the price of gold does not go up next year." Then $S = B_1 \cup B_2$. We would like to know whether our actual world belongs to B_1 or B_2, but unfortunately we don't know and there is no observation that we can perform at present which will give us the right answer with certainty.

We now describe the way in which an observation X can give us more information regarding next year's gold price. In this discussion we shall use some elementary facts regarding conditional probability and conditional expectation set forth in the mathematical appendix.

The event that the observation X yields the value x_r is a subset of S that we shall simply denote as x_r. The theorem of Bayes says

$$p(B_1 \mid x_r) = \frac{p(x_r \mid B_1)p(B_1)}{p(x_r)}$$

Dividing this equation by a similar equation for B_2:

$$\frac{p(B_1 \mid x_r)}{p(B_2 \mid x_r)} = \frac{p(x_r \mid B_1)}{p(x_r \mid B_2)} \left[\frac{p(B_1)}{p(B_2)} \right]$$

Now before we make the observation of X, its value is uncertain. Hence, we may consider $p(B_1 | X)$ as a random variable, and since the above equation holds for all values x_r of X we may write

$$\frac{p(B_1 | X)}{p(B_2 | X)} = \frac{p(X | B_1)}{p(X | B_2)} \left[\frac{p(B_1)}{p(B_2)} \right]$$

as an equation among random variables.

Suppose we perform X and observe x_r. Our subjective probability will no longer be $p(.)$. Instead, we incorporate this knowledge by adopting the conditional probability $p(. | x_r)$.

We want to say something like this: If B_1 is really the case, then we should expect the left-hand side of the above equation to be larger than $p(B_1)/p(B_2)$. Equivalently, we should expect that the first term on the right-hand side is greater than one. Define

$$R(X) = \frac{p(X | B_1)}{p(X | B_2)}$$

Letting E denote expectation, the supplement contains a simple proof of the following:

Theorem 6.3: $E(R(X) | B_1) \geq 1$, and equality holds if and only if

$$p(R(X) = 1 | B_1) = 1$$

Theorem 6.3 shows that if B_1 is the case, and if $R(X)$ is not unity given B_1, then after observing X we may expect to be more confident of B_1 than we were before performing the observation. The quantity $R(X)$ is sometimes called the likelihood ratio of B_1 to B_2 given X.

Theorem 6.3 does not use the fact that $p(B_1) + p(B_2) = 1$, and it can be generalized straightaway to larger partitions. Defining

$$R_{1j}(X) = \frac{p(X | B_1)}{p(X | B_j)} \qquad j = 1, \ldots, m$$

one shows that $E(R_{1j}(X) | B_1) \geq 1$ just as in the proof of Theorem 6.3.

Theorem 6.3 gives a nice account of learning from observations, but it does not answer the question "why observe."

Why indeed? Before giving an answer, two deeply trivial remarks must be made. First, in deriving the representation of preference in terms of expected utility we assumed that the subject's preferences are defined with respect to all mathematically possible acts. If he can actually *choose* any of these acts, then he can always choose the act yielding the maximum utility in every possible world. Such a person has no decision problem and has no reason to observe anything. Observation becomes interesting only if the set of *accessible* acts is strictly smaller than the set of all acts.

Second, we have seen that Savage's decision theory provides for changing subjective probabilities via conditionalization on the results of observation. It does not provide for *changing subjective utilities*. For the absolute stoic, the decision problem of life is not a question of choosing the accessible act with the highest expectation, it is rather a question of changing one's utilities in such a way that the act of doing nothing yields maximum pleasure. Such a person also has no reason to observe anything.

Hence, in explaining why observation is rational under certain circumstances, we must assume that not all acts are accessible, and that utilities cannot be changed. As before, we let F denote the set of all acts. A decision problem may be characterized as a subset F of **F**, where **F** denotes the accessible acts. We assume that F is finite. In light of Theorem 6.2 we may assume that all acts are utility-valued, and we write $E(f)$ to denote the expected utility of $f(E(f)$ corresponds to Uf in Theorem 6.2). We define the *value* $v(F)$ of F as follows:

$$v(F) = \max_{f \in F} E(f)$$

The point of performing an observation is that we can choose an act from F *after* inspecting the result of the observation. For observation X with possible values $x_1, \ldots, x_n$, we define the *value of F given* $X = x_i$, $v(F \mid x_i)$ as

$$v(F \mid x_i) = \max_{f \in F} E(f \mid x_i)$$

Here, the conditional expectation $E(f \mid x_i)$ is the expectation of f with respect to the conditional probability $p(. \mid X = x_i)$. As this holds for every possible value of X, we may define the *value of F given X*, $v(F \mid X)$ as

$$v(F \mid X) = \sum_{i=1}^{n} \max_{f \in F} E(f \mid x_i) p(x_i)$$

The point of performing observations can now be captured in a theorem with a two-line proof:

Theorem 6.4: For any observation X and any subset F of **F**:

$$v(F \mid X) \geqslant v(F)$$

Proof: Let f' be any act in F.

$$\sum_{i=1}^{n} \max_{f \in F} E(f \mid x_i) p(x_i) \geqslant \sum_{i=1}^{n} E(f' \mid x_i) p(x_i) = E(f')$$

Hence $v(F \mid X) \geqslant \max_{f \in F} E(f) = v(F)$. ∎

SUPPLEMENT[2]

In this supplement, proofs are given for the theorems mentioned in the text. The proof of Theorem 6.2 is a combination of the proof of Savage and the proof of a similar theorem by Ramsey (1931). Villegas (1964) has given a shorter proof of a theorem resembling Theorem 6.1. However, he makes assumptions that are substantially stronger than those of Savage, and that entail the existence of a

[2] I am grateful to Peter Wakker and Bram Meima for reading this supplement and providing many useful comments.

random variable $x: S \to [0, 1]$ whose distribution is unitorm.[3] While it is interesting to have an axiomatic basis for assuming a uniform random variable, Savage's weaker assumptions are more relevant for the problem of representing uncertainty. The proof of Theorem 6.1 economizes on that of Savage by not analyzing the properties of fineness and tightness separately.

The first section introduces notation and formalizes the axioms for rational preference. Theorems 6.1, 6.2, and 6.3 are proved in the following sections.

Definitions and Notation

C is a finite set with at least two elements, and S is a nonempty set. $F = C^S$ is the set of C-valued functions on S. 2^S denotes the set of all subsets of S.[4] For $A \in 2^S$, A' denotes the complement of A, and $A - B = A \cap B'$. $A \supset B$ means that B is a subset of A, and $A \subset B$ means that B is a superset of A. A binary relation r on a set A is called a *weak order* if r is *transitive* [$(a, b) \in r$ and $(b, c) \in r$ imply $(a, c) \in r$] and *connected* [for all $a, b \in A$, $(a, b) \in r$ or $(b, a) \in r$].

Definition. A relation $\leqslant .$ on 2^S is called a *qualitative probability* if

- $\leqslant .$ is a weak order.

- For all $B, C, D \in 2^S$, if $B \cap D = C \cap D = \varnothing$, then $B \leqslant . C$ if and only if $B \cup D \leqslant . C \cup D$.

- For all $B \in 2^S$, $\varnothing \leqslant . B \leqslant . S$, and not $\varnothing \geqslant . S$.

If $A \geqslant . B$ and $A \leqslant . B$, then we write $A = . B$. If $A \geqslant . B$ and not $A \leqslant . B$, then we write $A > . B$. A qualitative probability $\geqslant .$ satisfies the *principle of refinement* if for all $A, B \in 2^S$ such that $A > . B$, there exists a finite partition $\{C_i\}_{i=1,...,n}$ of S, such that $A > . B \cup C_i$, $i = 1, ..., n$.

For any $f \in F$, $A \in 2^S$ we let $f \mid A$ denote the restriction of f to A. Let $\geqslant$ be a relation on F; we write $f \mid A \geqslant g \mid A$ ("f is *conditionally preferred to g given A*") if for all $h \in F$

$$f . \mid A \cup h \mid A' \geqslant g \mid A \cup h \mid A'$$

$f \approx g$ denotes "$f \geqslant g$ and $f \leqslant g$". $f > g$ denotes "$f \geqslant g$ and not $g \geqslant f$". For $c, d \in C$ we write $c \geqslant d$ if the constant function on S with value c dominates the constant function on S with value d. $B \in 2^S$ is called *null* if for all $f, g \in F$, $f \mid B \approx g \mid B$. For $A, B \in 2^S$, we write $A \geqslant B$ if for all $c, d \in C$ with $c > d$, and for all $f, g \in F$ satisfying

$$f \mid A = g \mid B = c \qquad f \mid A' = g \mid B' = d$$

we have $f \geqslant g$.

Definition. A relation $\geqslant$ on F is a *rational preference* if

- $\geqslant$ is a weak order.

---

[3] The reader should be especially cautious in applying theorem 4.1 of Villegas (1964), as the notion of isomorphism used there is not the one common to probabilists.

[4] 2^S denotes the set of maps from S to $\{0, 1\}$; each such map corresponds to one subset of S.

- For all $f, g \in F$, for all $A \in 2^S$, $f \mid A \geqslant g \mid A$ or $f \mid A \leqslant g \mid A$ (or both) (*the sure thing principle*).
- For all $A, B \in 2^S$, $A \geqslant B$ or $B \geqslant A$ (or both) (the *principle of definition*).
- If $f(s) \geqslant g(s)$ for all $s \in S$, then $f \geqslant g$; if in addition $f \mid B > g \mid B$ for some nonnull $B \in 2^S$, then $f > g$ (the *principle of dominance*).
- For some $c, d \in C$, $c > d$.
- If $f > g$, then there exists a partition $\{A_i\}_{i=1,\dots,n}$ of S, such that for $j = 1, \dots, n$ and for all g_j satisfying

$$g_j \mid A_i = g \mid A_i \qquad \text{for } i \neq j$$

$f > g_j$ (called the *strengthened principle of refinement*).

Remark. It is easy to show that strengthened refinement implies refinement and that a rational preference induces a qualitative by setting $\geqslant \; = \; \geqslant$.. Moreover, this qualitative probability satisfies the principle of refinement, and the set S is infinite.

Proof of Theorem 6.1

We assume that $\leqslant$. is a qualitative probability satisfying the principle of refinement.

Lemma 6.1. For all $B, C, D, \in 2^S$,

(a) $B \leqslant . C$, $C \cap D = \emptyset$ imply $B \cup D \leqslant . C \cup D$.

(b) $B \cup D \leqslant . C \cup D$, $B \cap D = \emptyset$ imply $B \leqslant . C$.

(c) $B_1 \leqslant . C_1$, $B_2 \leqslant . C_2$, $C_1 \cap C_2 = \emptyset$ imply $B_1 \cup B_2 \leqslant . C_1 \cup C_2$.

(d) $B_1 \cup B_2 \leqslant . C_1 \cup C_2$, $B_1 \cap B_2 = \emptyset$ imply $B_1 \leqslant . C_1$, or $B_2 \leqslant . C_2$.

(e) $B_1 < . C_1$ $B_1 \cup B_2 \geqslant . C_1 \cup C_2$, $C_1 \cap C_2 = \emptyset$ imply $B_2 > . C_2$.

Proof. We show only (a) and (c) to give the idea behind the proofs.

(a) Let $D^\sim = D - B$. Then by the definition of qualitative probability

$$B \cup D^\sim \leqslant . C \cup D^\sim$$

Since $\emptyset \leqslant . D \cap B$, and both $\emptyset$ and $D \cap B$ are disjunct from $C \cup D^\sim$, we may "add" $C \cup D^\sim$ to both sides to get

$$C \cup D^\sim \leqslant . C \cup D$$

The result now follows from the transitivity of the qualitative probability relation.

(c) Let $Q = B_2 \cap C_1$, $B_2^\sim = B_2 - Q$, $C_1^\sim = C_1 - Q$. By (a)

$$B_1 \cup B_2^\sim \leqslant . C_1 \cup B_2^\sim = C_1^\sim \cup B_2 \leqslant . C_1^\sim \cup C_2$$

Now "add" Q to each side and apply (a). ∎

Lemma 6.2. If for all $E \in 2^S$ such that $E \cap B = \emptyset$ and $E > . \emptyset$ we have $B \cup E \geqslant . C$, then $B \geqslant . C$.

Proof. Suppose that $B < . C$. We find $E > . \emptyset$, with $E \cap B = \emptyset$ such that

$B \cup E <. C$. By the principle of refinement, there exists a partition $\{A_i\}$ such that $B \cup A_i <. C$, for all i. Since $B <. S$, there must be an i such that $B' \cap A_i >. \varnothing$. Put $E = B' \cap A_i$. ∎

Lemma 6.3. For all $B \in 2^S$ there exist subsets B_1 and B_2 of B such that $B_1 \cap B_2 = \varnothing$, $B_1 =. B_2$, and $B_1 \cup B_2 = B$.

Proof. The lemma is trivial if $B =. \varnothing$. Assume $B >. \varnothing$. We show that for all n there exists a threefold partition of B; $B = D_n \cup G_n \cup C_n$ such that

$$C_n \cup G_n \geqslant. D_n$$

$$D_n \cup G_n \geqslant. C_n$$

$$C_{n+1} \supset C_n$$

$$D_{n+1} \supset D_n$$

$$G_n - G_{n+1} \geqslant. G_{n+1}$$

Choose $C_1 \subset B$, $\varnothing <. C_1 <. B$. If $C_1 =. B - C_1$ we are finished. Assume that $C_1 <. B - C_1$. Choose $D_1 \subset B - C_1$; $D_1 <. C_1$, and set $G_1 = (B_1 - C_1) - D_1$. There exists an $E \subset G_1$ such that $C_1 \cup E <. B - C_1$, and E_1, E_2, such that $E_1 \cap E_2 = \varnothing$, $E = E_1 \cup E_2$, $\varnothing <. E_1 <. E_2$. By Lemma 6.1e

$$C_1 \cup E_1 <. (B - C_1) - E_1$$

With the refinement principle we can find a partition $\{A_i\}_{i=1,\ldots,n}$ of G_1 such that $A_1 = E_1$ and

$$A_1 >. A_2 \geqslant. \cdots \geqslant. A_n$$

Let r be the least j such that

$$\bigcup_{i=1}^{j} A_i \geqslant. \bigcup_{i=j+1}^{n} A_i$$

By construction $r > 2$, and

$$\bigcup_{i=2}^{r} A_i \leqslant. \bigcup_{i=r+1}^{n} A_i \cup A_1$$

Put

$$G_2 = \bigcup_{i=2}^{r} A_i \qquad D_2 = D_1 \cup \bigcup_{i=r+1}^{n} A_i \qquad C_2 = C_1 \cup A_1$$

It is easily verified that G_2, D_2, and C_2 satisfy the conditions on the threefold partition for $n = 2$. Iterating the above construction we construct C_n, G_n, and D_n for arbitrary n. We put

$$B_1 = \bigcup_{n=1}^{\infty} C_n \qquad B_2 = \bigcup_{n=1}^{\infty} D_n \cup \bigcap_{n=1}^{\infty} G_n$$

It is easy to verify that B_1 and B_2 partition B. We show that $B_1 =. B_2$ with the help of the following:

Claim. For all $H >. \emptyset$, there exists an integer n' such that for all $n > n'$, $G_n <. H$.

Proof. Let $\{A_i\}_{i=1,\dots,m}$ be a partition such that $A_i <. H$ for all i. Choose $n' = m$. For $n > n'$ suppose that $G_n \geqslant. H$; then $G_n >. A_i$ for all $i = 1, \dots, m$. By the properties of the threefold partitions and Lemma 6.1d we derive the contradiction:

$$G_n \cup (G_{n-1} - G_n) \cup \cdots \cup (G_1 - G_2) >. \cup A_i = S \quad \square$$

By construction $C_{m+i} \subset C_m \cup G_m$, for all i, hence

$$\bigcup_{i > m} C_i \subset C_m \cup G_m$$

It follows that $\cup_{i > m} C_i \leqslant. C_m \cup G_m$, or equivalently (Lemma 6.1b)

$$\bigcup_{i > m} C_i - C_m \leqslant. G_m$$

and similarly $\cup_{i > m} D_i - D_m \leqslant. G_m$. Suppose $B_2 <. B_1$. For sufficiently large m, n with $n > m$, we should have by two applications of Lemma 6.2:

$$B_2 \cup G_m <. B_1 - \left\{ \bigcup_{i > n} C_i - C_n \right\}$$

Since $B_2 \supset D_n$, $G_m \supset G_n$, this implies the contradiction

$$D_n \cup G_n <. B_1 - \left\{ \bigcup_{i > n} C_i - C_n \right\} = C_n$$

Suppose $B_1 <. B_2$. From the claim it follows that $\cap G_m =. \emptyset$. For sufficiently large m, n with $n > m$ we should have

$$B_1 \cup G_m <. B_2 - \left\{ \bigcup_{i > n} D_i - D_n \right\} = D_n$$

This implies the contradiction

$$C_n \cup G_n <. D_n \quad \blacksquare$$

Definition. $\{A_i\}_{i=1,\dots,m}$ is a *uniform partition* if it is a partition and if $A_i =. A_j$, i, $j = 1, \dots, m$.

Lemma 6.4. For all n, there exists a uniform partition of S into 2^n elements.

Proof. We show first that if

$$B =. C \qquad G =. H \qquad B \cap C =. G \cap H =. \emptyset \qquad B \cup C =. G \cup H$$

then $B =. G$. Suppose to the contrary that $B <. G$. Then $C <. G =. H$, and by Lemma 6.1e $B \cup C <. G \cup H$, a contradiction. Suppose we have partitioned S into 2^{n-1} elements that are qualitatively equally probable. Applying Lemma 6.3 to each element, it follows that we can partition S into 2^n elements of equal qualitative probability. The lemma now follows by induction. $\blacksquare$

Definition. For all $B \in 2^S$, let $B(n)$ denote the largest k such that

$$\bigcup_{i=1}^{k} A_i \leqslant. B$$

where $\{A_i\}$, $i = 1,\ldots,2^n$, is a uniform partition.

Definition. Let $p(B) = \lim_{n\to\infty} B(n)/2^n$.

Theorem 6.1. There is a unique quantitative probability that represents $\geqslant.$; that is, there is a unique probability $p: 2^S \to [0,1]$ such that for all A, $B \in 2^S$:

$$A \geqslant. B \qquad \text{if and only if} \qquad p(A) \geqslant p(B)$$

Proof. Let p be given by the above definition. The reader can easily verify that p is a probability. If $A \geqslant. B$ then $A(n) \geqslant B(n)$, hence $p(A) \geqslant p(B)$. Suppose that $A <. B$. By the refinement principle there exists a partition $\{C_i\}_{i=1,\ldots,m}$ such that $A \cup C_i <. B$ for $i = 1,\ldots,m$. Choose i such that $C_i - A >. \varnothing$. Again, by the refinement principle there exists a partition $\{D_i\}_{i=1,\ldots,m'}$, such that $D_j <. C_i - A, j = 1,\ldots,m'$. Hence, for $n > m'$, there exists a uniform partition $\{A_i\}$, $i = 1,\ldots,2^n$ such that $A_j <. C_i - A$ for $j = 1,\ldots,2^n$. It follows that

$$p(C_i - A) \geqslant 2^{-n}$$

and consequently,

$$p(A) < p(A) + 2^{-n} \leqslant p(A \cup C_i) \leqslant p(B)$$

Finally, let p' be another probability that represents $\leqslant..$ Since p and p' must agree on all elements of all uniform partitions, it follows that for all n and all $B \in 2^S$:

$$\frac{B(n)}{2^n} \leqslant p'(B) < \frac{B(n) + 1}{2^n}$$

and hence p and p' agree on 2^S. ∎

Remark 2. It is evident from the above, that for all $r \in [0,1]$, there is $A \in 2^S$ such that $p(A) = r$.

Proof of Theorem 6.2

We assume that $\leqslant$ is a rational preference, that $\leqslant.$ is the qualitative probability induced by $\leqslant$ (see Remark 1), and that p is the probability given by Theorem 6.1. We assume that the (finitely many) elements $c_1,\ldots,c_n$ of C are ordered such that

$$c_1 \geqslant c_2 \cdots \geqslant c_n$$

with at least one of the preferences strict.

Lemma 6.5. If $A_1 \cap A_2 =. \varnothing$ and

(a) $f|A_1 \approx (<)g|A_1$
(b) $f|A_2 \approx (<)g|A_2$; then

$$f|A_1 \cup A_2 \approx (<)g|A_1 \cup A_2$$

(These equations can be read as all equivalences or all inequivalences.)

Proof. By the sure thing principle, (a) and (b) imply, respectively,

$$g\,|\,A_1 \cup f\,|\,A_2 \approx (>) f\,|\,A_1 \cup f\,|\,A_2$$

$$g\,|\,A_1 \cup f\,|\,A_2 \approx (<) g\,|\,A_1 \cup g\,|\,A_2$$

and the Lemma follows by the transitivity of preference. ■

Lemma 6.6. For all $A \in 2^S$ there exists a set $\{A_r\,|\,r \in [0, 1]\}$ such that $p(A_r) = r$ and such that

$$A = A_{p(A)}$$

$$A_r \subset A \qquad \text{if} \qquad r \leqslant p(A)$$

$$A_r \supset A \qquad \text{if} \qquad r \geqslant p(A)$$

Lemma 6.7. For all c_i, $c_j \in C$ such that $c_i > c_j$;

$$c_i\,|\,A \cup c_j\,|\,A' \approx c_i\,|\,B \cup c_j\,|\,B'$$

if and only if

$$p(A) = p(B)$$

Corollary. If $c_i \approx c_1\,|\,A \cup c_n\,|\,A'$ and $p(A) = p(B)$, then

$$c_i \approx c_1\,|\,B \cup c_n\,|\,B'$$

Lemma 6.6 follows directly from Remark 2. Lemma 6.7 is little more than the principle of definition, in combination with Remark 1 and Theorem 6.1.

Lemma 6.8. For all $A \in 2^S$ with $p(A) > 0$, there exists a unique $r \in [0, 1]$ such that

$$c_i \approx c_1\,|\,A_r \cup c_n\,|\,A_r'$$

Proof. By the principle of dominance and Lemmata 6.6 and 6.7,

$$c_1\,|\,A_r \cup c_n\,|\,A_r' > c_1\,|\,A_s \cup c_n\,|\,A_s'$$

if and only if $r > s$. Let s denote the infimum of all r such that

$$c_i < c_1\,|\,A_r \cup c_n\,|\,A_r'$$

If

$$c_i < c_1\,|\,A_s \cup c_n\,|\,A_s'$$

then by the strengthened refinement principle we could alter the right-hand side such that for some $\alpha > 0$,

$$c_i < c_1\,|\,A_{s-\alpha} \cup c_n\,|\,(A_{s-\alpha})'$$

contradicting the fact that s is the infimum. A similar argument holds for ">," and we conclude

$$c_i \approx c_1\,|\,A_s \cup c_n\,|\,A_s'$$

and that s is the only number for which this holds. ■

By Lemma 6.7, this equation depends only on s and not on A, hence we write

$$c_i \approx c_1 \,|\, s \cup c_n \,|\, (1 - s)$$

Definition. For $r \in [0, 1]$, and $B \in 2^S$, $r.B$ denotes an arbitrary subset of B with probability $rp(B)$.

Lemma 6.9. Let $c_i \approx c_1 \,|\, r \cup c_n \,|\, (1 - r)$, and let $p(B) = \frac{1}{2}$, then

$$c_i \,|\, B \approx c_1 \,|\, r.B \cup c_n \,|\, B \cap (r.B)'$$

Proof. Suppose

$$c_i \,|\, B < c_1 \,|\, r.B \cup c_n \,|\, B \cap (r.B)'$$

Then we should have by the definition of conditional preference:

$$c_i \,|\, B \cup c_n \,|\, B' < c_1 \,|\, r.B \cup c_n \,|\, (r.B)'$$

Since $p(B) = p(B')$, Lemma 6.7 entails that

$$c_i \,|\, B' \cup c_n \,|\, B < c_1 \,|\, r.B' \cup c_n \,|\, (r.B')'$$

which, according to the definition of conditional preference, means

$$c_i \,|\, B' < c_1 \,|\, r.B' \cup c_n \,|\, B' \cap (r.B')'$$

Lemma 6.5 now entails that

$$c_i < c_1 \,|\, r \cup c_n \,|\, (1 - r)$$

which is a contradiction. ∎

Lemma 6.10. For all integers n, Lemma 6.9 holds for B with $p(B) = k/2^n$, $k = 1, \ldots, 2^n$.

Lemma 6.11. Suppose $c_i \approx c_1 \,|\, r \cup c_n \,|\, (1 - r)$; then for all $B \in 2^S$:

$$c_i \,|\, B \approx c_1 \,|\, r.B \cup c_n \,|\, B \cap (r.B)'$$

Proof. The lemma is trivial if $B = . \varnothing$. Assume that $p(B) > 0$. Suppose

$$c_i \,|\, B < c_1 \,|\, r.B \cup c_n \,|\, B \cap (r.B)'$$

By the strengthened principle of refinement there exist integers n, k, and $D \in 2^S$, $p(D) = k/2^n > p(B)$, $D \supset B$ such that

$$c_i \,|\, D \cup c_n \,|\, D' < c_1 \,|\, r.B \cup c_n \,|\, (r.B)'$$

By Lemma 6.10

$$c_i \,|\, D \cup c_n \,|\, D' \approx c_1 \,|\, r.D \cup c_n \,|\, (r.D)'$$

The set $r.D$ may be chosen such that $r.D \supset r.B$; the principle of dominance then entails that

$$c_1 \,|\, r.D \cup c_n \,|\, (r.D)' > c_1 \,|\, r.B \cup c_n \,|\, (r.B)'$$

Hence

$$c_i \,|\, D \cup c_n \,|\, D' > c_1 \,|\, r.B \cup c_n \,|\, (r.B)'$$

from which contradiction the lemma follows. ∎

Theorem 6.2. Let C, F, S, $\leqslant$, and $\leqslant$. be as above, and let R denote the real numbers. Let $U: C \rightarrow R$ and let Uf be defined as

$$Uf = \sum_{c \in C} p(f^{-1}(c))U(c)$$

Uf is the "expected utility of f with respect to p." Then

1. There exists a utility function $U: C \rightarrow R$ such that for all $f, g \in D$,

$$f \geqslant g \qquad \text{if and only if} \qquad Uf \geqslant Ug$$

2. If U' is another real-valued function on C satisfying (1), then there exist real constants a, b with $a > 0$, such that

$$U' = aU + b$$

Proof. As before, we put $c_1 \geqslant c_2 \geqslant \cdots \geqslant c_n$, with at least one preference strict.

1. Define $U(c_1) = 1$; $U(c_n) = 0$, $U(c_i) = r_i$ where

$$c_i \approx c_1 | r_i \cup c_n | (1 - r_i)$$

We show that $f \geqslant g$ if and only if $Uf \geqslant Ug$. We may write

$$f = c_1 | f^{-1}(c_1) \cup c_2 | f^{-1}(c_2) \cup \cdots \cup c_n | f^{-1}(c_n)$$

By Lemmata 6.6 and 6.10,

$$f \approx c_1 | A \cup c_n | A'$$

where

$$A = \bigcup_{i=1}^{n} f^{-1}(c_i).r_i$$

Similarly,

$$g \approx c_1 | B \cup c_n | B'$$

$$B = \bigcup_{i=1}^{n} g^{-1}(c_i).r_i$$

By dominance, Lemmata 6.6 and 6.7, $f \geqslant g$ if and only if $p(A) \geqslant p(B)$. The proof of (1) is completed by noting that

$$p(A) = \sum_{i=1}^{n} p(f^{-1}(c_i).r_i) = \sum_{i=1}^{n} p(f^{-1}(c_i))r_i = Uf$$

$$p(B) = Ug$$

2. Note first that if W satisfies (1), then so does W', where

$$W' = aW + b \qquad a, b \in R, a > 0$$

Let W satisfy (1). By appropriately choosing constants a and b if necessary, we can arrange that $W(c_1) = 1$, $W(c_n) = 0$. It suffices to show that $W(c_i) = r_i$, where r_i is defined above, $i = 2, \ldots, n - 1$. But since

$$c_i \approx c_1 \,|\, r_i \cup c_n \,|\, (1 - r_i),$$
$$W(c_i) = W(c_1 \,|\, r_i \cup c_n \,|\, (1 - r_i)) = W(c_1) r_i + W(c_n)(1 - r_i) = r_i \quad \blacksquare$$

Proof of Theorem 6.3

Definition. A real function Y on **R** is convex (concave) on (a, b) if for all $x, y \in (a, b)$ and for all $\alpha \in (0, 1)$

$$Y(\alpha x + (1 - \alpha)y) \leqslant (\geqslant)\alpha Y(x) + (1 - \alpha)Y(y).$$

Jensen's Inequality. If a real function Y is convex (concave) and bounded on an interval I, and if $f(s) \in I$ for all $s \in S$, then

$$Y(Ef) \leqslant (\geqslant)E(Y(f)),$$

with equality if and only if, with probability one, Y is linear on the range of f.

Where B_1 and B_2 partition S, with $p(B_1) > 0$, $p(B_2) > 0$, we define

$$R(X) = \frac{p(X \,|\, B_1)}{p(X \,|\, B_2)}$$

The possible values of X are $x_1, \ldots, x_n$.

Theorem 6.3. $E(R(X) \,|\, B_1) \geqslant 1$ with equality holding if and only if

$$p(R(X) = 1 \,|\, B_1) = 1$$

Proof. We first note that

$$E(R(X)^{-1} \,|\, B_1) = \sum_{i=1}^{n} \left(\frac{p(x_i \,|\, B_2)}{p(x_i \,|\, B_1)} \right) p(x_i \,|\, B_1) = 1$$

Since $\log(x)$ is concave and $-\log(x)$ convex, Jensen's inequality implies

$$\log[E(R(X) \,|\, B_1)] \geqslant E(\log R(X) \,|\, B_1) = E(-\log R(X)^{-1} \,|\, B_1)$$
$$\geqslant -\log E(R(X^{-1}) \,|\, B_1) = -\log 1 = 0$$

It follows that $E(R(X) \,|\, B_1) \geqslant 1$, and from Jensen's inequality we know that equality holds if and only if $R(X)$ is constant with probability 1, as log is nowhere linear. The latter condition can hold only if

$$p(x_i \,|\, B_1) = p(x_i \,|\, B_2)$$

for all i such that $p(x_i \,|\, B_1) > 0$. $\blacksquare$

7

Relative Frequencies and Exchangeability

This chapter studies the relation between subjective probability and relative frequency. Not only is this essential for understanding subjective probability, it is also important for the Bayesian models for using expert opinion. In Theorem 6.3 we have seen how we *expect* to learn from experience, and in Theorem 6.4 how we *expect* to profit from observations. However, these theorems do not explain why we like to observe repeated events and like to use their relative frequencies as probabilities.

Although the relation between subjective probability and relative frequency has been well understood since the 1930s, the more popular literature has been very slow in appreciating it, and there is still much confusion. Perhaps the most important piece of disinformation concerns an alleged antagonism between subjective probabilities and relative frequencies. There is to be sure a certain opposition between the subjectivist and frequentist *interpretation* of probability, as seen in the Appendix. Unfortunately, this often gets understood as an opposition between subjective probabilities and relative frequencies per se, as if the latter were different from and "more objective" than the former. This is nonsense. Subjective probabilities can be, and often are, limiting relative frequencies. In particular, this happens when a subject's belief state leads him to regard the past as relevant for the future in a special way. Thanks to the work of Bruno De Finetti, we can give a precise mathematical account of this type of relevance: past and future together should form an *exchangeable sequence*, prior to observation.

Equally regrettable is the fact that most books on Bayesian decision theory scarcely mention the word "exchangeability" and certainly do not accord it the position it deserves. The reason for this may lie in some beguiling mathematics. De Finetti's famous "representation theorem" shows that an exchangeable sequence of events can be uniquely represented as lotteries over certain "archetypal" sequences: A given exchangeable sequence corresponds to drawing an archetype from a given distribution over all archetypes. In the case of an *infinite* sequence of exchangeable events, these archetypal sequences are just the old Bernoulli independent coin-tossing processes, which probabilists know and love. Thus, in the infinite case, exchangeability entails *conditional independence*: Once the archetype is specified, the sequence may be regarded as independent tosses with a (not necessarily fair) coin. The proof of the theorem in the infinite case involves advanced techniques in probability theory and cannot be presented to the practicing decision analyst.

Conditional independence is very appealing to probabilists, but it is not the essence of exchangeability. For finite exchangeable sequences, the archetypes are not Bernoulli processes, but rather urn-drawing processes without replacement. For finite exchangeable sequences conditional independence does not hold, and learning by experience can be modeled as learning the composition of an urn by successively drawing out all of its balls. This is a better and more intuitive model than learning the "probability of heads" by tossing the coin infinitely many times. For De Finetti, there is no such thing as a "probability of heads," and this is why De Finetti placed great emphasis on the finite case of the "representation theorem." The extent to which mathematicians have failed to grasp the philosophical importance of exchangeability is reflected in Feller (1971, p. 229). After giving an elegant proof of the representation theorem for infinite sequences, he remarks that the theorem "fails" for finite sequences.

This chapter studies learning from experience in the case of finite sequences, shows why probabilities sometimes track relative frequencies, and states and proves the finite version of De Finetti's representation theorem. In a mathematical supplement to this chapter the infinite case is proved. This proof allows the reader without advanced probability knowledge to follow the argument, if he is prepared to accept the moment convergence theorem mentioned in the Appendix.

THE FREQUENTIST ACCOUNT OF LEARNING BY EXPERIENCE

It is useful to begin by developing the naive frequentist account of estimating probabilities by relative frequencies. Consider an event E. The naive frequentist says, if you don't know $p(E)$, then perform a long sequence of independent trials of E. According to the weak law of large number, it is practically certain that the observed frequency of E in this sequence will be very close to $p(E)$. Such trials are called *Bernoulli trials* for E, and the sequence of trials is called a *Bernoulli sequence*. Frequentists can give no hard operational criteria for determining whether a sequence is a Bernoulli sequence.

Let us observe two features of this account:

1. Probability exists. By this we mean that the person reasoning in the above manner assumes that there is some objective "thing," $p(E)$.
2. If this objective thing is initially unknown, it can be discovered with probability approaching certainty by constructing sufficiently long Bernoulli sequences.

De Finetti begins his *Theory of Probability* (1974) with the boldface exclamation **"Probability does not exist."** He means that probability is not part of the objective world, cannot be unknown, and therefore cannot be "discovered:"

The old (frequency) definition cannot, in fact, be stripped of its so to speak, "metaphysical" character: One would be obliged to suppose that beyond the probability distributions corresponding to our judgment, there must be another, unknown, corresponding to something real, and that the different hypotheses about the unknown distribution—according to which the various trials would no longer be dependent, but independent—would constitute *events* whose probability one could consider. From our point of view these statements are completely devoid of sense, and no one has given them a justification which seems satisfactory, even in relation to a different point of view (De Finetti, 1937).

Someone flipping a coin repeatedly, prior to betting on its outcomes, may explain his behavior by rehearsing the frequentist account. This does not prove that the "thing" which he is looking for actually exists. Saying a prayer does not prove the existence of God. Nonetheless, a subjectivist should give some account of what this person *is* doing. If a state of partial belief is represented by a subjective probability measure, under what partial beliefs would this person's behavior make sense?

EXPECTED FREQUENCY

In Chapter 6 we showed on the basis of expectation why it makes sense to perform observations, though we said nothing about how the results of observations should affect our beliefs. We can also say something quite general about the relation between probabilities and *expected* relative frequencies.

Theorem 7.1. Let $A_1, \ldots A_n$ be events whose probabilities of occurrence are $p_1, \ldots, p_n$. Let $w_i^{(n)}$ denote the probability that exactly i of these events occur; $i = 0, \ldots, n$, and let $\bar{p} = (1/n)(p_1 + \cdots + p_n)$. Then:

$$\bar{p} = \frac{1}{n} \sum_{i=1}^{n} i w_i^{(n)}$$

Proof. The proof is trivial. Let 1_i denote the indicator function of $A_i, i = 1, \ldots, n$. The right-hand side is just the expected relative frequency of occurrence, and by the linearity of expectation this is equal to

$$E\left(\frac{1}{n}\right) \sum_{i=1}^{n} 1_i = \frac{1}{n} \sum_{i=1}^{n} E(1_i) = \bar{p} \quad \blacksquare$$

Theorem 7.1 says that the average probability is equal to the expected relative frequency. It makes no assumption on the probabilities of the A_i, and says nothing about changing our beliefs on the basis of relative frequencies.

EXCHANGEABILITY

Learning from experience can take many forms. We are interested in a particular form, learning from observed frequencies. In what sort of belief states are we inclined to learn from observed relative frequencies? Roughly speaking, we must believe that the past is similar to the future. Let $\{A_i\}_{i=1,\ldots}$ denote the sequence of events {"it rains on day i"}, $i = 1, \ldots$. Consider a particular event, say A_{100}. Would we be prepared, before observing any days, to say that after observing A_1–A_{99}, our probability for A_{100} should be approximately equal to the relative frequency of occurrence of A_1–A_{99}? That depends on whether we know anything special about A_{100}, relative to the other events. For example suppose the sequence starts on June 1, and suppose we know that it hardly rains at all in the summer, but that September is the wettest month of the year. Then $p(A_{100})$ may *change* as a result of observing A_1–A_{99}, but it will not incline toward the relative frequency of rain in the summer.

Suppose we do not have any such knowledge. If we think about it, we can express this lack of knowledge mathematically as follows: Let 1_i denote the indicator of A_i; then for any i, $i = 1, \ldots, 99$ our probability for any outcome sequence for $1_1, \ldots, 1_{100}$ is unaffected by interchanging the outcomes of 1_i and 1_{100}. Instead of interchanging the outcomes, we can interchange the variables, keeping the outcomes fixed. Let Q denote an arbitrary sequence of 100 0s and 1s. This means

$$p((1_1, \ldots, 1_{100}) = Q) = p((1_1, \ldots, 1_{i-1}, 1_{100}, 1_{i+1}, \ldots, 1_{99}, 1_i) = Q)$$

Definition. The sequence of indicator functions $\{1_i\}_{i=1,\ldots,n}$ is exchangeable with respect to p if for every permutation π of $1, \ldots, n$ and every sequence Q of 0s and 1s of length n

$$p((1_1, \ldots, 1_n) = Q) = p((1_{\pi(1)}, \ldots, 1_{\pi(n)}) = Q)$$

The sequence $\{1_i\}_{i=1,\ldots,\infty}$ is exchangeable with respect to p if for every n, the sequence $\{1_i\}_{i=1,\ldots,n}$ is exchangeable with respect to p.

When no confusion can arise, we shall speak of exchangeable events, without directly referring to the probability with respect to which they are exchangeable. In order to study the relation between probability and relative frequency, let $\{1_i\}_{i=1,\ldots,\infty}$ be an infinite sequence of indicators for the exchangeable events $\{A_i\}_{i=1,\ldots,\infty}$, let $Q \in \{0,1\}^n$ be an outcome of the first n indicators, and let $w_r^{(n)}$ denote the probability that exactly r of the first n events occur. We assume all such terms are different from zero. Finally, we write $p(Q)$ for $p((1_1, \ldots, 1_n) = Q)$. The probability of Q may depend on the numbers of 0s and 1s in Q, but it may not depend on where they occur. Hence

$$p(Q) = \frac{w_r^{(n)}}{\binom{n}{r}}$$

Suppose we observe the values of $1_1, \ldots, 1_n$, and find the sequence Q with r 1s and $n-r$ 0s. After this observation our probability for A_{n+1} is not $p(A_{n+1})$ but $p(A_{n+1}|Q)$. We now calculate this latter quantity.

$$p(A_{n+1}|Q) = \frac{p(A_{n+1} \cap Q)}{p(Q)} = \frac{w_{r+1}^{(n+1)} \binom{n}{r}}{w_r^{(n)} \binom{n+1}{r+1}} \tag{7.1}$$

In order to reduce this, note

$$\frac{w_r^{(n)}}{\binom{n}{r}} = \frac{w_r^{(n+1)}}{\binom{n+1}{r}} + \frac{w_{r+1}^{(n+1)}}{\binom{n+1}{r+1}}$$

In prose, the probability of a particular n-sequence with r 1s equals that of a particular $n + 1$-sequence with r 1s plus that of an $n + 1$-sequence with $(r + 1)$ 1s. Putting $s = n - r$ and simplyfing, we find

$$w_r^{(n)} = \frac{w_r^{(n+1)}(s + 1) + w_{r+1}^{(n+1)}(r + 1)}{n + 1}$$

and substituting this in (7.1):

$$p(A_{n+1}|Q) = \frac{(r + 1)w_{r+1}^{(n+1)}}{(s + 1)w_r^{(n+1)} + (r + 1)w_{r+1}^{(n+1)}}$$

$$= \frac{r + 1}{n + 2 + (s + 1)[(w_r^{(n+1)}/w_{r+1}^{(n+1)}) - 1]}$$

Suppose that for all n and r,

$$w_r^{(n+1)} = w_{r+1}^{(n+1)} \tag{7.2}$$

It then follows that

$$p(A_{n+1}|Q) = \frac{r + 1}{n + 2} \tag{7.3}$$

This is the famous "Laplace rule" for betting on "success on the next trial" given r successes on n previous trials. As r and n get large, this approaches the observed relative frequency of occurrence r/n. In general, the ratio $w_r^{(n+1)}/w_{r+1}^{(n+1)}$ determines how fast we will "learn from experience." We record this as

Theorem 7.2. Let $\{A_i\}_{i=1,\ldots,\infty}$ be exchangeable, and suppose for all $n = 1, \ldots, \infty$ and all $r = 0, 1 \cdots n$, $w_r^{(n)} > 0$. Suppose there is a constant K such that for all n:

$$\max_{r=0,\ldots,n} n \left| \frac{w_r^{(n+1)}}{w_{r+1}^{(n+1)}} - 1 \right| \leqslant K$$

then if Q is an outcome sequence for $A_{1,\ldots,\infty}$, Q_n the initial segment of Q of length n, and $\#Q_n$ the relative frequency of 1s in Q_n:

$$|p(A_{n+1}|Q_n) - \#Q_n| \to 0 \qquad \text{as } n \to \infty$$

The conditions of Theorem 7.2 do not always apply. If the events are independent (a special case of exchangeability) then $p(A_{n+1}|Q) = P(A_{n+1})$ and nothing is learned from the observation Q.

DE FINETTI'S FINITE REPRESENTATION THEOREM

De Finetti's theorem gives us a way of characterizing all finite exchangeable sequences of events. For a sequence of M exchangeable events, the numbers j/M, $j = 0, \ldots, M$ represent the possible "limiting" relative frequencies. In certain exchangeable M-sequences, a given "limiting" relative frequency is certain to occur, namely when the sequence consists of drawing black or white balls from an urn of known composition without replacement. The theorem says that every exchangeable M-sequence can be represented in the following way: First choose an urn according to a distribution $\{w_j^{(M)}\}_{j=0,\ldots,M}$ over the possible urn compositions j/M, then draw balls out of the selected urn. An exchangeable sequence is thus uniquely characterized by a distribution $\{w_j^{(M)}\}_{j=0,\ldots,M}$. In the case of an infinite sequence of exchangeable events, the theorem says exactly the same thing, except that the urns are now infinite. Drawing balls from an infinite urn is equivalent to flipping a coin, hence the sequences in which a given limiting relative frequency is certain to occur are just Bernoulli sequences. To avoid switching back and forth between events and their indicators, we shall speak of the outcome of an event being 1 if the event occurs, and 0 if it does not occur.

Theorem 7.3. (De Finetti, 1937, 1974). An exchangeable sequence of length M is uniquely determined by the probabilities $w_j^{(M)}$, $j = 0, \ldots, M$.

Proof. Let $A_1, \ldots, A_M$ be exchangeable. For any outcome sequence Q of length n, $n \leqslant M$, containing r 1s,

$$p(Q) = \frac{w_r^{(n)}}{\binom{n}{r}}$$

Write for brevity $w_n = w_n^{(n)}$. The key to the proof is the following:

Claim

$$\frac{w_r^{(n)}}{\binom{n}{r}} = w_r - \binom{n-r}{1} w_{r+1} + \binom{n-r}{2} w_{r+2} \cdots (-1)^{n-r} w_n \qquad (7.4)$$

Proof. The left-hand side is the probability of a sequence of length n beginning with r 1s followed by $n - r$ 0s. Since the probability of an event is the expectation of its indicator, the probability of such a sequence is given by

$$E(1_1 \cdots 1_r (1 - 1_{r+1}) \cdots (1 - 1_n))$$

$$= E\left(1_1 \cdots 1_r \left[-\sum_{i=r+1}^{n} 1_i + \sum_{\substack{i<j \\ i,j=r+1}}^{n} 1_i 1_j \cdots (-1)^{n-r} 1_{r+1} \cdots 1_n \right]\right)$$

There are $n - r$ terms in the first summand, $\binom{n-r}{2}$ terms in the second summand, etc. Multiplying through and using the linearity of expectation proves the claim. $\square$

From the claim, it follows that the probability of every event can be calculated in terms of the w_n's, $n = 1, \ldots, M$. It suffices to show that these can be calculated in terms of $w_j^{(M)}$, $j = 0, \ldots, M$. Writing (7.4) with $w_{M-1}^{(M)}$ on the left-hand side, we can express w_{M-1} in terms of $w_{M-1}^{(M)}$ and w_M. Writing then $w_{M-2}^{(M)}$ on the left-hand side, we can solve for w_{M-2} in terms of $w_{M-2}^{(M)}$, $w_{M-1}^{(M)}$, and w_M. Continuing in this way, we express all the w_n's in terms of $w_j^{(M)}$, $j = 1, \ldots, M$. ∎

In order to specify an exchangeable process of length M, it suffices to specify the probability that j of the M events occur, for $j = 0, \ldots, M$. Hence, any distribution over the numbers $0, \ldots, M$ determines a unique exchangeable process. Among such processes, one class deserves special mention. These are the processes that assign probability 1 to one value of j, and probability 0 to all the others. Such processes can be thought of as drawing balls of two different colors from an urn without replacement. A black ball on the ith draw corresponds, say, to $1_i = 1$, and a white ball on the ith draw to $1_i = 0$. If the urn is known to contain M balls, j of which are black, then $w_j^{(M)} = 1$ and $w_i^{(M)} = 0$ for $i \neq j$. For $M \geqslant n$, $j \geqslant r$, the probability of finding exactly r black balls on n drawings is given by the *hypergeometric distribution*:

$$w_r^{(n)} = \frac{\binom{n}{r}\binom{M-n}{j-r}}{\binom{M}{j}} \tag{7.5}$$

A general exchangeable process of length M may be modeled as drawings from an urn containing M balls of which the number of black balls is unknown. The probability $w_j^{(M)}$ may be thought of as a probability that the urn contains j black balls. It follows that the probabilities $w_r^{(n)}$ can be written as follows:

$$w_r^{(n)} = \sum_{j \geqslant r} p(r \text{ black balls on } n \text{ draws} \,|\, \text{urn has } j \text{ black balls})$$

$$p(\text{urn has } j \text{ black balls}) \tag{7.6}$$

Hence, we see that a finite exchangeable process is a "conditional urn-drawing process." Urn drawings are not independent since the result of one draw alters the probabilities for succeeding draws. Before drawing any balls, the probability of black on the last draw is j/M, if j of the M balls are black. After having drawn $M - 1$ balls, the probability of black on the last draw is either 0 or 1, depending on whether all the black balls have already been drawn.

Filling in from (7.5) in (7.6), we have

$$w_r^{(n)} = \sum_{j=r}^{M} \frac{\binom{n}{r}\binom{M-n}{j-r}}{\binom{M}{j}} w_j^{(M)} \tag{7.7}$$

This gives us an easier way to calculate $w_r^{(n)}$ in terms of the $w_j^{(M)}$'s.

Up to now we have assumed that our inference problem was to determine the probability of occurrence of the rth event, conditional on observations of $1_1, \ldots, 1_{r-1}$. We may also consider the problem of trying to guess the true composition of an urn with M balls, on the basis of observational data. The following notation may be suggestive. Let j/M denote the hypothesis that j of the urn's M balls are black, and let (n, r) denote the data that r black balls have been observed on n draws (note that the data does not specify an outcome sequence, only the number observed and the number of black balls). Bayes' theorem says

$$p(j/M|(n, r)) = \frac{p((n, r)|j/M)p(j/M)}{p(n, r)} = \frac{\binom{n}{r}\binom{M-n}{j-r}w_j^{(M)}}{\binom{M}{j}w_r^{(n)}}$$

A sequence of M tosses with a coin is also an exchangeable sequence, hence it must be representable by a unique distribution over $\{j/M\}_{j=0,...,M}$. The reader may verify that putting

$$w_j^{(M)} = \binom{M}{j}p^j(1 - p)^{M-j}$$

in (7.7) yields [applying the binomial Theorem (A.8) in the appendix to $1 = (p + 1 - p)^{m-n}$],

$$w_r^{(n)} = \binom{n}{r}p^r(1 - p)^{n-r}$$

corresponding to the probability of seeing exactly r "heads" in n tosses with a coin for which the probability of heads is p.

SUPPLEMENT

In this supplement we compare the weak laws of large numbers for the frequentist and subjective interpretations, and give a proof of De Finetti's representation theorem for infinite sequences of exchangeable events. Given the moment convergence theorem of the Appendix, this proof is perhaps the simplest; however, there are simple proofs that use only Helly's theorem (Heath and Sudderth, 1975). The mathematics underlying the applications of De Finetti's theorem in Chapters 11 and 13 are developed here, and further generalizations are indicated.

The Weak Laws of Large Numbers for Exchangeable Events

Definition. A set of random variables $\{X_n\}_{n=1,...,\infty}$ converge in probability (or in measure) if for all $\beta > 0$

$$p(|X_n - X_m| > \beta) \to 0 \qquad \text{as } n, m \to \infty$$

$\{X_n\}$ converge in probability to a random variable X if for all $\beta > 0$

$$p(|X_n - X| > \beta) \to 0 \qquad \text{as } n \to \infty$$

If $\{X_n\}$ converge in probability, then there exists a random variable X such that

$\{X_n\}$ converges to X in probability, and X is unique up to a set of probability zero (Tucker, 1967, p 102).

Let $1_1, \ldots$ be an infinite sequence of indicators for events $A_1, \ldots$. Define

$$X_n = \frac{1}{n} \sum_{i=1}^{n} 1_i$$

If the events A_i form a Bernoulli sequence with parameter p, then the weak law of large numbers (see the Appendix) says that X_n converge in probability to the constant p. This theorem forms the foundation for the frequency interpretation of probability. A corresponding role in the subjective interpretation is played by the *weak law of large numbers for exchangeable events*, according to which the X_n converge in probability, but not necessarily to a constant.

Theorem 7.4. If $A_1, \ldots$ are exchangeable, then X_n converge in probability.

Proof. Fix $\beta > 0$, and let $k > h$. Since

$$E(X_h - X_k)^2 \geqslant \int\limits_{(X_h - X_k)^2 \geqslant \beta^2} |X_h - X_k|^2 \, dp \geqslant p((X_h - X_k)^2 \geqslant \beta^2)\beta^2$$

it suffices to show $E(X_h - X_k)^2 \to 0$.

$$E(X_h - X_k)^2 = E\left(\frac{1}{h^2} \sum_{i,j=1}^{h} 1_i 1_j + \frac{1}{k^2} \sum_{i,j=1}^{k} 1_i 1_j - \frac{2}{hk} \sum_{i=1}^{h} \sum_{j=1}^{k} 1_i 1_j\right)$$

Since $1_i 1_i = 1_i$, $E(1_i 1_i) = w_1$. If $i \neq j$, $E(1_i 1_j) = w_2$. The coefficient of w_1 on the right-hand side above is

$$\frac{h}{h^2} + \frac{k}{k^2} - \frac{2h}{hk} = \frac{1}{h} - \frac{1}{k}$$

Similarly, the coefficient of w_2 is

$$\frac{h(h-1)}{h^2} + \frac{k(k-1)}{k^2} - \frac{2h(k-1)}{hk} = \frac{1}{k} - \frac{1}{h}$$

Hence

$$E(X_h - X_k)^2 = \left(\frac{1}{h} - \frac{1}{k}\right)(w_1 - w_2) \to 0 \qquad \text{as } h, k \to \infty \quad \blacksquare$$

De Finetti's Representation Theorem for Infinite Sequences of Events

Let p_h be the probability on j/h, $j = 0, \ldots, h$, induced by X_h, such that $p_h(j/h) = p(X_h = j/h)$. We may consider p_h as a countably additive measure on the unit interval, setting

$$p_h(B) = \sum_{j=0}^{h} p(X_h = j/h) 1_{B \cap (j/h)}$$

for any measurable set B. De Finetti's theorem says essentially that the measures p_h converge as $h \to \infty$, and characterizes the limit measure. For the notion of convergence of probability measures, we refer to the moment convergence theorem of the Appendix.

Theorem 7.5. (De Finetti). Let $A_1, \ldots$ be exchangeable with respect to p, with X_h and p_h defined as above, and with w_n as in (7.4). Then as $h \to \infty$, the measures p_h converge to a measure on the unit interval and w_n is the nth moment of this measure.

Proof. By the moment convergence theorem (A.6) of the Appendix, it suffices to show that $E(X_h)^n \to w_n$ as $h \to \infty$. Recall the multinomial expansion:

$$(1_1 + \cdots + 1_h)^n = \sum_{\substack{r_i \geq 0 \\ r_1 + \cdots + r_h = n}} \frac{n!}{r_1! \cdots r_h!} 1_1^{r_1} \cdots 1_h^{r_h}.$$

By exchangeability, the expectation for each term in the summand is determined by the number of r_i's that are different from 0. For each term in which k of the r_i's differ from 0, the expectation of the product of indicators is w_k. We divide both sides by h^n, take expectations on both sides and estimate the coefficients of w_k, $k = 1, \ldots, n$. There are $h!/k!(h - k)!$ ways of choosing k of the $1_1, \ldots, 1_h$. For $k = n$ the coefficient of w_n is

$$\left(\frac{1}{h^n}\right)\binom{h}{n} n! = \frac{h(h-1)\cdots(h-n+1)}{h^n} \to 1 \qquad \text{as } h \to \infty$$

For $k < n$, the coefficient of w_k will be smaller than

$$\left(\frac{1}{h^n}\right)\binom{h}{k} n! = \frac{h(h-1)\cdots(h-k+1)n!}{h^n k!} \to 0 \qquad \text{as } h \to \infty$$

It follows that $E(X_h)^n \to w_n$ as $h \to \infty$ ∎

We denote the probability measure on the unit interval given by Theorem 7.5 as p. Since w_n is the nth moment of p, we may write

$$w_n = \int_0^1 x^n \, dp(x)$$

Recall the binomial expansion:

$$(1 - x)^n = \sum_{k=0}^n (-1)^k \binom{n}{k} x^k$$

Let Q be an outcome sequence of length n containing exactly r 1s. Using the above facts in combination with Equation (7.4), we see that

$$p(Q) = \frac{w_r^{(n)}}{\binom{n}{r}} = \int_0^1 x^r (1 - x)^{n-r} \, dp(x) \tag{7.8}$$

We can interpret this as follows. If the events in question form a Bernoulli sequence with parameter x, then the probability of Q would be $x^r(1 - x)^{n-r}$. The limiting relative frequency in this case is certain to be x. When the events are exchangeable, the probability of Q is given by mixing Bernoulli sequences according to the probability measure dp on $[0, 1]$. This is the analog of Equation (7.6) for infinite sequences of exchangeable events.

Equivalent Observations

The key to using De Finetti's theorem in applications involves finding prior probability measures dp for which the integral in (7.8) can be evaluated easily. For exchangeable events, the *beta densities* are most useful in this respect. For positive integers a and b define

$$B(a, b) = \frac{(a - 1)!(b - 1)!}{(a + b - 1)!}$$

then the beta density with parameters a and b, a, $b > 0$, is defined as

$$f(x) = \frac{x^{a-1}(1 - x)^{b-1}}{B(a, b)} \qquad x \in [0, 1] \tag{7.9}$$

Substituting $dp(x) = f(x)\, dx$ in (7.8), we find

$$p(Q) = \frac{B(r + a, n - r + b)}{B(a, b)} \tag{7.10}$$

The right-hand side of (7.10) admits a convenient interpretation. Let $\tilde{Q}$ be a sequence of $(a+b-2)$ outcomes containing exactly $(a-1)$ 1s and $(b-1)$ 0s, and let $Q*\tilde{Q}$ be the sequence gotten by concatenating Q and $\tilde{Q}$. Let us adopt the uniform prior density $dp(x) = dx$. Then the conditional probability of $Q*\tilde{Q}$ given $\tilde{Q}$ is

$$p(Q*\tilde{Q}\,|\,\tilde{Q}) = \frac{p(Q*\tilde{Q})}{p(\tilde{Q})} = \frac{B(r + a, n - r + b)}{B(a, b)} \tag{7.11}$$

Hence, the parameters a and b in (7.9) can be interpreted in terms of *equivalent observations*. That is, the probability $p(Q)$ in (7.10) is the probability that we would have if we started with the uniform prior density, observed $(a - 1)$ 1s and $(b - 1)$ 0s, and conditionalized on these observations.

Setting $r = 1$, $n = r$ in (7.8) and substituting (7.9) for $dp(x)$ we derive the expectation of the beta density with parameters a, b:

$$E(x) = \frac{B(a + 1, b)}{B(a, b)} = \frac{a}{a + b} \tag{7.12}$$

Under the interpretation of $(a - 1)$ and $(b - 1)$ as equivalent observations, this reduces to the Laplace rule in Equation (7.3). The variance of the beta density with parameters a, b is calculated as

$$V(x) = E(x^2) - [E(x)]^2 = \frac{ab}{(a + b)^2(a + b + 1)} \tag{7.13}$$

Events may be thought of as random variables taking two possible values, 0 and 1. The results of this chapter can be generalized for random variables taking a finite number of possible values. Instead of urns with two colors of balls, we then must consider drawings from urns whose balls can have any of a given finite number of colors. Instead of Bernoulli, or coin-tossing processes, we have to consider multinomial processes. These may be thought of as a sequences of rolls with a die having a given finite number of faces. The approach to relative frequencies and the representation of exchangeable sequences go through *mutatis mutandis*.

Let $y_1, y_2, \ldots$ be an infinite sequence of exchangeable random variables taking outcomes in the set $\{1, \ldots, k\}$, and let Q be a sequence of outcomes of length n containing exactly r_i occurrences of outcome i, $i = 1, \ldots, k$. Let $S(k) = \{(\pi_1 \cdots \pi_k) | \pi_i \geqslant 0, \, \Sigma\pi_i = 1\}$, then the appropriate generalization of De Finetti's theorem entails that

$$p(Q) = \int_{\pi \in S(k)} \prod \pi_i^{r_i} \, dp(\pi) \qquad (7.14)$$

$p(Q)$ is a mixture of independent "dice-tossing" processes. The set of *Dirichlet densities* over this set forms an analog to the beta densities. The Dirichlet density with parameters $a_1 \cdots a_k$ is given by (see Box and Tiao, 1973, chap. 2)

$$f(\pi) = \frac{\prod \pi_i^{a_i - 1} \, d\pi}{D(a_1 \cdots a_k)} \qquad (7.15)$$

where $a_i > 0$, $i = 1, \ldots, k$, and

$$D(a_1 \cdots a_k) = \frac{\prod_{i=1}^{k} (a_i - 1)!}{(\Sigma a_i - 1)!} \qquad (7.16)$$

The interpretation of $a_i - 1$ as the number of equivalent observations of outcome i holds just as in the case of the beta densities. It is well to emphasize that the notion of equivalent observation is defined with respect to the uniform prior distribution on $S(k)$, and hence depends on k. If we start with the uniform distribution on $S(k)$, and marginalize to a distribution on $S(k - 1)$ by identifying the outcomes k and $k - 1$, the result is not the uniform distribution on $S(k - 1)$. The impact of this remark can be brought home as follows. Substituting (7.15) in (7.14) we calculate

$$p(y_{n+1} = j | Q) = \frac{(r_j + a_j)}{\Sigma(r_i + a_i)} \qquad (7.17)$$

If $n = 0$, so that $r_i = 0$, $i = 1, \ldots, k$; then this becomes

$$p(y_1 = j) = \frac{a_j}{\Sigma a_i}. \qquad (7.18)$$

Compare (7.18) with (7.12). The numerator is the number of equivalent observations of outcome j plus 1, whereas the denominator is the total number of equivalent observations plus *the number k of alternatives*. If $k > 2$, then $\Sigma a_i >$ (number of equivalent observations $= j$) + (number of equivalent observations $\neq j$) + 2.

Extension to l_p Symmetric Measures

The results can be further extended to continuous random variables. A precise statement requires more mathematical preliminaries (see Hewitt and Savage, 1955, for a good discussion), but we can give the following informal statement. Let $X_1, \ldots$ be an infinite sequence of exchangeable random variables taking values on the real line. For any set $\{E_i\}_{i=1,\ldots,n}$ of (possibly infinite) intervals, let $P(E_1, \ldots, E_n)$ denote the probability that X_i takes a value lying in E_i, $i = 1, \ldots, n$. Then there exists a unique probability measure μ over the set of probabilities on the real line such that for all n, and all $E_1, \ldots, E_n$,

$$p(E_1, \ldots, E_n) = \int \prod_{i=1}^{n} \tilde{p}(E_i) \, d\mu(\tilde{p})$$

The De Finetti theorem is less helpful in this case, as the set of probabilities on the real line over which μ is defined is quite large. The problem is that exchangeability is too weak for studying infinite sequences of continuous random variables. Intuitively, exchangeability means that the probability of a sequence of outcomes depends only on the relative frequencies with which the outcomes occur in the sequence. If the outcomes can be chosen from the real line, then there are too many possible outcomes to see repetitions in finite sequences.

Interesting extensions of De Finetti's theorem to the case of real-valued random variables involve strengthening the assumptions. A probability is exchangeable if it is invariant under the action of the finite permutation group. We strengthen the assumptions by requiring invariance under the action of a group that includes the finite permutation group as a subgroup. Theorems can be proved relating strengthened invariance properties with restrictions on the measure μ above. We give one recent example. Let $X_1, \ldots$ be a sequence of random variables, and assume for every k, the density f_k of the random vector $(X_1, \ldots, X_k)$ exists, is continuous, and has the form

$$f_k(x_1, \ldots, x_k) = f_k(\Sigma |x_i|^p)$$

In other words, f_k depends only on the l_p norm of the point $(x_1, \ldots, x_k)$. Note that the process $X_1, \ldots$ is exchangeable, as the finite permutation group is a subgroup of the group of "rotations under the l_p norm." Then for a unique measure μ on $(0, \infty)$, f_k can be written as (Berman, 1980; for a simpler proof see Cooke and Misiewicz, 1988)

$$f_k(\Sigma |x_i|^p) = \int_0^\infty \prod_{i=1}^{k} e^{s|x_i|^p} \left(\frac{p}{2}\right)^k s^{k/p} T\left(\frac{1}{p}\right)^k \mu \, ds$$

where the gamma function Γ is defined as

$$\Gamma(t) = \int_0^\infty x^{t-1} e^{-x} \, dx$$

As a consequence, if f_k is invariant under normal rotations, that is, if $p = 2$, then f_k is a mixture of independent normal densities with mean zero (this latter result was first established by Schoenberg, 1938).

For a mathematical discussion of recent results in exchangeability, see Aldous (1985), and for applications, Cooke, Misiewicz and Mendel (1990).

8
Elicitation and Scoring

In Chapter 6 it was demonstrated that a rational agent should represent his uncertainty as subjective probability. This alas does not mean that everyone is able to produce meaningful probability numbers on demand. Considerable thought must be given to the manner in which probabilities are elicited.

Scoring refers to the process of assigning a numerical value to a subjective probability. It was initially regarded as a tool for eliciting probabilities, but was recognized early on as a potential tool for evaluating subjective probability assessments (Winkler, 1969). The choice of a scoring procedure necessarily affects the form in which the probabilities are expressed; hence the subjects of elicitation and scoring cannot be separated. With regard to scoring there are two questions to be considered:

1. Does the score reward those features that we would like subjective probability assessments to display?
2. Does a score introduce a reward structure that distorts the assessment of subjective probabilities?

It is perhaps remarkable that we can find scores for which both questions can be answered satisfactorally. The present chapter is concerned with the first question: What are the features we should like subjective probability assessments to display, and how can we measure them? The scoring concepts introduced in this chapter will evolve into weights for evaluating and combining probability assessments in the next chapter.

Despite a widespread interest in elicitation in the professional journals, relatively little attention has been paid to the problem of eliciting *expert* probabilities. The research centers, on the other hand, have recently taken an active interest in this question (von Winterfeld, 1989; ESRRDA, 1989). In the United States, these developments are related to the emergence of NUREG-1150, a 34-volume report using expert judgment to assess the frequency of serious accidents at nuclear reactors (see Chap. 2). In response to sharp criticism of a draft report, a massive investment was made in eliciting expert subjective probabilities (Wheeler et al., 1989). Recommendations (von Winterfeld, 1989) deriving from this experience involve bringing the experts together, defining and decomposing the issues, explaining the methods, training in probability assessment (at least 1 day), eliciting and finally resolving disagreements. The experts' probability assessments are aggregated by simple averaging. There is no attempt to evaluate the quality of the probabilities obtained, and no attention is paid to scoring. The method of aggregation is therefore the simplest conceivable. From the examples discussed in Chapter 15, it emerges that simply averaging experts' probabilities is suboptimal.

Even if the suboptimal method of aggregation were compensated by the effort in elicitation, the methods evolving from NUREG-1150 would still be quite infeasible for most practical applications. For practical reasons, if for no other, the issues of scoring and elicitation must be coupled.

The first section of this chapter takes up methods of elicitation, reviews some standard techniques, and describes a method employed at the European Space Agency. The next section broaches the issue of scoring from the viewpoint of the first question above. Methodological problems inherent in certain standard scoring methods are discussed. A final section is devoted to practical guidelines.

ELICITATION

Indirect Methods; Betting Rates

The most popular elicitation technique among theoreticians is surely that of betting rates, introduced independently by Ramsey (1931) and De Finetti (1937). Perhaps because of their somewhat artificial character, it is hard to find any application of betting rates to expert probability elicitation. However, because of their theoretical and historical interest, they will be treated briefly below.

Suppose we are interested in a subject's degree of belief in event A. We give the subject a lottery ticket of the form

Receive $100 if A occurs

We are now interested in the value that the subject attaches to this ticket. For a positive amount of money x less than $100, we inquire if he is willing to trade this ticket for x. Let x_A be the smallest amount for which he is willing to make such a trade. We may then say that he is indifferent between the ticket and receiving x_A with certainty. x_A is called the *certainty equivalent* for the above lottery.

If the subject is a rational agent, then his preferences should be representable as expectations. If his preferences for two options are equal, his expected utility associated with each option should be the same. If we assume that his utility for money is linear in money, then his expected utility of the lottery ticket is

Expected utility of lottery $= 100Kp(A)$

for some constant K. His expected utility of x_A is simply Kx_A. Setting these two expectations equal, we derive

$$p(A) = \frac{x_A}{100}$$

This is a good example of measurement as it is practiced in the physical sciences. For example, we can measure the momentum of a charged particle by observing its deflection in a magnetic field. In associating its deflection with momentum, we assume the particle obeys the laws of physics (e.g., the conservation of momentum), and that its mass is constant during the measurement. Similarly, in the above scheme, we assume that the agent obeys the laws of rational preference (e.g., Savage's axioms), and assume that his utility function for money is linear.

Unfortunately, the scant empirical evidence suggests that the utility of money is highly nonlinear. Galanter (1962) asked a group of subjects the following question: "Suppose we give you x dollars; reflect on how happy you are. How much should we have given you in order to have made you twice as happy?" Galanter gives the results presented in Table 8.1 for the three values of x: $10, $100, $1000.

If utility were linear in money, then U($10) = 10K. It follows that to double the utility of $10, the subject should require $20, etc. Galanter reports that a good fit for the above data is given by the function $U(\$x) = 3.71x^{.43}$. The experiment was also attempted using losses of money to become "twice as unhappy," but subjects found such questions "too difficult."

Without wishing to overestimate the value of the numbers in Table 8.1; they do tend to confirm a rather self-evident intuition. People may be willing to run certain risks to win one million dollars, but this does not mean they would run the same risk twice to win two million.

The nonlinearity of the value of money would become especially disturbing if the subject were asked to estimate very small or very large probabilities. Suppose an expert believes that the probability of a core melt in a nuclear reactor was 10^{-5} per reactor year. If we give him the lottery ticket

Receive $100 if reactor X suffers a core melt in year Y.

and if his utility were linear in money, this lottery would have the value of one-tenth of a cent for him. In order to make the lottery meaningful, we would have to raise the amount in the lottery ticket to, say, $100,000. The expert now has to compare winning $100,000 if the core melt occurs, to some amount on the order of $10.

Direct Methods

The simplest method of measuring a person's degrees of belied is simply to ask him what his degree of belief is. While this method is surely the most common, it is equally surely the worst, especially for persons who are not familiar with the notion of probability. Most people have poor intuitions regarding numerical probabilities.

Table 8.1 Money Required to Double Happiness

| Given | Twice as Happy | |
	Mean	Median
$10	$53	$45
100	537.50	350
1000	10,220	5000

Source: Galanter, 1962.

Evidence for this among experts may be gleaned from the "electric lawn mower"
data in Table 2.7. More than 8% of probabilities assessed by the experts were either
0 or 1. Given the events in question, these assessments are clearly absurd.

A better method has been proposed by Lindley (1970). The idea is that states
of uncertainty can be compared with regard to intensity. Hence, it makes sense to
assess the probability of an event A by comparing the intensity of one's uncertainty
regarding A to that of some other event B, to which numbers may easily be
assigned. Consider a lottery basket containing 1000 tickets, each with a number
between 1 and 1000. Suppose we ask a person:

> For which number N, $0 \leqslant N \leqslant 1000$, is your uncertainty regarding the
> occurrence of A equal to your uncertainty for drawing a ticket from this basket
> with a number less than or equal to N?

The number N, which he gives, divided by 1000, may be taken as a measure of his
degree of belief in the event A. A slight variation on this method involves spinning a
"probability wheel" (De Groot, 1970), similar to a roulette wheel.

Other direct methods are the discrete tests and quantile tests for calibration
discussed in Chapter 4. Although these are treated in the following section, it is
appropriate to recall them briefly here. In the discrete tests, a subject assigns an
uncertain event to one of a designated number of probability bins. The probability
associated with a probability bin is taken as an estimate of the subject's probability
for the event in question. In assessing the distributions of continuous variables, the
quantile test involves asking the subject to state certain fixed quantiles, of his
distribution. Although these tests were designed to measure calibration, they also
serve to elicit probabilities.

Parametric Elicitation

In some applications, the nature of the quantities whose distributions are assessed
may be such as to suggest a particular class of probability distributions. In such
cases more elegant elicitation procedures can be derived. One such procedure is
described below. This specific implementation presupposes that the experts'
distributions are approximately lognormal; however, the idea may be applied to
any class of distributions determined by two parameters. This procedure was
developed for the European Space Agency (Preyssl and Cooke, 1989) for the
assessment of failure frequencies.

Preliminary discussions with experts at the European Space Agency indicated
a preference for breaking the elicitation down into two steps:

Step 1: Indicate a best estimate for the failure frequency in question.
Step 2: Indicate "how certain" one is of the best estimate.

Furthermore, given the traditions in the aerospace sector (see Chap. 2), there was a
preference for qualitative as well as quantitative elicitation procedures. The
elicitation was broken down into two steps as follows:

Step 1: The expert is asked for his median estimate of the failure frequency in
         question. His answer is M.

Step 2: The expert is asked how surprised he/she would be if the true value turned out to be a factor 10 or more higher. The answer is a number r, $0 < r < 1$, reflecting the probability that the true value should exceed the median by a factor of 10 or more. (For median estimates greater than 0.1, the expert is simply asked to state his upper 95% confidence bound directly.)

The numbers M and r determine a unique lognormal distribution. The techniques described in the Appendix can be applied to find the experts' 5% and 95% confidence bounds. The 5% and 95% confidence bounds are then $M/k_{.95}$, $Mk_{.95}$, respectively, where

$$k_{.95} \approx \left(\frac{\exp(-0.658)}{y_{1-r}} \right)$$

and y_{1-r} is the $(1 - r)$th quantile or the standard normal distribution.

It was found that experts became comfortable with this procedure very quickly, and particularly liked splitting the assessment task into a "best estimate" and a "degree of uncertainty" task.

SCORING

Roughly speaking, scoring is a numerical evaluation of probability assessments on the basis of observations. Although scoring rules were originally introduced as a technique for *eliciting* probabilities, it will be obvious that scoring rules are of great importance for evaluating and combining expert opinions. A subject is asked for his probability distribution for an uncertain quantity and is told that he will be scored on the basis of his assessment and the observed value of that quantity. In this section we are concerned with scoring as a way of rewarding those properties of expert subjective probability assessments that we value positively. Two properties will be discussed below, *calibration* and *entropy*. It will be useful to introduce these notions by discussing an ideal measurement situation.

An Ideal Measurement

It is useful at first to discuss an ideal setup for measuring calibration and entropy. Suppose that we are required to estimate the mean time to failure in days of a new system component that cannot be subjected to destructive experimental tests. Our only way of obtaining quantitative data is to ask the opinion of experts acquainted with similar kinds of components. Given the amount of uncertainty inherent in predictions of this sort, the experts may feel uncomfortable about giving point predictions, and may prefer to communicate something about the range of their uncertainty. The best they could do in this respect would be to give their subjective probability mass functions (or density functions in the case of continuous variables) for the quantity in question. In other words, they could provide a histogram over the positive integers such that the mass above the integer i is proportional to their subjective probability that the mean time to failure is i days.

The mean time to failure will eventually become known, and when it is known, we may want to pose the question how good was this expert's assessment. Calibration and entropy are relevant to performing this kind of evaluation. We assume that subjective probability mass functions over a finite number M of integers can be solicited from each of several experts, for a large number of uncertain quantities.

The entropy associated with a probability mass function P over the integers $i = 1, \ldots M$ is:

$$H(P) = - \sum_{i=1}^{M} P(i) \ln P(i) \tag{8.1}$$

where $P(i)$ is the probability assigned to the integer i. $H(P)$ is a good measure of the degree to which the mass is "spread out." Its maximal value $\ln M$ is attained if $P(i) = 1/M$ for all i, and its minimal value 0 is attained if $P(i) = 1$ for some i. Obviously, low entropy is a desideratum in expert probabilistic assessment. Other things being equal, we should prefer the advice of the expert whose probability functions have the lowest entropy. Other things are usually not equal.

To get an idea how a calibration score could be defined, suppose for the sake of argument that an expert gives the same probability mass function P for a large number n of physically unrelated uncertain quantities. By observing the true values for all these quantities we generate a sample distribution S with $S(i)$ equal to the number of times the value i is observed, divided, by n.

It might appear reasonable to say that the expert is miscalibrated if $S \neq P$. Upon reflection, however, this is easily seen to be quite unreasonable. Suppose the true values represent independent samples from a random variable with distribution P. P certainly "corresponds to reality" (by assumption), but in general we will not have $S = P$, as statistical fluctuations will cause P and S to differ. In line with the intuitive definition of calibration, we might say that the expert was well-calibrated if S and P agree in the long run. The problem with this, as Keynes was fond of saying, is that in the long run we are all dead. This definition gives us no way of measuring calibration for finite samples. We shall see shortly that "$S = P$ in the long run" is a necessary but not sufficient condition for calibration.

Still speaking roughly, we want to say that the expert is well-calibrated if the true values of the uncertain quantities can be regarded as independent samples of a random variable with distribution P. This entails that the discrepancy between S and P should be no more than what one might expect in the case of independent multinomial variables with distribution P. We therefore propose to interpret the statement

The expert is well-calibrated.

as the statistical hypothesis:

Cal(P) := the uncertain quantities are independent and identically distributed with distribution P

We want to define a calibration score as the degree to which the data supports the hypothesis Cal(P). A procedure for doing this is described below.

The "discrepancy between S and P" can be measured by the relative information of S with respect to P, $I(S, P)$:

$$I(S, P) = \sum_{i=1}^{M} S(i) \ln \frac{S(i)}{P(i)} \qquad (8.2)$$

$I(S, P)$ may be taken as a measure of surprise, which someone would experience if he believed P and subsequently learned S. $I(S, P) = 0$ if and only if $P = S$, and larger values correspond to greater surprise. Obviously, large values of $I(S, P)$ are critical for Cal(P). We interpret the "degree to which the data support the hypothesis Cal(P)" as the probability under Cal(P) of observing a discrepancy in a sample distribution S' at least as large as $I(S, P)$, on n observations:

$$\text{Prob}\{I(S', P) \geqslant I(S, P) \,|\, \text{Cal}(P), n \text{ observations}\} \qquad (8.3)$$

This probability can be used to define statistical tests in the classical sense. Of particular interest is the following fact (see Appendix): If P is concentrated on a finite number M of integers that include all observed values, then as the number n of observations gets large $2nI(S, P)$ becomes χ^2-distributed with $M - 1$ degrees of freedom (see Hoel, 1971). The natural logarithm must be used in Equation (8.2). Expanding the logarithms in (8.2) via a Taylor series and retaining the dominant terms yield the familiar χ^2 statistic for testing goodness of fit between the sample distribution S and the "theoretical distribution" P.

We call the above conditional probability the expert's calibration score for the n observations, and we propose to use this quantity to measure calibration in expert assessments.

We can now understand why asymptotic convergence of S to P is not sufficient for good calibration. Suppose that P is concentrated on six values so that the number of degrees of freedom of the χ^2 distribution is five, and suppose that as the number of observations n goes to infinity, the expert's calibration score converges to 1%. From a χ^2 table we conclude that $2nI(S, P)$ converges to 15. This entails that $I(S, P)$ converges to 0, and hence that S converges to P. However, for all n greater than some n_0, the hypothesis Cal(P) would be rejected at the 5% significance level.

The basic principle of a classical approach to expert evaluation can now be outlined: Good experts should have good entropy scores and good calibration scores. This theory is normative in the sense that it prescribes how experts should perform. In Chapter 10 we present evidence that experienced probability assessors do indeed perform better as a group with respect to both scores than inexperienced assessors, for items relating to their field.

Implementation: Quantile Tests

The approach presented above is not yet very practical since it requires a large number of quantities for which the expert gives the same probability distribution. If a set of random variables X_i have invertible cumulative distribution functions $F_i, i = 1, 2, \ldots,$ it is not difficult to find transformations under which the variables become identically distributed. It suffices to consider the transformed variables

$F_i(X_i)$, as these all have the uniform distribution on the unit interval.[1] However, it is seldom convenient to elicit an entire distribution function for every random variable.

The quantile tests encountered in Chapter 4 may be seen as providing practical implementations of the ideas in the preceding section. The implementation involves casting the assessments in a form that yields sets of variables with identical distributions.

Instead of eliciting the entire mass function from an assessor, we elicit various quantiles from his mass function. The rth quantile of a mass function P over the integers is by definition the smallest value i such that

$$\sum_{j < i} P(j) \geq r$$

In the experiment described in Chapter 10, the 1%, 25%, 50%, 75%, and 99% quantiles were elicited. For each uncertain item, a multinomial variable is introduced with probability vector $p = p_1, \ldots, p_6$. p_1 denotes the probability that the true value of the original value is less than or equal to the 1% quantile, p_2 the probability that this value falls between the 1% and the 25% quantiles, etc. Obviously,

$$p_1 = p_6 = 1\% \qquad p_2 = p_5 = 24\% \qquad p_3 = p_4 = 25\%$$

Observing between which quantiles the true values fall, a sample distribution $s = s_1, \ldots, s_6$ is generated. s_1 represents the number of true values falling beneath the 1% quantile, divided by the total number of observations, etc.

The method of scoring calibration for the ideal measurement applies equally well to the probability vectors p and s. For n observations, the scoring variable for calibration is

$$\text{Prob}\{I(s', p) \geq I(s, p) \mid \text{Cal}(P), n \text{ observations}\} \qquad (8.4)$$

Since $2nI(s, p)$ is asymptotically χ^2 distributed with five degrees of freedom under hypothesis $\text{Cal}(P)$ on n observations, the above probability can be approximated by consulting an appropriate χ^2 table. The approximation is good if $np_i \geq 4$, $i = 1, \ldots, 6$.

Measuring entropy in quantile tests introduces some complications. We are not interested in the entropy of the distribution p, but rather in the entropy in the original distribution P. The original distribution P has not been elicited, and we only know some quantiles from P.

There are many ways of constructing an approximation P' to P using the information contained in the elicited quantiles. In an experiment using quantile tests discussed in Chapter 10, we first define an "intrinsic range" for each quantity, such that we can reasonably put $P'(i) = 0$ for i outside this range. We then construct P' by "smearing out" the probability mass evenly between the various quantiles. The example shown in Figure 8.1 makes this clear.

Aside from being simple, the mass function P' has something else to recommend it. Of all mass functions over the given intrinsic range whose 5%, 50%, and 95% quantiles agree with those of P, P' has the largest entropy (Kullback,

---

[1] We have $F_i(x) = \text{Prob}(X_i \leq x)$. Hence $\text{Prob}(F_i(X_i) \leq r) = \text{Prob}(X_i \leq F_i^{-1}(r)) = F_i(F_i^{-1}(r)) = r$.

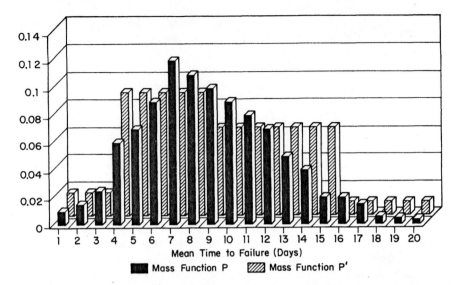

Figure 8.1 A mass function P approximated by a mass function P′ in which the mass between the 0% quantile (0 days) and the 5% quantile (3 days) has been evenly smeared out, and similarly for the 50% quantile (9 days), the 95% quantile (15 days), and the 100% quantile (20 days). P′ is the minimal information approximation (relative to the uniform measure) to P agreeing with P at the 5%, 50% and 95% quantiles.

1959). Hence, if we are interested in scoring entropy, and we must choose an approximation to P on the basis of quantile information, then P' is the natural choice. We can use the entropy $H(P')$ of P' as an approximation to $H(P)$.

The above technique is appropriate when P is a mass function over a finite number of integers. If the range of the uncertain quantity is really continuous then P should be replaced by a probability density function over the real numbers. In this case additional subtleties arise that are signalled in the Appendix. If $f(x)$ represents the probability density at the point x, then the entropy of the distribution associated with f is not well-defined. However, the relative information of f with respect to the uniform distribution over the intrinsic range is well-defined and can be used in this context (see the Appendix).[2] Of course, one must remember that sharply peaked distributions correspond to low entropy and to high relative information, with respect to the uniform distribution.

Let us assume that mass functions of the form P' have been constructed for suitable intrinsic ranges for all uncertain quantities. Now another problem arises.

---

[2] For mass functions S and P over the integers $1, \ldots, M$, the relative information of S with respect to P, $I(S, P)$, is

$$I(S, P) = \sum_{i=1}^{M} S(i) \ln \frac{S(i)}{P(i)}$$

For density functions the corresponding expression is

$$I(S, P) = \int S(x) \ln \frac{S(x)}{P(x)} dx$$

When quantile tests are constructed, the quantities involved may have different physical units (e.g., days, kilograms, percentages). Moreover, the intrinsic ranges of possible values may be quite different.

If the uncertain quantities all had the same intrinsic range, then we could reasonably adopt the entropy in the joint distribution over all quantities as an entropy score. If the quantities are independent, this joint entropy is simply the sum of the entropies for each quantity.

If the uncertain quantities do not have the same intrinsic range, then there are good arguments for not using the "joint entropy" as an entropy score. For example, if one of the uncertain quantities can take a billion possible values, then the maximal entropy for this quantity is ln 1,000,000,000 = 20.7. If the other quantities can take only one of a hundred possible values, the maximal entropy for these quantities is ln 100 = 4.6. Simply adding the scores may therefore give inordinate weight to quantities with intrinsically larger ranges. In particular, if we rank the experts according to "joint entropy," then we may well find that the entropy rank is largely determined by the rank on the variable with the largest intrinsic range. If we do not wish the intrinsic ranges of the uncertain quantities to influence the entropy score, then the joint entropy score is not appropriate.

The natural way to eliminate the influence of different intrinsic ranges is simply to normalize the intrinsic ranges to unity. We illustrate this with a calculation. Calling the lower and upper limits of the intrinsic range the 0 and 100% quantiles, respectively, let m_j denote the number of units between the jth and the $(j - 1)$th quantiles. In the example described above, j runs from 1 to 6. Clearly, $P'(i) = p_j/m_j$, when i is greater than the $(j - 1)$th quantile and less than or equal to the jth quantile. Letting Rg denote the intrinsic range,

$$H(P') = - \sum_{i \in \mathrm{Rg}} P'(i) \ln P'(i)$$

$$= - \sum_{j=1}^{6} p_j \ln \frac{p_j}{m_j}$$

$$= H(p) + \sum_{j=1}^{6} p_j \ln m_j$$

If we now rescale the intrinsic range by replacing m_j in the above expression with m_j/m, where $m = \Sigma m_j$ is the total number of units in the original range, then $H(P')$ is transformed into

$$H(P') - \ln m$$

Since $\ln m$ is the maximal value of $H(P')$, the above expression is always less than or equal to 0. From the Appendix we recognize the above expression as the negative of the relative information of P' with respect to the uniform distribution over the intrinsic range. Letting U denote this uniform distribution, it follows that we can represent entropy under rescaling by calculating the quantity $I(P', U)$. $I(P', U)$ is always nonnegative, and low values of $H(P')$ correspond, under rescaling, to high values of $I(P', U)$.

Implementation: Discrete Tests

Figures 2.2 and 2.4 represent results of discrete tests. These can be viewed as another way of adapting the ideal measurement to the demands of the real world. This is the method most commonly encountered in the psychometric literature.

In a typical application, a subject is given a number of statements and asked to determine whether the statements are true or false, and asked to state his subjective probability that his answer is correct (for variations, see Lichtenstein, Fischoff, and Phillips, 1982). The probabilities are coarse grained into discrete "probability bins," usually 50%, 60%, ..., 90%. In this way we generate sets of identically distributed variables, one set for each bin. For each probability bin, we can apply the ideas introduced in the discussion of the ideal measurement. The only problem is how to combine the scores for all bins simultaneously.

A calibration score can be introduced as follows. If p_i is the probability corresponding with the ith bin, then we associate the distribution $P_i = \{p_i, 1 - p_i\}$ with bin i, $i = 1, \ldots, B$. A sample distribution $S_i = \{s_i, 1 - s_i\}$ can also be associated with bin i, by considering the proportion s_i of items in the ith bin that are correct. Let n_i denote the number of items in bin i.

What does it mean to be well-calibrated on this type of test? In accordance with the previous discussion, we must identify good calibration with a statistical hypothesis. We shall say that a subject is well-calibrated for a discrete test if he is well-calibrated for each bin in the sense defined previously, and if the items in distinct bins are independent.

If the subject is well-calibrated and if the items in each bin are independent, then we should expect the relative information score for each bin to fluctuate asymptotically as a chi square variable with one degree of freedom. If the items are independent across bins, then the scores for the various bins should fluctuate independently. A complication arises here if the total number of items n is fixed beforehand, as this can introduce a dependency in the fluctuations of the scores for each bin. However, it is not difficult to show (Cooke, Mendel, and Thys, 1988) that the fluctuations become independent as the number of items in each bin gets large. This entails that the quantity

$$RI := \sum_{i=1}^{B} 2n_i I(S_i, P_i) \tag{8.5}$$

is the sum of asymptotically independent asymptotic chi square variables with one degree of freedom. It follows that RI itself is asymptotic chi square with B degrees of freedom, where B is the number of bins. Hence, RI can be used to define a likelihood function, which in turn can be used as a calibration score, just as in the case of quantile tests.

Concerning the entropy score, we note that in a discrete test all variables have the same intrinsic range, namely "correct" and "not-correct." The *response entropy*

$$H(P) := \sum_{i=1}^{B} n_i H(P_i) \tag{8.6}$$

is the entropy in the joint distribution for all items when the items are distributed according to the calibration hypothesis. The term "response" serves to remind us that $H(P)$ measures the entropy in what the subject *says*. This is to be contrasted with the *sample entropy*:

$$H(S) := \sum_{i=1}^{B} n_i H(S_i) \qquad (8.7)$$

The sample entropy measures the entropy in the subject's *performance*, assuming that all items are independent. Whereas the response entropy refers to the distribution associated with the calibration hypothesis, the sample entropy does not correspond to a distribution which anyone believes, or would like to believe, barring the exceptional case that the joint distributions P and S coincide.

An example will illustrate the meaning of the response entropy. Suppose for 1 year two experts are asked each day to give their probability of rain on the next day. The probability bins run from 10% to 90%. Both experts know that the yearly "base rate" of rain is 20%. The first expert simply predicts rain with 20% probability each day and may expect to be well-calibrated. The second expert distinguishes between days in which he thinks rain is more or less likely. As he is also aware of the base rate, he will assign days to probability bins such that

$$\sum_{i=1}^{B} \left(\frac{n_i}{n} \right) p_i = 20\%$$

Let $H(E_j)$ denote the entropy score of expert j, $j = 1, 2$. Considering $H(P_i)$ as a function of p_i, we note that $H(\cdot)$ is concave, so by Jensen's inequality (see supplement to Chap. 6):

$$H(E_1) = nH(20\%) > \sum_{i=1}^{B} n_i H(P_i) = H(E_2)$$

Under the calibration hypothesis, the entropy of the second expert's responses is less than that of the first.

Comparison of Quantile and Discrete Tests

Although the quantile and discrete tests are quite different in form, the theory underlying the scoring of calibration and entropy is the same. This becomes clear when these tests are recognized as concessions with respect to the ideal measurement situation. Both tests construct identically distributed variables from variables which are not identically distributed.

Roughly speaking, quantile tests are better for measuring calibration. The scoring variable $I(s, p)$ in Equation (8.4) is based on variables that really are distributed according to p, and the chi square approximation for (8.4) will require relatively few observations, as the total sample of n variables need not be disaggregated into sets having different distributions. The scoring variable, Equation (8.5), on the other hand, involves distributions P_i that do not really correspond to the subject's probabilities, as these have been coarse grained. In addition, the total sample has been disaggregated into B sets corresponding to the B probability bins. A larger total sample will generally be required to justify the chi square approximation to the distribution of R in (8.5). In an experiment involving discrete tests, discussed in Chapter 10, the distribution of R had to be calculated directly. This is a cumbersome procedure, even on a computer.

For the measurement of entropy, the discrete tests have two advantages over the quantile tests. First, the variables all have the same intrinsic ranges, and second, the distribution P is not elicited in the quantile test and must be approximated by some distribution P'. Note that the notion of sample entropy is not meaningful for quantile tests.

METHODOLOGICAL PROBLEMS

In designing a psychometric experiment to measure calibration, the first problem is to define a suitable scoring variable. With regard to quantile tests no scoring variable has been proposed in the literature. The only numerical representation of calibration is the "surprise index" mentioned in Chapter 4.

With regard to discrete tests the situation is different. In the more recent literature, the "weighted Euclidean distance" score is used. Using the notation of the previous section, with B probability bins, bin i associated with the distribution $P_i = (p_i, 1 - p_i)$, this score becomes

$$\sum_{i=1}^{B} \frac{(p_i - s_i)^2 n_i}{n} \tag{8.8}$$

where, as before, n_i is the number of events in bin i, n is the total number of items, and s_i is the relative frequency of occurrence in bin i. Murphy (1973) extracted this score as the "calibration term" of the Brier (1950) score. There is no evident way of extending (8.8) to quantile tests; thereby suggesting that calibration for such tests is not meaningful. Equation (8.8) is very different from the calibration score derived in Equation (8.5) of the previous section:

$$RI = \sum_{i=1}^{B} 2n_i I(S_i, P_i) \tag{8.5}$$

Equation (8.5) is related to the familiar χ^2 variable for testing goodness of fit in statistics. Indeed, taking the Taylor expansion of the logarithm in (8.5) and retaining the dominant terms, we find that (8.5) can be approximated by

$$\sum_{i=1}^{B} n_i \left[\frac{(p_i - s_i)^2}{p_i(1 - p_i)} \right] \tag{8.9}$$

As pointed out in Lichtenstein, Fischoff, and Phillips (1982), the sampling distribution of the score (8.8) is not known. This means that when the results of an experiment are reported in terms of a scoring variable (8.8), we have no way of distinguishing miscalibration from mere statistical fluctuations. In practice, people seem to rely on gathering a large amount of data. How much data is necessary? Of course, without performing a statistical analysis it is impossible to say. To get a rough idea of the fluctuations, suppose a subject is given 15 uncertain events, and assigns them all to the 90% bin. Suppose that 73% of these (i.e., 11), in fact, occur. Assuming the events are independent with probability of occurrence 90%, the probability of seeing 11 occurrences or fewer is a little greater than 5%. Hence, we could not reject the hypothesis that the subject was well-calibrated at the 5% significance level. If the subject distributes the 15 events more or less evenly over five bins, it may be very difficult to reject the calibration hypothesis at this significance level.

Intuitively, more responses should be used. One way of obtaining more responses, of course, is to give the subject more uncertain items. As this method places demands on both the subject and on the experimenter, the method often used in practice is to give the same items to more subjects. This latter method is employed when one is interested in a set of subjects who have been selected according to some common characteristics, or have been "treated" in some way.

However, this way of obtaining more responses is highly suspect. If two subjects place the same event in the same bin, then these two responses are not independent, but are completely coupled. The event either occurs for both subjects, or does not occur for both. It cannot occur for one and fail to occur for the other. Hence, these two responses cannot fluctuate independently. In general, we should expect that different subjects should have some tendency to place the same events in the same bins, especially if the subjects have been preselected or treated in some way. Hence, the number of *independent* events in the various bins need not increase very rapidly as the number of subjects assessing the same events increases. In any case, the number of independent responses in any bin cannot exceed the total number of events. Experiments involving a small number of variables (on the order of 10) and a large number of subjects (on the order of 50–100) are not uncommon in the literature (see, e.g., Lichtenstein and Fishhoff, 1977). Experiments of this type always neglect to report the multiplicities of the events in the various bins. It remains to be demonstrated whether such experiments yield statistically significant results.

PRACTICAL GUIDELINES

It is appropriate to conclude this chapter with several prosaic, but vitally important, guidelines for eliciting opinions from experts. The guidelines apply equally to all elicitation techniques.

The questions must be clear.

It is very difficult to formulate clear unambiguous questions. This is particularly true with regard to unlikely failure events. An expert will become quickly, and justifiably, irritated if he gets the idea that the analyst doesn't know what he wants.

Prepare an attractive format for the questions and graphic format for the answers.

An attractive format is simply good psychology. Experience shows that experts dislike writing numbers for subjective probabilities. They much prefer to place a check on a scale, or place an "x" in a box, etc. The numerical interpretation must be clearly visible. This will greatly speed up the elicitation.

Perform a dry run.

Once the questions have been selected and a format chosen, a dry run must be performed on a small number of experts. The author has yet to perform a dry run that did not result in significant improvements.

An analysis must be present during the elicitation.

This is absolutely essential. Some items will inevitably require clarification, as it is impossible to anticipate all possible unintended interpretations. Moreover, the presence of an analyst shows that the study has sufficient priority for the expert to take it seriously.

Prepare a brief explanation of the elicitation format, and of the model for processing the responses.

According to the principles formulated in Part I, experts have a right to know how their answers will be processed. A short explanation should be given, and supporting material should be made available if desired. A clear, concise explanation of the format should be carefully prepared, and a few practice items should be walked through before starting the elicitation.

Avoid coaching.

The analyst is not an expert, the expert is the expert. The expert must be convinced that the analyst is interested in the expert's opinion.

The elicitation session should not exceed 1 hour.

The elicitation process is more taxing for the expert than for the analyst. If more time is required a second session should be arranged.

9

Scoring Rules for Evaluating and Weighing Assessments

The previous chapter discussed scoring as a means of defining desirable properties of expert probability assessments and designing elicitation techniques that facilitate the measurement of these properties. The present chapter discusses scoring as a reward structure.

Scoring rules are implicitly brought into play whenever an assessment is associated with a reward structure. Experts, being human, naturally experience "influence" in matters relating to their expertise as a form of reward. Therefore, when expert assessments are combined, the method of combination is a scoring rule and introduces a regard structure.

Scoring rules were initially introduced as tools for eliciting probability assessments. A scoring rule is called (*strictly*) *proper* if a subject receives his maximal expected score if (and only if) his stated assessment corresponds to his true opinion. When probabilities are elicited under a reward structure that is not a proper scoring rule, the subject is being encouraged to state an assessment at variance with his true opinion. In this chapter we are interested in scoring rules not as tools for eliciting, but as tools for *evaluating* expert assessments. Important contributions to the theory of scoring rules can be found in Shuford, Albert, and Massengil, (1966), Winkler and Murphy (1968), Stael von Holstein (1970), Matheson and Winkler (1976), Friedman (1983), De Groot and Bayarri (1987), and Savage (1971). The latter reference provides the most comprehensive discussion of propriety (the property of being proper). De Groot and Fienberg (1983, 1986) and Winkler (1986) discuss "goodness" of probability appraisers in relation to scoring rules.

The problem of scoring expert assessments is intimately related to the problem of assigning weights to experts' assessments and forming weighted combinations of their opinions. A system of weights is a reward structure and hence a score. The "classical model" discussed in Part III forms weighted combinations of expert opinions, where the weights are themselves (asymptotically) strictly proper scoring rules. The main goal of the present chapter is to lay the mathematical foundation for these developments by showing that calibration and entropy can be combined in a proper reward structure. Although the scores introduced here are called "weights," by way of anticipation, they are treated here simply as scoring variables. The problem of combining expert assessments is deferred to Part III.

The developments in the later sections of this chapter are somewhat more technical than the previous chapters, and the proofs appeal to graduate level probability theory. Although the more technical proofs are placed in a supplement, some sections will still be a bit rough on nonmathematicians. Such readers can skip these sections, if they are willing to take the assertions in Chapter 12 at face value. The first section initializes the discussion by considering improper scoring rules, and the second section reviews the literature in this field.

IMPROPER SCORING RULES

It is important to appreciate that the most natural way of scoring probability assessors is also grossly improper. Consider an uncertain quantity taking values in the set $\{o_1,\ldots,o_m\}$. An expert is asked to state his distribution over the possible outcomes. His response will be a probability vector $p = p_1,\ldots,p_m$ satisfying

$$\sum p_i = 1 \qquad p_i \geqslant 0 \qquad i = 1,\ldots,m$$

The most natural way of scoring the expert in this situation is to assign a score proportional to p_i if outcome i occurs. In other words, his score $R(p,i)$ can be written as a function of his assessment p and the observed outcome i as follows:

$$R(p,i) = Kp_i$$

for some constant K. Suppose his true belief is represented by the probability vector q. Let E_q denote expectation with respect to q. $R(p,i)$ is *strictly proper* for q if the maximum of $E_q(R(p,i))$ for all probability vectors p is attained if and only if $p = q$. More generally,

$R(p,i)$ is *strictly proper* (positive sensed) if for all probability vectors q,

$$\operatorname*{argmax}_{p} E_q(R(p,i)) \text{ is unique and equals } q$$

The "argmax" operator returns the set of arguments that maximize the function on which it operates; saying that the argmax is unique means that argmax is a singleton set. For negatively sensed scoring rules, ((argmin" replaces "argmax."

Let us see whether $R(p,i)$ is strictly proper. We have

$$E_q(R(p,i)) = \sum_{i=1}^{m} q_i R(p,i) = \sum_{i=1}^{m} Kq_i p_i$$

Suppose $m = 2$, $K = 1$, $q_1 = \frac{1}{4}$, $q_2 = \frac{3}{4}$. If the expert sets $q = p$, his expected score is $\frac{5}{8}$; however, if he puts $p_2 = 1$, then his expected score becomes $\frac{3}{4}$. In general, a maximal expected score will be attained by setting $p_i = 1$ if q_i is larger than $q_j, j \neq i$. In short, this scoring rule encourages experts to state overconfident assessments.

Even more disquieting is the fact that precisely this scoring rule emerges when a "decision maker" uses experts in a natural Bayesian setting, as suggested in perhaps the first paper on this subject (Roberts, 1965; this example was first discussed in Winkler, 1969). Suppose a decision maker chooses his probability measure p to be a weighted sum of the measures p_1 and p_2 of experts 1 and 2:

$$p(\cdot) = w_1 p_1(\cdot) + w_2 p_2(\cdot) \qquad w_1 + w_2 = 1 \qquad w_i \geqslant 0, \, i = 1, 2$$

After observing event A, the decision maker's posterior probability for B is given by

$$p(B \mid A) = \frac{p(A \text{ and } B)}{p(A)} = \frac{w_1 p_1(B \text{ and } A) + w_2 p_2(B \text{ and } A)}{w_1 p_1(A) + w_2 p_2(A)}$$

However, $p(B \mid A)$ can also be written as a weighted combination of the experts' posterior probabilities, according to the "updated weights" w_i':

$$p(B \mid A) = w_1' p_1(B \mid A) + w_2' p_2(B \mid A)$$

Writing $p_i(B \text{ and } A)$ as $p_i(B \mid A)p_i(A)$, and equating the right-hand sides of the last two equations, it is easy to check that

$$\frac{w_1'}{w_2'} = \frac{w_1 p_1(A)}{w_2 p_2(A)}$$

Assuming that expert 1 experiences the ratio of updated weights on the left-hand side as a score, and assuming that he can influence neither the original weights nor expert 2's assessment, we see that expert 1's score is proportional to his probability $p_1(A)$. The same holds for expert 2. DeGroot and Bayarri (1987) have recently studied this phenomenon and found that the situation improves somewhat if the experts incorporate beliefs regarding the assessments of other experts.

SCORING RULES FOR INDIVIDUAL VARIABLES

The scoring rule discussed above assigns a score to each individual assessment on the basis of an observed realization. It is natural to focus on such rules when we are interested in elicitation. However, if we wish to evaluate sets of assessments and assign them weights, then we might consider rules that assign a score to a set of assessments on the basis of a set of realizations. These scores need not be the result of adding scores of individual variables. It turns out that scores of the latter type are better suited for evaluating assessments. These are called *scoring rules for average probabilities* for reasons that become clear in the following section. Before turning to these, it is important to appreciate the limitations of scoring individual variables.

Consider an uncertain quantity with possible outcomes $o_1, \ldots, o_m$. Let $p = p_1, \ldots, p_m$ be a probability vector for these outcomes, and let $R(p, i)$ be the score for assessment p upon observing i. The best known strictly proper scoring rules[1]

---

[1] The strict propriety of the first two rules can be proved by writing $p_m = 1 - p_1 - p_2, \, -p_{m-1}$ and setting the derivative with respect to p_i of the expected score equal to 0. For the third score, the result follows easily from the nonnegativeness of the relative information (see the Appendix).

are (Winkler and Murphy, 1968):

$$R(p, i) = 2p_i - \sum p_j^2 \qquad \text{(quadratic scoring rule)}$$

$$R(p, i) = \frac{p_i}{(\Sigma p_j^2)^{1/2}} \qquad \text{(spherical scoring rule)}$$

$$R(p, i) = \ln p_i \qquad \text{(logarithmic scoring rule)}$$

Under the quadratic and spherical scoring rules the scores depend on the assessments of outcomes that did not, in fact, occur (for n greater than 2). A scoring rule has been called *relevant* if the scores depend only on the assessments of the observed outcomes. Following Winkler (1969), relevance is felt to be an important property for a theory of weights.

Shuford, Albert, and Massengil (1966) prove the surprising result that the logarithmic score (plus a constant) is the only strictly proper scoring rule that is relevant in this sense, when n is greater than 2 (MacCarthy, (1956), cited in Savage, 1971, attributes this insight to Andrew Gleason).

Hence, if a system of weights is to be relevant and strictly proper, the choice is determined up to the logarithmic scoring rule plus a constant. However, the logarithmic rule has its drawbacks. First of all, its values range from 0 to $-\infty$, and it is positively sensed (higher values are better). In order to get positive, positively sensed weights we must make the weights proportional to $K + R$ for some large constant K. For sufficiently small p_i, however, the weight becomes negative. This may seem like a mere theoretical inconvenience; however, risk analysis, a potential area of application, typically deals with low probabilities. Indeed, failure probabilities as low as 1.51×10^{-79} have been calculated (Honglin and Duo, 1985).

Friedman (1983) advances several further arguments against the logarithmic score. He contends that it unduly penalizes underestimating small probabilities and is not "effective" with respect to any metric, where a score is said to be effective with respect to a metric if reducing the distance (in the sense of the metric) between the elicited and the believed distributions increases the expected score.

De Groot and Fienberg (1986) give an additive decomposition of proper scoring rules into a "calibration term" and a "refinement term." This generalizes an earlier result of Murphy (1973), who speaks of "resolution" instead of "refinement" (as "refinement" applies only for well-calibrated assessors, Murphy's term is retained here). The following formulation of their result provides a useful perspective for the ensuing discussion.

Consider N uncertain quantities with identical, finite ranges (e.g., "win," "lose," and "draw"). An expert assigns each quantity to one of B "probability bins." Each bin is associated with a probability vector $p_{i,.}$ over the range of possible outcomes, $i = 1, \ldots, B$ (the notation $p_{i,.}$ indicates that $p_{i,.}$ is a vector with a component for each possible outcome). The vector $p_{i,.}$ is assumed to agree with the expert's distribution for all items in the ith probability bin. Let $s_{i,.}$ be the sample distribution for the items in the ith bin, and let n_i denote the number of items placed in the ith bin. Let R be a strictly proper scoring rule, $\mathbf{p} = p_{1,.}, \ldots, p_{B,.}$, $\mathbf{s} = s_{1,.}, \ldots, s_{B,.}$, $\mathbf{n} = n_1, \ldots, n_B$, and let $R(\mathbf{p}, \mathbf{n}, \mathbf{s})$ denote the total score gotten by summing the scores for all uncertain quantities. If t and r are distributions over the possible outcomes, let $E_t(R(r)) = \Sigma t_i R(r, i)$ denote the expected score for one variable with distribution t when the elicited distribution is r. De Groot and Fienberg's decomposition can then be expressed as

$$R(\mathbf{p}, \mathbf{n}, \mathbf{s}) = \sum_{i=1}^{B} n_i[E_{s_{i,.}}R(p_{i,.}) - E_{s_{i,.}}R(s_{i,.})] + \sum_{i=1}^{B} n_i E_{s_{i,.}}R(s_{i,.}) \qquad (9.1)$$

$$\underbrace{\phantom{\sum_{i=1}^{B} n_i[E_{s_{i,.}}R(p_{i,.}) - E_{s_{i,.}}R(s_{i,.})]}}_{\text{Calibration term}} \qquad \underbrace{\phantom{\sum_{i=1}^{B} n_i E_{s_{i,.}}R(s_{i,.})}}_{\text{Resolution term}}$$

Note that the right-hand side is a random variable, despite appearances, as the sample distributions with respect to which the expectations are taken, are random variables. The first term under the summation is the "calibration term" and the second is the "resolution term." Since R is strictly proper, the calibration term equals zero if and only if $p_{i,.} = s_{i,.}$ for all i. Note that the choice of calibration and resolution terms is completely determined by R. The calibration term in and of itself is not strictly proper.

It is interesting to examine this decomposition for the special case that R is the logarithmic scoring rule. Let A denote the finite set of possible outcomes, and assume that each outcome has positive probability under the distributions $p_{i,.}$ over A, $i = 1, \ldots, B$. The decomposition (9.1) of the logarithmic score becomes

$$R(\mathbf{p}, \mathbf{n}, \mathbf{s}) = -\sum_{i=1}^{B} [(n_i I(s_{i,.}, p_{i,.}) + n_i H(s_{i,.})] \qquad (9.2)$$

The resolution term is simply the entropy in the joint sample distributions, when these are considered independent. Calibration is measured by the relative information for each bin, weighted by the number of items in each bin. Equation (9.2) will emerge as a special case of scoring rules for average probabilities. The calibration term by itself will be shown to be strictly proper, and it will prove possible to assign a multiplicative entropy penalty when entropy is associated with the assessed distributions (propriety in the latter case is asymptotic).

The principal disadvantage in using a scoring variable like (9.1) or (9.2) to derive weights (which De Groot and Fienberg do *not* propose) is the following. The resulting scores cannot be meaningfully interpreted without knowing the number of quantities involved and their overall sample distribution.

For example, suppose we score two experts assessing different sets of quantities with (9.2). Suppose the first expert assesses only one quantity, assigns one of the possible outcomes probability 1, and suppose this outcome is observed. His score is then maximal, namely 0. Suppose the second expert does the same, but for 1000 quantities. His score will also be 0. On the basis of their respective scores, we cannot distinguish the performance of these two experts, and would have to assign them equal weights. Intuitively, however, the second expert should receive a greater weight, as his performance is more convincing (the first expert might have gotten lucky). Dividing R by the number N of uncertain quantities (which De Groot and Fienberg in fact do) would not help.

The point is this: A scoring rule, being a function of the values of uncertain quantities, is a random variable, and interpreting the values of the score requires knowledge of the score's distribution.

Moreover, if S denotes the total sample distribution, then the maximal value of the resolution term in (9.2) is $NH(S)$. The resolution terms for different sets of uncertain quantities with different sample distributions therefore cannot be compared.

In practice we shall often want to pool the advice of different experts who have assessed different quantities in the past. In light of the above remarks, there would be no way of combining experts' opinions via scores derived from different sets of test variables. Even if the scores did pertain to the same variables, it would be impossible to assess the importance of the differences in scores without some knowledge of the distribution of the scoring variable.

The theory developed below involves scoring variables that are not gotten by summing scores for individual variables. This generalization provides considerably more latitude in choosing (relevant) proper scoring rules. We shall find that the calibration term in (9.2) is strictly proper in an appropriate sense of the word, and has a known distribution under common assumptions. Weights can then be derived on the basis of the significance level of the calibration term, and these weights are shown to be asymptotically strictly proper. Similar remarks apply to the entropy penalty $\Sigma n_i H(s_{i,.})$.

It is significant and perhaps surprising that a strictly proper calibration score exists whose sample distribution under customary assumptions is known. The theory presented below generates a large class of strictly proper scores, but only one has been found with this property.

PROPER SCORING RULES FOR AVERAGE PROBABILITIES

In this section we develop a theory of strictly proper scoring rules for average probabilities and prove a useful representation theorem. $(\Omega, \mathscr{F})$ will denote an arbitrary measurable space. It is assumed that all probability measures are countably additive and that all random variables on Ω are $\mathscr{F}$ measurable. $\mathbf{R}$ and $\mathbf{N}$ denote the real numbers and the integers, respectively. For $A \in \mathscr{F}$, 1_A denotes the indicator function of A. The following notation will be adopted:

$O = \{o_1, \ldots, o_m\}$	Set of outcomes
$X = X_1, \ldots,$	$X: \Omega \to O^\infty$; X is $\mathscr{F}$-measurable
$M(O)$	Set of nondegenerate probability vectors over O; $p \in M(O)$, $p = p_1, \ldots, p_m$, $\Sigma p_i = 1$; $p_i > 0$, $i = 1, \ldots, m$
$k_i^{(N)}$	$\Sigma_{j=1}^N 1_{\{X_j = o_i\}}$
$s_i^{(N)}$	$k_i^{(N)}/N$, $i = 1, \ldots, m$
$s^{(N)}$	$(s_1^{(N)}, \ldots, s_m^{(N)})$
$M(X)$	The set of nondegenerate distributions for X: if $P \in M(X)$, then $0 < P(s^{(N)})$, for all $s^{(N)}$; we assume that if $P \in M(X)$, then P is defined on $\mathscr{F}$. The X_i need not be independent under P.
Q	Assessor's probability for X; $Q \in M(X)$
$q_i^{(N)}$	$(1/N) \sum_{j=1}^N Q(X_j = o_i)$
$q^{(N)}$	$(q_1^{(N)}, \ldots, q_m^{(N)})$; $q^{(N)} \in M(O)$

We suppose that an assessor is asked to state his/her average over $X_1, \ldots, X_N$ of the probabilities of occurrence for the outcomes $\{o_1, \ldots, o_m\}$. From Theorem 7.1, we know this is equivalent to asking for the expected relative frequencies of occurrences of these outcomes. In general he/she will respond with a probability vector $p \in M(O)$. We are interested in scoring rules that reward the expert for responding truthfully, that is, ones which reward the expert for responding with $p = q^{(N)}$. We shall first investigate scoring rules having this property for all N, and subsequently study rules that have this property in some asymptotic sense. E_Q denotes expectation with respect to Q. We assume $Q \in M(X)$.

Definition. A *scoring rule for average probabilities* is a real-valued function R: $M(O) \times \mathbf{N} \times \Omega \to \mathbf{R}$.

Notational convention. When $R(p, N, \omega)$ depends on ω only through the sample distribution $s^{(N)}$, we shall write $R(p, N, s^{(N)})$. As we shall be exclusively concerned with scoring rules of this form, we drop the superscript (N), and simply write $R(p, N, s)$.

Definition. For M a subset of $M(X)$, $R(p, N, s)$ is *positive (negative) sensed* and *M-strictly proper* if, for all $Q \in M$,

$$\operatorname*{argmax}_{p \in M(O)} (\operatorname{argmin}) E_Q R(p, N, s) \quad \text{is unique and equals} \quad q^{(N)}$$

The argmax (argmin) is taken over all nondegenerate probability vectors over the outcome set O. $R(p, N, s)$ is called strictly proper if it is $M(X)$-strictly proper. Strict propriety is stronger for scoring rules for average probabilities than for scoring rules for individual variables, as the set $M(X)$ from which the assessor's probability is taken is larger than the set $M(O)$ from which the "response distribution" is drawn.

Theorem 9.1. With the notation as above, let $R(p, N, s)$ be differentiable in p. Then the following are equivalent:

for all $Q \in M(X)$,

$$\nabla_p E_Q(R(p, N, s))|_{p = q^{(N)}} = 0. \tag{9.3}$$

For $i, j, k \in \{1, \ldots, m - 1\}$, there exist integrable functions g_i and g_{ikj} such that

$$\frac{\partial}{\partial p_i} R(p, N, s) = g_i(p, N)(p_i - s_i) + \sum_{k < j} g_{ikj}(p, N)(s_k p_j - p_k s_j). \tag{9.4}$$

The proof is found in the supplement to this chapter.

Remark 1. If R is differentiable in p, then (9.3) is a necessary condition for strict propriety. If R is strictly convex or strictly concave in p, then (9.3) is also sufficient for strict propriety (with the rule's sense chosen appropriately).

Remark 2. Scoring rules of the sort described in this section are applicable to the quantile tests for calibration. In such tests the expert is asked to state quantiles corresponding to the probabilities $0 = f_0 < f_1 < f_2 < \cdots < f_{M-1} < f_M = 1$ of continuously distributed variables $X'_1, \ldots, X'_N$. The variable X_i is then said to take outcome j if $X'_{i, j-1} < X'_i \leqslant X'_{i, j}$, where $X'_{i, j}$ denotes the jth quantile of X'_i. If the expert is asked to state his probabilities for these outcomes then a strictly proper rule for average probabilities encourages him to respond with the probabilities $f_j - f_{j-1}, j = 1, \ldots, M$.

Remark 3. If the outcome set is $\{0, 1\}$, then the variables $X_1, \ldots, X_N$ are indicator functions for uncertain events. In this case the terms $(s_k p_j - p_k s_j)$, $k \neq j$, vanish and (9.4) takes the form:

$$\frac{\partial}{\partial p_i} R(p, N, s) = g_i(p, N)(p_i - s_i) \tag{9.4a}$$

Remark 4. Under the conditions of Remark 3, it follows from Theorem 9.1 that if R satisfies Equation (9.4a), then

$$R(p, N, s) = -s \int_0^p g(x) \, dx + \int_0^p x g(x) \, dx$$

A "dual" equation appears in Savage (1971):

$$G(p, x) = -p \int_x^1 f(y) \, dy + \int_x^1 y f(y) \, dy$$

$G(p, x)$ is interpreted as the income of a subject who states price x when his true price is p, for commodities of which the experimenter will buy $f(y)$ units at price y, $0 \leqslant y \leqslant 1$.

Examples

The following proposition shows that the relative information score

$$R(p, N, s) = I(s^{(N)}, p) = \sum_{i=1}^m s_i^{(N)} \ln \frac{s_i^{(N)}}{p_i}$$

is strictly proper.

Proposition 9.1. Let $s, p \in M(O)$.

(i) $I(s, p)$ is a convex function of p; and putting $p_m = 1 - \sum_{i=1}^{m-1} p_i$;

$$\frac{\partial}{\partial p_i} I(s, p) = \frac{p_i - s_i + \sum_{j=1}^{m-1}(s_i p_j - s_j p_i)}{p_i(1 - p_1 - \cdots - p_{m-1})} \qquad i = 1, \ldots, m-1$$

(ii) For $m = 2$, writing $I(s, p)$ as a function of s_1, p_1, and dropping the subscripts,

$$\frac{\partial}{\partial s} I(s, p) = \ln \frac{s - sp}{p - sp} = \frac{s - p}{p - p^2} + o(s - p) \qquad s - p \to 0$$

where $f(y) = o(y)$ means that $f(y)/y \to 0$ as y approaches some limit (in this case 0).

The proof is found in the supplement.

The last statement in Proposition 9.1 shows that $I(p, s)$ is *not* a strictly proper scoring rule.

Another possible choice for R is

$$R(p, N, s) = \sum_{i=1}^{m-1} c_i (s_i - p_i)^2$$

where the c_i are positive constants. From Remark 1 and Theorem 9.1 it is easy to verify that this is a strictly proper scoring rule. Moreover, it corresponds to a "quadratic loss function," where R is the loss incurred when one takes "action" p while s is the "true value." Indeed, scoring rules for average probabilities can be regarded as loss functions for the random variable s. The set of negatively sensed proper scoring rules for average probabilities may be regarded as the set of loss functions for the random variable s that are minimized for the action p equal to the expected value of s.

The following three propositions extend the formalism of Theorem 9.1 to cover the case of an arbitrary finite number B of "probability bins." Σ_b denotes summation over the bins $b = 1, \ldots, B$. We revive the notation of the previous section:

$p_{b,.}$	Probability vector associated with bin b; $p_{b,.} \in M(O)$
$s_{b,.}$	Sample distribution associated with bin b
$\mathbf{p}$	$p_{1,.}, \ldots, p_{B,.}$
$\mathbf{s}$	$s_{1,.}, \ldots, s_{B,.}$
n_b	Number of variables assigned to bin b
$\mathbf{n} = n_1, \ldots, n_B$	$\mathbf{n}$ is called the occupation vector
N	$\Sigma_b n_b$

The assessor performs his assessment by assigning each X_i, $i = 1, \ldots$ to one of the B bins, and assigning each bin a probability vector. The assessor is asked to state his/her average probability vector for each bin. The vector of sample distributions $\mathbf{s}$ will be understood to depend on $\mathbf{n}$, but we suppress this dependence in the notation and write $R(\mathbf{p}, \mathbf{n}, \mathbf{s}) := \Sigma_b R_b(p_{b,.}, n_b, s_{b,.})$. Taking "argmax" over the set of non-degenerate vectors $\mathbf{p}$, the definition of propriety extends to $R(\mathbf{p}, \mathbf{n}, \mathbf{s})$ in the obvious way.

Proposition 9.2. If $R(\mathbf{p}, \mathbf{n}, \mathbf{s}) := \Sigma_b R_b(p_{b,.}, n_b, s_{b,.})$ for scoring rules R_b for average probabilities having the same sense, then R is strictly proper if and only if R_b is strictly proper, $b = 1, \ldots, B$.

Proof. This follows immediately from the additivity of expectation:

$$E(R) = \Sigma_b E(R_b) \quad \blacksquare$$

In particular, this proposition applies if the R_b's are all the same scoring rule. However, taking $B = 1$, then, in general,

$$R(\mathbf{p}, \mathbf{n}, \mathbf{s}) \neq \sum_{i=1}^{N} R(p, 1, x_i)$$

This can readily be verified for the relative information score, taking $N = 2$, $M = \{0, 1\}$, $x_1 = 1$, $x_2 = 0$, $p = \frac{1}{2}$. The right-hand side is $2 \ln 2$, whereas the left-hand side is 0. This emphasizes the difference between using scoring rules for average probabilities, as against using scoring rules for individual variables and adding the scores.

Proposition 9.3. If $R(\mathbf{p}, \mathbf{n}, \mathbf{s})$ is a strictly proper scoring rule, then so is

$$R^{\sim}(\mathbf{p}, \mathbf{n}, \mathbf{s}) = R(\mathbf{p}, \mathbf{n}, \mathbf{s}) + f(\mathbf{s}, \mathbf{n})$$

where f is an arbitrary real function on the sample and occupation vectors.

Proof. Assume that R is positively sensed. Since $f(\mathbf{s}, \mathbf{n})$ does not depend on $\mathbf{p}$,

$$\sup_{\mathbf{p}} ER^{\sim}(\mathbf{p}, \mathbf{n}, \mathbf{s}) = \sup_{\mathbf{p}} ER(\mathbf{p}, \mathbf{n}, \mathbf{s}) + Ef(\mathbf{s}, \mathbf{n}) = ER(\mathbf{q}, \mathbf{n}, \mathbf{s}) + Ef(\mathbf{s}, \mathbf{n})$$

$$= ER^{\sim}(\mathbf{q}, \mathbf{n}, \mathbf{s}) \quad \blacksquare$$

ASYMPTOTIC PROPERTIES

In this section we revert to the formalism of Theorem 9.1, involving one probability bin, and we study asymptotic properties of scoring rules for average probabilities. A strong and weak form of asymptotic propriety are distinguished. Results with the weak form are easier to prove, and seem sufficient for applications. As before, M denotes a subset of $M(X)$. The definitions will be formulated for positively sensed rules, for negatively sensed rules, "argmin" replaces "argmax."

Definition. $R(p, N, s)$ is *strongly asymptotic M-strictly proper* if for all $Q \in M$,

$$\operatorname*{argmax}_{p} E_Q R(p, N, s) = p_N \text{ for some (not necessarily unique) } p_N, \text{ and if } q^{(N)} \to r$$

as $N \to \infty$, then also $p_N \to r$ as $N \to \infty$.

$R(p, N, s)$ is *weakly asymptotic M-strictly proper* if for all $Q \in M$, whenever

$$q^{(N)} \to r \quad \text{as } N \to \infty \quad \text{and} \quad r' \in M(0) \quad r' \neq r$$

then there exists $N' \in \mathbf{N}$ such that for all $N > N'$;

$$E_Q R(r, N, s) > E_Q R(r', N, s)$$

The difference between strong and weak asymptotic propriety may be characterized as follows. For strong asymptotic propriety we first take the argmax over possible assessments, and take the limit as $N \to \infty$ of these argmax's. In weak asymptotic propriety we first choose an assessment r'. If r' is not equal to the limit of the average probabilities then it is possible to choose an N' in $\mathbf{N}$ such that for all $N > N'$, the expected score under r' is worse than the expected score using the limits of the average probabilities.

We now fix M to be the set of nondegenerate product probabilities for X; that is, $P \in M$ implies $P = p^\infty$ for some $p \in M(0)$. Under P the variables $X_1, \ldots,$ become independent and identically distributed with distribution p. Our goal in this section is to show that a simple Fisher-type test of significance and a weighted test of significance are weakly asymptotic M-strictly proper scoring rules for average probabilities.

Before turning to this, we first examine the asymptotic distributions under Q, for $Q \in M$, of the strictly proper scoring rules for average probabilities introduced in the previous section. We define

$$RI(p, N, s) = 2NI(s, p)$$

Statisticians recognize $2NI(s, q)$ as the log likelihood ratio; it has an asymptotic chi-square distribution with $m - 1$ degrees of freedom under Q. From Proposition 1, RI is a strictly proper scoring rule for average probabilities, for all $Q \in M(X)$. Moreover, if we have B probability bins, $B > 1$, then we can simply add the scores $RI(q, N, s)$ for each bin. The resulting sum will have an asymptotic chi-square distribution with $B(m - 1)$ degrees of freedom (we must also assume the variables in different bins are independent under Q).

If we expand the logarithm in RI and retain only the dominant term, we arrive at

$$\sum_{i=1}^{m} \frac{N(p_i - s_i)^2}{p_i}$$

which is the familiar chi-square variable for testing goodness of fit. This has the same asymptotic properties as RI, but is not a strictly proper scoring rule for average probabilities. The terms p_i in the denominators cause the gradient in Equation (9.3) to have a term in $(p_i - s_i)^2$.

For $B = 1$, $m = 2$, the quadratic score has tractable properties. Put $p = p_1$, $s = s_1$. Under $P = p^\infty$ the variable

$$QU(p, N, s) = (p - s)^2$$

approaches a squared normal variable with mean 0 and variance $(p - p^2)/N$. QU is also strictly proper. If QU is standardized by dividing by the variance, the result will no longer be strictly proper. Without standardization, the distribution of sums of scores QU for $B > 1$ will not be tractable.

If we desire a test statistic that is also a strictly proper scoring rule for average probabilities, we are well-advised to confine attention to RI. We define a simple and a weighted test of significance using RI. $\chi^2_{(m-1)}$ denotes the cumulative chi-square distribution function with $m - 1$ degrees of freedom.

$$w_t(p, N, s) = \begin{cases} 1 & \text{if } RI(p, N, s) \leqslant t \\ 0 & \text{if } RI(p, N, s) > t \end{cases}$$

$$W_t(p, N, s) = [1 - \chi^2_{(m-1)}(RI(p, N, s))]w_t(p, N, s)$$

The score w_t reflects simple significance testing. As we know the (asymptotic) distribution of $RI(p, N, s)$ under suitable hypotheses, we can choose t such that the probability that $RI(p, N, s) > t$ is less than some fixed number α. t is then called the *critical value* and α the *significance level* for testing the hypothesis that the N observations are independently drawn from the distribution p. W_t distinguishes "degrees of acceptance" according to the probability under Q of seeing a relative information score in N observations at least as large as $RI(p, N, s)$. The following three propositions are proved in the supplement to this chapter.

Proposition 9.4. For $t \in (0, \infty)$, the score w_t is weakly asymptotic M-strictly proper.

Proposition 9.5. If $t \in (0, \infty)$, then W_t is weakly asymptotically M-strictly proper.

Propositions 9.4 and 9.5 are central for the "classical model" introduced in Chapter 12. They show that simple and graduated significance testing constitutes an asymptotically strictly proper reward structure under suitable conditions. Weighing expert probability assessors with these scores is similar to treating the experts as classical statistical hypotheses. This explains the designation "classical model." However, weighing expert assessments differs in two important respects from testing hypotheses. First, good expert assessments must not only "pass statistical tests," but must have high information, or equivalently, low entropy. Second, the choice of significance level, or equivalently, the choice of t, is not determined by the same considerations which apply in hypothesis testing. We return to this last point at the end of this chapter.

Proposition 9.6. For all $t \in (0, \infty)$, and for any function $f: M(0) \times \mathbf{N} \to [a, b]$ with $0 < a < b < \infty$, the scores

$$w_t(p, N, s)f(p, N) \qquad W_t(p, N, s)f(p, N)$$

are weakly asymptotic M-strictly proper.

Remark 1. The definition of f in this proposition is quite arbitrary, so long as it does not depend on Ω and so long as its range is bounded and bounded away from 0. Referring to Remark 2 of Theorem 9.1, f might depend on the distributions of the variables X_i' instead of $M(0)$.

Remark 2. Propositions 9.4, 9.5, and 9.6 will also go through if RI is replaced in the definitions of w_t and W_t by the sum relative information score introduced in Formula (8.5):

$$2 \sum_{b=1}^{B} n_b I(s_{b,.} p_{b.,})$$

The number of degrees of freedom must be changed to $B(m - 1)$, and the class M must be altered appropriately.

The proofs of Propositions 9.4, 9.5, and 9.6 do not use the propriety of the score RI. Had we used any other "goodness of fit" statistic to define w and W, these proofs would still go through. The use of RI reflects a preference for a statistic which is itself strictly proper.

A MENU OF WEIGHTS

In this section we consider weights for combining probability distributions, based on the theory of scoring rules. This section will use the notation of Propositions 9.2 and 9.3, that is, we consider the case where the variables X_i take values in the set of outcomes $0 = \{o_1, \ldots, o_m\}$, and the subject assigns each variable to one of B "probability bins" with assessed distribution $p_{b,.} \in M(0)$, $b = 1, \ldots, B$. As noted in Remark 2 to Theorem 9.1, quantile tests can constitute a special case with $B = 1$.

Definition. For a given finite set of variables, a *weight for an expert assessment p* is a nonnegative, positively sensed scoring rule for the average probabilities. A *system of weights for a finite set of experts* (perhaps assessing different variables) is a normalized set of weights for each expert, if one of the experts' weights is positive; otherwise the system of weights is identically zero.

The above definition explicitly accounts for the eventuality that all experts might receive zero weight. Based on the discussions of the previous sections, we formulate four desiderata for a system of weights. Such weights should

1. Reward low entropy and good calibration
2. Be relevant
3. Be asymptotically strictly proper (under suitable assumptions)
4. Be meaningful, prior to normalization, independent of the specific indicator variables from which they are derived

The last desideratum is somewhat vague, but is understood to entail that the unnormalized weights for experts assessing different variables can be meaningfully compared.

The requirement of asymptotic strict propriety requires explanation. Let us assume that an expert experiences "influence on the beliefs of the decision maker" as a form of reward. Let us represent the decision maker's beliefs as a distribution P_{dm}, which can be expressed as some function G of experts' weights w_e and assessments P_e, $e = 1, \ldots, E$:

$$P_{dm} = G(w_1, \ldots, w_E; P_1, \ldots, P_E)$$

Expert e's influence is found by taking the partial derivative of P_{dm} with respect to that argument which e can control, namely P_e. Maximizing expected influence in this sense is not always the same as maximizing the expected value of w_e. However, if G is a weighted average:

$$P_{dm} = \sum_{e=1}^{E} \frac{w_e P_e}{K} \qquad K = \sum_{e=1}^{E} w_e$$

then

$$\frac{\partial P_{dm}}{\partial P_e} = \frac{w_e}{K}$$

When giving his assessment, expert e will not generally know the weights of the other experts; he may not even know who they are or how many there are. Therefore the normalization constant K is effectively independent of the variable P_e that e can manipulate. Maximizing the expected influence $\partial P_{dm}/\partial P_e$ is effectively equivalent to maximizing the (unnormalized) weight w_e. In Chapter 11 we shall see that the above form for G is the only serious candidate.

Requirement 2 is satisfied by scoring rules for average probabilities in the following sense: The scores do not depend on the probabilities of outcomes that might have been observed but were not. Of course, the average of probabilities is not itself the probability of an outcome that can be observed. However, if the average of the probabilities of independent events converges to a limit, then the observed relative frequencies converge with probability 1 to the same limit.[2]

The scores w_t and W_t introduced in the previous section may be considered as weights which incorporate the notion of significance testing. They reward good calibration, but they are not sensitive to entropy.

Before discussing weights that are sensitive to entropy, we must distinguish the two notions of entropy.

$$H(\mathbf{n}, \mathbf{s}) = \frac{1}{N} \sum_b n_b H(s_{b,.}) \quad \text{(sample entropy)} \tag{9.5}$$

$$H(\mathbf{n}, \mathbf{p}) = \frac{1}{N} \sum_b n_b H(p_{b,.}) \quad \text{(response entropy)} \tag{9.6}$$

where

$$H(p_{b,.}) = -\sum_{i=1}^{m} p_{b,i} \ln(p_{b,i})$$

and similarly for $H(s_{b,.})$. In conjunction with Remark 2 to Theorem 9.1, we note that the distinction between sample and response entropy is not meaningful for quantile tests.

Proposition 9.3 allows us to add an arbitrary function of the sample distribution to a proper scoring rule. The maximal value of the sample entropy $H(\mathbf{n}, \mathbf{s})$ is $\ln m$. We can define positive, positively sensed weights taking values in $[0, 1]$ as follows:

$$w_t + \ln m - H(\mathbf{n}, \mathbf{s}) \tag{9.7}$$

$$W_t + \ln m - H(\mathbf{n}, \mathbf{s}) \tag{9.8}$$

These weights bear a resemblance to the De Groot-Fienberg decomposition of the logarithmic scoring rule. Two criticisms may be directed against them. First, these weights can be zero only if w_t and W_t are zero *and* if $H(\mathbf{n}, \mathbf{s})$ assumes the value $\ln m$. Hence, a poorly calibrated expert can still receive substantial weight. Moreover, the maximal value of $\ln m$ can only be obtained if exactly half of the test events occur. Second, the term $H(\mathbf{n}, \mathbf{s})$ is also a random variable. Different numerical values for $H(\mathbf{n}, \mathbf{s})$ cannot be meaningfully compared without taking the distribution of $H(\mathbf{n}, \mathbf{s})$ into account.

These objections can at least be partially met. Instead of considering the sample entropy $H(s_{b,.})$ for each bin, we could consider the relative information

---

[2] Since according to Kolmogorov's sufficient condition for the strong law of large numbers (Tucker, 1967, p. 124), for indicator functions 1_i,

$$p\left(\lim_{n \to \infty} \frac{1}{n} \sum_{i=1}^{n} (1_i - E(1_i)) = 0\right) = 1$$

Hence, if the average probabilities $(1/n)\Sigma E(1_i)$ converge to $\tilde{p}$ as $n \to \infty$, then with probability 1

$$\lim_{n \to \infty} \frac{1}{n} \sum_{i=1}^{n} 1_i = \tilde{p}$$

$I(s_{b,\cdot}, S)$, where S denotes the total sample distribution over all variables. If the variables are independent and distributed according to S, the quantity

$$D(n, s) = \sum_b 2n_b \, I(s_{b,\cdot}, S)$$

approaches χ_B^2 in distribution as n goes to infinity. Large values of $D(n, s)$ indicate a "high resolution of the base rate," which would be unlikely to result from statistical fluctuations under the distribution S. $\chi_B^2(D)$ will be termed the *base rate resolution index*. The following thus represent improvements over the weights (9.7) and (9.8):

$$w_t + \chi_B^2(D) \tag{9.9}$$

$$W_t + \chi_B^2(D) \tag{9.10}$$

The weights (9.7) to (9.10) are easily seen to be asymptotically strictly proper. For large n, $1 - \chi_B^2(D)$ represents the significance level at which we would reject the hypothesis that the expert had assigned test variables to bins randomly. These weights still have the property that poorly calibrated experts can receive substantial weight.

Weights using a multiplicative entropy penalty based on the response entropy can avoid this problem. A suitable form for such weights is

$$\frac{w_t}{H(\mathbf{n}, \mathbf{p})} \tag{9.11}$$

$$\frac{W_t}{H(\mathbf{n}, \mathbf{p})} \tag{9.12}$$

If the experts' weights are derived from assessment of *different* variables in the past, then comparing their response entropies might not be meaningful. For example, suppose one expert has assessed the probability of rain in The Netherlands, where the weather is quite unpredictable, and another has assessed the probability of rain in Saudi Arabia. The latter would have a lower response entropy simply because his assessments would all be near 0. If these experts' assessments were to be combined via weights derived from their past assessments, then their response entropies should not be used. In such situations the term $1/H(\mathbf{n}, \mathbf{p})$ should be replaced by

$$\frac{1}{N} \sum_b n_b \, I(p_b, S) \tag{9.13}$$

where S is the overall sample distribution. Weight (9.13) is the average information in the expert's assessments relative to the base rate. To satisfy the conditions of Proposition 9.6, (13) must be bounded, and bounded away from 0, and the overall sample distribution must be treated as a constant.

Weights (9.11), (9.12), and (9.13) are 0 whenever the expert's calibration score exceeds the critical value. Moreover, it is possible to argue that the response entropy is a more appropriate index for lack of information than the sample entropy. The weighted combinations refer indeed to the assessed probabilities $p_{b,\cdot}$.

and not to the sample probabilities $s_{b,}$. If $\mathbf{s} = \mathbf{p}$, then these two coincide. However, an experiment discussed in Chapter 10 will illustrate that the sample and response entropies can differ, even for well-calibrated assessors. The weak asymptotic propriety of these weights is proved in Proposition 9.6.

In the case of quantile tests, the term $1/H(\mathbf{n}, \mathbf{p})$ would simply be a constant and should be replaced in (9.11) and (9.12) by some other appropriately bounded function of the assessment. Recalling the discussion in Chapter 8, the average relative information of the assessor's (approximated) density functions with respect to the uniform distribution can be used. The details of this will be postponed until Chapter 12.

HEURISTICS OF WEIGHING

We conclude with some heuristic remarks on choosing a score for weighing expert opinion. First, the unnormalized weights for each expert depend only on the assessments of each expert and the realizations. When a decision maker calls various experts together for advice on a particular variable, he could compute weights based on prior assessments of *different* variables. When experts are combined, their weights must be normalized to add to unity. However, the decision maker must ensure that the experts are all calibrated on the same effective number of variables. In the language of hypothesis testing, this ensures that the experts are subjected to significance tests of equal power. The technique of equalizing the power of the calibration tests will be discussed in Chapter 12.

It seems obvious that we do not want to assign a high weight to experts who are very poorly calibrated, regardless how low their entropy is, and regardless which information measure is used. Entropy should be used to distinguish between experts who are more or less equally well-calibrated. The weight (9.12) does in fact behave in this way. The calibration scores W_t will typically range over several orders of magnitude, while the entropy scores typically remain within a factor 3. Because of the form of the weight (9.12), when weights of different experts are normalized, the calibration term will dominate, unless all experts are more or less equally calibrated. The weight (9.10) does not have this property. We also note, referring to the discussion of the previous chapter, that there are other measures of "lack of information" that can be substituted for the function f in Proposition 9.6. Any other measure should be chosen in such a way that the calibration term will dominate in weight (9.12).

In testing hypotheses, it is usual to say that a hypothesis is "rejected" when the test statistic exceeds its critical value. This way of speaking, of course, is not appropriate in dealing with expert probability assessments. In hypothesis testing, choice of the significance level α means choosing to reject true hypotheses with frequency α. In combining expert opinions, we are not trying to find the "true" expert and "reject" all the others as "false." Indeed, if we collect enough data, then we will surely succeed in rejecting all experts, as no one will be *perfectly* calibrated. The significance level α is chosen so as to optimize the decision maker's "virtual weight," as explained in Chapter 12.

SUPPLEMENT

This supplement gives the proofs for various results cited in the text.

Proof of Theorem 9.1

Theorem 1. With the notation as above, let $R(p, N, s)$ be differentiable in p. Then the following are equivalent:

For all $Q \in M(X)$, $\nabla_p E_Q(R(p, N, s))|_{p=q^{(N)}} = 0$. (9.3)

For $i, j, k \in \{1, \ldots, m-1\}$, there exist integrable functions g_i, and g_{ikj}, such that:

$$\left(\frac{\partial}{\partial p_i}\right) R(p, N, s) = g_i(p, N)(p_i - s_i) + \sum_{k<j} g_{ikj}(p, N)(s_k p_j - p_k s_j).$$ (9.4)

The proof uses three lemmata, the first was introduced as theorem (7.1).

Lemma 9.1. Let $A_1, \ldots, A_N \in \mathscr{F}$ with indicator functions $1_1, \ldots, 1_N$. Then

$$\frac{1}{N} \sum_{i=1}^{N} Q(1_j = 1) = \frac{1}{N} \sum_{k=1}^{N} kQ(\text{exactly } k \text{ of } A_1, \ldots, A_N \text{ occur})$$

Lemma 9.2. Let A be an $L \times n$ matrix with rank L, $L < n$, and let $b \in \mathbf{R}^L$, $b \neq 0$. Let

$$V = \{x \in \mathbf{R}^n \mid Ax = b, x_i > 0, i = 1, \ldots, n\}.$$

Suppose $V \neq \varnothing$, then the subspace generated by V has dimension $n - L + 1$.

Proof. Let Y be the subspace of vectors orthogonal to each $x \in V$. It suffices to show that Dim $Y = L - 1$. Let $Z = \{z \in \mathbf{R}^n \mid Az = 0\}$. Then Dim $Z = n - L$. Pick $x^{(0)} \in V$. We can find a basis $e^{(1)}, \ldots, e^{(n-L)}$ of Z such that $x_i^{(0)} + e_i^{(l)} > 0, i = 1, \ldots, n$, and thus $x^{(0)} + e^{(l)} \in V, l = 1, \ldots, n - L$. For all $y \in Y$:

$$0 = \sum x_i^{(0)} y_i + \sum e_i^{(l)} y_i$$

or

$$\sum x_i^{(0)} y_i = -\sum e_i^{(l)} y_i \qquad l = 1, \ldots, n - L$$

The left-hand side does not depend on l, or on the choice of basis, hence both sides must vanish, and y must be orthogonal to $x^{(0)}$ and to all $z \in Z$. Since $b \neq 0, x^{(0)} \notin Z$, and y is orthogonal to $n - L + 1$ linearly independent vectors. The proof is completed by noting that any solution to the linear system $Ax = b$ may be written $x = x^{(0)} + z$ for some $z \in Z$. Hence $y \in Y$ if and only if y is orthogonal to $x^{(0)}$ and to $e^{(l)}, l = 1, \ldots, n - L$, and Dim $Y = L - 1$. ∎

Lemma 9.3. Let $R(p, N, s)$ satisfy Equation (9.3), and fix N. Let

$$W = \left\{k \in \mathbf{N}^{m-1} \mid k_i \geq 0, i = 1, \ldots, m-1, \sum_{i=1}^{m-1} k_i \leq N\right\}$$

For $Q \in M(X)$, let

$$Q(k_{i_1}, \ldots, k_{i_L}) = Q\{\text{outcome } i_j \text{ occurs } k_{i_j} \text{ times in } X_1, \ldots, X_N | j = 1, \ldots, L\}.$$

Use $k \in W$ to index the coordinates of $\mathbf{R}^{|W|}$. Then, by writing

$$Q_k = Q(k) \qquad k \in W$$

we may consider each element of $M(X)$ as a vector in $\mathbf{R}^{|W|}$. We write $R(p, N, s) = R(p, N, k/N)$, where $k \in W$. For each differentiable scoring rule, for each $p \in M(0)$ and each $i = 1, \ldots, m - 1$ we may consider $R(p, i)$ with coordinates

$$R(p, i)_k = \frac{\partial}{\partial p_i} R\left(p, N, \frac{k}{N}\right) \qquad k \in W$$

as an element of $\mathbf{R}^{|W|}$. For $p \in M(0)$ let

$$A(p) = \text{the subspace of } \mathbf{R}^{|W|} \text{ generated by } \{Q \in M(X) | q^{(N)} = p\}.$$

In other words, $A(p)$ is the subspace generated by probability vectors Q whose vector of average probabilities equals p. Let

$$B(p, i) = \text{the subspace of } \mathbf{R}^{|W|} \text{ generated by the vectors } R(p, i) \text{ where the scoring rule } R \text{ is differentiable in } p \text{ and satisfies (9.4).}$$

Then for all $p \in M(0)$, and $i = 1, \ldots, m - 1$,

$$\text{Dim } A(p)^{\perp} \leqslant \text{Dim } B(p, i)$$

where $A(p)^{\perp}$ denotes the subspace of vectors in $\mathbf{R}^{|W|}$ orthogonal to $A(p)$.

Proof. We first show that Dim $A(p)^{\perp} = m - 1$. Letting Σ' denote summation over W, the vectors $Q \in A(p)$ are in the positive cone and satisfy the m linear equations:

$$\Sigma'Q(k) = 1 \qquad \Sigma'k_iQ(k) = Np_i \qquad i = 1, \ldots, m - 1$$

The following consideration shows that these equations are independent: Since for $Q \in M(X)$, $q_i^{(N)} > 0$, $i = 1, \ldots, m$, we can always find two probability vectors Q, Q' whose vectors of average probabilities $q^{(N)}$ and $q'^{(N)}$ disagree in just one coordinate $j, j \in \{1, \ldots, m - 1\}$. By Lemma 9.2, the dimension of $A(p)$ equals $|W| - (m - 1)$, hence the dimension of $A(p)^{\perp}$ equals $m - 1$.

We show that Dim $B(p, i) \geqslant m - 1$. Fix i. If $m = 2$, Dim $A(p)^{\perp} = 1$, the functions g_{if_j} are all zero, and it is trivial to show that $B(p, i)$ has dimension 1. Assume $m \geqslant 3$. It suffices to find $m - 1$ linearly independent vectors in $B(p, i)$. In fact, it suffices to find $m - 1$ vectors whose components on $m - 1$ coordinates $k^{(1)}, \ldots, k^{(m-1)}$ are linearly independent, where

$$k_h^{(j)} = N\delta_{jh} \qquad \delta_{jh} = 1 \text{ if } j = h \text{ and } = 0 \text{ otherwise}$$

It suffices to find scoring rules $R^{(1)}, \ldots, R^{(m-1)}$ satisfying Equation (9.4) such that the $(m - 1)$ by $(m - 1)$ matrix Y with

$$y_{jh} = \frac{\partial}{\partial p_i} R^{(j)}\left(p, N, \frac{k^{(h)}}{N}\right)$$

has full rank. We choose

$$R^{(i)}(p, N, s) = \tfrac{1}{2}p_i^2 - p_i s_i$$

$$R^{(j)}(p, N, s) = -s_j \ln p_j - s_i \ln p_i - (1 - s_i - s_j) \ln(1 - p_i - p_j), \; j \neq i$$

It is easy to verify that

$$\frac{\partial}{\partial p_i} R^{(i)} = p_i - s_i$$

$$\frac{\partial}{\partial p_i} R^{(j)} = \frac{p_i - s_i + s_i p_j - p_i s_j}{p_i(1 - p_i - p_j)}$$

$$\frac{\partial}{\partial p_j} R^{(j)} = \frac{p_j - s_j + s_i p_j - p_i s_j}{p_j(1 - p_i - p_j)}$$

other derivatives being zero. This shows that the scores $R^{(j)}$ satisfy (9.4); $j = 1, \ldots, m - 1$. Multiplying row j by $p_i(1 - p_i - p_j)$ for $j \neq i$ the matrix Y has the form

$$
\begin{array}{c}
k^{(1)} \quad\quad \ldots\ldots\ldots\ldots\ldots\ldots \quad k^{(i)} \quad\quad \ldots\ldots\ldots \quad k^{(m-1)}
\end{array}
$$

$$
\begin{array}{ccccccc}
0 & p_i & p_i & p_i & \cdots & p_1 + p_i - 1 & \cdots & p_i \\
p_i & 0 & p_i & p_i & \cdots & p_2 + p_i - 1 & \cdots & p_i \\
p_i & p_i & 0 & p_i & \cdots & p_3 + p_i - 1 & \cdots & p_i \\
\cdots & \vdots & \vdots & \vdots & \cdots & \vdots & \cdots & \cdots \\
p_i & p_i & p_i & p_i & \cdots & p_i - 1 & \cdots & p_i \\
\cdots & \vdots & \vdots & \vdots & & \vdots & & \vdots \\
p_i & p_i & p_i & p_i & \cdots & p_{m-1} + p_i - 1 & \cdots & 0
\end{array}
\quad i\text{ th row}
$$

That Y has full rank can be seen by subtracting the ith row of the above matrix from each of the other rows. The result is

$$
\begin{array}{ccccccccc}
-p_i & 0 & 0 & 0 & \cdots & p_1 & 0 & 0 & \cdots & 0 \\
0 & -p_i & 0 & 0 & \cdots & p_2 & 0 & 0 & \cdots & 0 \\
0 & 0 & -p_i & 0 & \cdots & p_3 & 0 & 0 & \cdots & 0 \\
\vdots & \vdots & \vdots & \vdots & & \vdots & \vdots & \vdots & & \vdots \\
p_i & p_i & p_i & p_i & \cdots & p_i - 1 & p_i & p_i & \cdots & p_i \\
0 & 0 & 0 & 0 & \cdots & p_{i+1} & -p_i & 0 & \cdots & 0 \\
\vdots & \vdots & \vdots & \vdots & & \vdots & \vdots & \vdots & & \vdots \\
0 & 0 & 0 & 0 & \cdots & p_{m-1} & 0 & 0 & \cdots & -p_i
\end{array}
$$

These rows are linearly dependent if and only if

$$p_1 + p_2 + \cdots + p_{i-1} + p_{i+1} + \cdots + p_{m-1} = 1 - p_i$$

However, this condition cannot hold if $p_m > 0$, which is the case if $p \in M(0)$. It follows that Y has full rank, and the proof is completed. ∎

We now prove Theorem 9.1. We fix N and adopt the notation of Lemma 9.3.

Equation (9.4) implies (9.3):

$$
\frac{\partial}{\partial p_i} E_Q R(p, N, s) = \sum' Q(k) \frac{\partial}{\partial p_i} R\left(p, N, \frac{k}{N}\right)
$$

$$
= \sum' Q(k) \left\{ g_i(p, N) \left[p_i - \left(\frac{k_i}{N}\right) \right] \right.
$$

$$
+ \sum_{f<j} g_{ifj}(p, N) \left[\left(\frac{k_f}{N}\right) p_j - p_f \left(\frac{k_j}{N}\right) \right] \right\}
$$

$$
= g_i(p, N) \sum_{k_j=0}^{N} Q(k_i) \left[p_i - \left(\frac{k_i}{N}\right) \right]
$$

$$
+ \sum_{f<j} g_{ifj}(p, N) \sum_{k_f, k_j=0}^{N} Q(k_f, k_j) \left[\left(\frac{k_f}{N}\right) p_j - p_f \left(\frac{k_j}{N}\right) \right] \quad (9.5)
$$

If $p = q^{(N)}$, the last expression equals 0 by Lemma 9.1.

Equation (9.3) implies (9.4): Let R be a differentiable scoring rule satisfying (9.3). Then for all $Q \in M(X)$

$$
(\partial/\partial q_i)(E_Q R(q, N, s)) = \sum' Q(k)(\partial/\partial q_i) R(q, N, k/N) = 0.
$$

It follows that for all $q \in M(0)$ and $i = 1, \ldots, m-1$, $R(q, i) \in A(q)^\perp$. From Equation (9.5) it follows that $B(q, i)$ is contained in $A(q)^\perp$. From Lemma 9.3 it now follows that $B(q, i) = A(q)^\perp$, hence $R(q, i) \in B(q, i)$. Since this holds for all $q \in M(0)$, $i-1, \ldots, m-1$, and since R is differentiable, it follows that R has the form of Equation (9.4). ∎

Proof of Proposition 9.1

The statements regarding the partial derivatives can be verified by direct calculation. To verify that $I(s, p)$ is convex, it suffices to verify for $r \in (0, 1)$, $p_1, p_2 \in M(0)$; $p = rp_1 + (1-r)p_2$, that $rI(s, p_1) + (1-r)I(s, p_2) \geq I(s, p)$. We have

$$
I(s, p) = \sum_{i=1}^{m} s_i \ln s_i - \sum_{i=1}^{m} s_i \ln p_i
$$

The first term is the negative entropy of s and is always nonpositive. It suffices to verify that

$$
\sum_{i=1}^{m} s_i [r \ln p_{1,i} + (1-r) \ln p_{2,i}] \leq \sum_{i=1}^{m} s_i \ln p_i
$$

which indeed follows immediately from the concavity of the function $\ln x$.

To verify the estimate in the last equation of (ii), write $s = p + \Theta p$. Then

$$
\frac{s}{p} = 1 + \Theta \qquad \Theta = \frac{s-p}{p} \qquad \frac{1-s}{1-p} = 1 - \frac{\Theta p}{1-p}
$$

Using the Taylor expansion, valid for $x \in (-1, 1)$,

$$\ln(1 + x) = x - \frac{x^2}{2} + \frac{x^3}{3} \cdots$$

$$\ln \frac{s}{p} - \ln \frac{1-s}{1-p} = \Theta + \frac{\Theta p}{1-p} + o(\Theta)$$

$$= \frac{s-p}{p-p^2} + o(\Theta) \quad \blacksquare$$

Proof of Proposition 9.4

Choose $Q \in M$. Then $\lim_{N \to \infty} E_Q w_t(q, N, s) = \chi^2_{m-1}(t) > 0$. Choose $r \in M(0)$ with $r \neq q$. By the strong law of large numbers, $s \to q$ Q-almost surely, and by Egoroff's theorem, for every $d > 0$, the convergence is uniform on a set of Q-probability greater than $1 - d$. Choose $d < \chi^2_{m-1}(t)$. For some $l_d > 0$ we can find $N_d \in N$ such that on this set, for all $N > N_d$, $I(s, r) > l_d$. For $N > \max\{N_d, t/l_d\}$, on this set

$$2NI(s, r) > t$$

Hence, for sufficiently large N, $Q\{w_t(r, N, s) = 0\} > 1 - d$ and

$$E_Q w_t(r, N, s) < d < \chi^2_{m-1}(t) \quad \blacksquare$$

This argument also shows that $E_Q w_t(r, N, s) \to 0$ as $N \to \infty$. The proof of Proposition 9.5 uses the following lemma:

Lemma 9.4. For any (right continuous) cumulative distribution function F and any $z \in R$,

$$0 \leqslant \int_{-\infty}^{z} F(x) \, dF(x) - \frac{F(z)^2}{2} \leqslant \max_x [F(x) - F_-(x)]$$

where $F_-(x) = \sup_{y < x} F(y)$.

Proof. Let 1_A denote the indicator function of the set A. Since F is bounded, F is integrable with respect to dF and the Fubini theorem may be applied.

$$\int_{-\infty}^{z} F(x) \, dF(x) = \int\!\!\int 1_{(-\infty, z]}(x) 1_{(-\infty, x]}(y) \, dF(y) \, dF(x)$$

$$= \int\!\!\int 1_{[y, z]}(x) 1_{(-\infty, z]}(y) \, dF(x) \, dF(y)$$

$$= \int [F(z) - F(y) + F(y) - F_-(y)] 1_{(-\infty, z]}(y) \, dF(y)$$

$$= F(z)^2 - \int_{-\infty}^{z} F(y) \, dF(y) + \int_{-\infty}^{z} [F(y) - F_-(y)] \, dF(y)$$

The two inequalities now follow from the fact that

$$0 \leqslant \int (F(y) - F_-(y)) \, dF(y) \leqslant \max_x (F(x) - F_-(x)) \quad \blacksquare$$

Corollary.

$$\int_{-\infty}^{z} F(x)\, dF(x) \geq \frac{F(z)^2}{2}$$

with equality holding for all z if and only if F is continuous.

Proof of Proposition 9.5

Choose $Q \in M$. Let Q_N denote the expert's cumulative distribution function for $RI(q, N, s)$.

$$E_Q W_t(q, N, s) = \int_0^t [1 - \chi_{m-1}^2(x)]\, dQ_N(x)$$

$$= E_Q w_t(q, N, s) - \int_0^t \chi_{m-1}^2(x)\, dQ_N(x).$$

$Q_N \to \chi_{m-1}^2$. χ_{m-1}^2 is continuous and bounded, so the Helly Bray theorem in conjunction with the corollary to Lemma 9.4 yield, as $N \to \infty$,

$$\int_0^t \chi_{m-1}^2(x)\, dQ_N(x) \to \int_0^t \chi_{m-1}^2(x)\, d\chi_{m-1}^2(x)$$

$$= \frac{\chi_{m-1}^2(t)^2}{2}$$

Hence, $\lim_{N \to \infty} E_Q W_t(q, N, s) = \chi_{m-1}^2(t) - \chi_{m-1}^2(t)^2/2 > 0$.

Suppose $r \in M(0)$ with $r \neq q$. Since $1 - \chi_{m-1}^2(t) < 1$, it follows from the proof of Proposition 9.4 that

$$\lim_{N \to \infty} E_Q W_t(r, N, s) = 0 \quad \blacksquare$$

Remark. Note that the continuity of χ_{m-1}^2 is essential in the above proof. If we replaced χ_{m-1}^2 in the definition of W_t by a noncontinuous distribution, for example, Q_N, then Proposition 9.5 would yield only a crude estimate of $E_Q W_t$. This illustrates the advantage of studying propriety from the asymptotic perspective. Note also that $t \neq \infty$ is essential.

Proof of Proposition 9.6

We treat only the first score, as the argument for the second is similar. Choose $Q \in M$. Suppose $r \in M(0)$, $r \neq q$. We must show that for all sufficiently large N

$$\frac{E_Q w_t(q, N, s)}{E_Q w_t(r, N, s)} > \frac{f(r, N)}{f(q, N)}$$

The right-hand side is bounded from above by b/a, and this bound does not depend on N. From the proof of Proposition 9.4, the left-hand side goes to ∞ as $N \to \infty$ $\quad \blacksquare$

10

Two Experiments with Calibration and Entropy

This chapter discusses two experiments recently carried out at the Delft University of Technology. Both involve the measurement of calibration and entropy with experts. One experiment (Cooke, Mendel, and Thys, 1988) was designed in such a way that the performance of "experienced" and "inexperienced" experts could be compared both on general knowledge items and on items relating to their common field of expertise. The experts were all mechanical engineers, and "expertise" in this test refers to *technical* expertise. The experiment used the quantile tests discussed in Chapter 8. The second (Bhola et al., 1991) also used more and less experienced experts, but the sense of expertise might be described as *managerial* rather than technical. It involved assessments by project leaders of the probabilities that their project proposals would be realized. In a follow-up experiment the evolution of calibration scores can be tracked.

In the first test, the experienced subjects outperformed the inexperienced subjects, while in the second test the reverse occurred. This will warn against any simple minded conclusions relating performance with experience. However, tentative conclusions can be drawn from each test.

The first section of this chapter reviews the psychometric literature on calibration and knowledge. The subsequent sections describe and analyze the experiments.

CALIBRATION AND KNOWLEDGE; BRIEF REVIEW OF THE LITERATURE

Several attempts have been made in the past to relate calibration to "knowledge." Adams and Adams (1961), in one of the earliest studies, found no correlation between knowledge and calibration for subjects taking a final examination. In this case "knowledge" was determined by the number of exam questions answered correctly. Sieber (1974) found similar results. Lichtenstein and Fischhoff (1977) found a negative correlation between calibration and "difficulty" on general knowledge items. Moreover, they found that calibration first improves and then declines with increasing knowledge.

It is important to note that these results concern general knowledge items. Another study (Lichtenstein, Fischhoff, and Phillips, 1982) looked at calibration and knowledge on tests involving almanac items. The subject is asked to determine whether a given statement is true or false, and state his subjective probability associated with his answer. These tests follow the format of the discrete tests discussed in Chapter 8, except that the lowest probability allowed is 50%. It was found that calibration improved as the number of true statements in the test was increased.

From this brief review, it is clear that the results are difficult to interpret. In particular, there is no well-defined psychometric variable corresponding to "knowledge." Moreover, the scoring variables used to measure calibration in many of these studies and the design of the experiments raise additional problems for interpretation, as discussed in Chapter 8.

There are two senses of knowledge that should be distinguished in relation to calibration. On the one hand there is "objective knowledge" in the sense of training and experience. Although we possess no universal measure for knowledge in this sense, we can easily define experimental groups that differ with respect to knowledge in this sense. A second sense of knowledge might be called "subjective knowledge" or "confidence." In a properly designed experiment, we can measure (lack of) subjective knowledge as entropy in subjective probability distributions.

The relationship between knowledge in these various senses and calibration is subtle. The first experiment discussed below produced the following results. "Objective knowledge" was significantly correlated with good calibration and with "subjective knowledge"; whereas subjective knowledge and calibration showed a strong negative correlation. More precisely, the group of experienced experts performed significantly better than the inexperienced group with respect to both calibration and entropy, on items related to their common field of expertise. On general knowledge items the groups did not differ significantly. Moreover, in both groups, and on all items, there was a significant negative correlation between good calibration and low entropy.

In the second experiment involving project managers, there was a negative correlation between experience and performance in calibration, although these results are somewhat more ambiguous. The less experienced managers tended to assess probabilities of a different kind of project proposal.

THE MECHANICAL ENGINEER EXPERIMENT

An experiment was recently conducted at a Dutch training facility for operators of large technological systems (Cooke, Mendel, and Thys, 1988). The purpose of the experiment was to investigate whether experts with more practical experience exhibit better calibration and better entropy, and if so, whether this generalizes to other areas outside their field of expertise.

The Subjects

The experimental subjects fell into two groups. One group, the inexperienced operators, was in the last year of a 5-year training program, roughly equivalent to a bachelor of science program at an American university. Their field of study was mechanical engineering. All these subjects were between 20 and 25 years of age, and all had completed a course in statistics.

The second group, the experienced operators, had all completed the training program. Their average was 36 years, and they had on the average 15 years of practical experience. Some of them were teachers at the training facility. Twenty-two inexperienced and twelve experienced subjects took both general knowledge and expertise-specific, or technical, calibration tests. Three additional experienced subjects took only the general knowledge test. All subjects were male.

The Tests

The tests were modeled on the quantile calibration tests of Alpert and Raiffa (1982). Some of the general knowledge items were taken literally from this test, and others were adapted to the situation in Holland. The following are examples of the uncertain quantities from the technical tests:

The maximal efficiency of the Tyne RM1A gas turbine
The maximum admissible intake temperature for gas in the Olympus power turbine

Each test contained 10 uncertain quantities, and for each quantity the 1%, 25%, 50%, 75%, and 99% quantiles were elicited.

Scoring and Results

Calibration was scored in the manner set forth in Chapter 8. The intrinsic range for each quantity was the lowest 1% and the highest 99% quantile elicited for that quantity, plus 10% overshoot and below. For each subject the relative information $I(s, p)$ of the sample with respect to the "theoretical" distribution p was determined. As the number of uncertain quantities was the same for all subjects, $I(s, p)$ was used as an index for calibration. The entire set of subjects is rank-ordered for each of the two tests, rank 1 corresponding to best calibration [i.e., lowest value of $I(s, p)$]. The rank results are presented in Table 10.1. Fractional ranks correspond to ties.

The Wilcoxon two-sample test is used to determine whether the experienced operators are significantly higher ranked with respect to calibration in the general knowledge and the technical tests (Hodges and Lehmann, 1970). This was indeed the case for the technical test (significance level 0.012), but not for the general knowledge test.

For each uncertain quantity, we also score the entire group of subjects with respect to entropy, as described in Chapter 8. For each quantity, each expert's density was approximated by a minimum information fit (P' in the notation of Chapter 8) over the intrinsic range consistent with the assessed quantiles. Since the quantities involved different physical dimensions, we considered the relative

Table 10.1. Calibration and Information Scores on the General Knowledge and Technical Tests

	Calibration rank		Information rank	
Subject number	General knowledge	Technical	General knowledge	Technical
1	8	2	13	23
2	27	31	1	7
3	3	25	31	4
4	6.5	6	24	10
5	17	1	22	24
6	25.5	4	18	33
7	11	7	36	22
8	22	15	15	5
9	12	11	11	16
10	29	3	6	1
11	19.5	22	27	3
12	25.5	20	28	2
13	31	—	5	—
14	23	—	21	—
15	18	—	26	—
16	4	12	19	30
17	24	26	10	27
18	9	13	17	26
19	1	5	25	28
20	32	30	29	8
21	5	23	32	9
22	19.5	8	34	34
23	2	19	33	19
24	10	24	35	15
25	21	17.5	8	20
26	33	16	37	32
27	14.5	21	30	13
28	34	17.5	14	18
29	30	32	23	12
30	35	29	12	17
31	16	27	2	25
32	14.5	14	20	29
33	36	33	3	6
34	6.5	9	9	31
35	37	10	7	14
36	28	34	16	11
37	13	28	4	21

Subjects 1–15 were experienced operators.
Subjects 16–37 were inexperienced operators.

information $I(P', U)$ of P' with respect to the uniform density on the intrinsic range U. Low values of $I(P', U)$ correspond to highly entropic, or highly uninformative, distributions. The values for $I(P', U)$ are then added for each subject and the subjects are ranked. Rank 1 corresponds to the "best" entropy score [i.e., largest value for the sum of the terms $I(P', U)$]. The results are presented in Table 10.1. The

Wilcoxon two-sample test for determining whether experienced subjects had significantly better entropy scores yields results almost identical to those for calibration. On the technical test the experienced operators had significantly more information (significance level 0.022), but not on the general knowledge test.

The Spearman rank correlation test was used to test the null hypothesis "good calibration is not correlated with bad entropy" (Siegel, 1956). For the experienced group on both tests and for the whole group on the general knowledge test there was significant correlation at the 5% level. For the inexperienced group on both tests and for the whole group on the technical test there was significant correlation at the 1% level.

From Figure 10.1, we see that one (experienced) subject was extremely well calibrated (rank 3) and very informative (i.e., low entropy, entropy rank 1) on the technical test. This subject emerges as a very good expert. Interestingly, there was no "good expert" for the general knowledge items. The above-mentioned individual was ranked twenty-ninth and sixth for calibration and entropy, respectively, on the general knowledge test.

A chi square table may be used to determine the calibration score formula Eq. 8.4)] for these subjects. Since there are 10 items in each test and five degrees of freedom in the "theoretical distributions," a calibration index greater than 0.6 would be significant at the 5% level. If we regard the uncertain quantities as independent, we could not reasonably believe that scores higher than 0.6 were produced by statistical fluctuations. Of the 34 subjects participating in the technical test, 31 would be regarded as miscalibrated at the 5% level. For the general knowledge, these figures are 37 and 33, respectively. For the experienced group 9 of the 12 would be "rejected" at the 5% level on the technical test, and 14 of the 15 on the general knowledge test. Since the expected number of observations in the tails of the theoretical distributions is quite small (0.1 for each tail), the chi square approximation is not very reliable. For example, the theoretical probability of finding more than one event in the tails is 0.017, but two observations in either of the tails contribute only 0.599 to the calibration scores. Hence, the chi square approximation is "charitable."

As with other tests reported in the literature, these calibration results can be described as poor. One way of judging this is to compare the number of rejected assessors with the number which would be rejected if P' were consistently replaced by the uniform distribution over the intrinsic ranges. On the technical test, only two experienced operators and three inexperienced operators would be rejected at the 5% significance level in this case. This means, roughly speaking, that most of the subjects behave more as if they were asked for the fractiles 17%, 33%, 50%, 67%, and 83%.

The "surprise index" is the percentage of the true values falling in the tails of the elicited distributions. On the general knowledge tests the surprise indices for experienced and inexperienced operators were 43% and 45%, respectively. On the technical test these were 30% and 43%.

A number of remarks are in order.

There is a natural antagonism between good calibration and low entropy. Choosing a "tighter" function P will produce a lower entropy score, but will make it more "likely" that the true values will fall in the tails, producing a low

i denotes inexperienced operator; e denotes experienced operator.

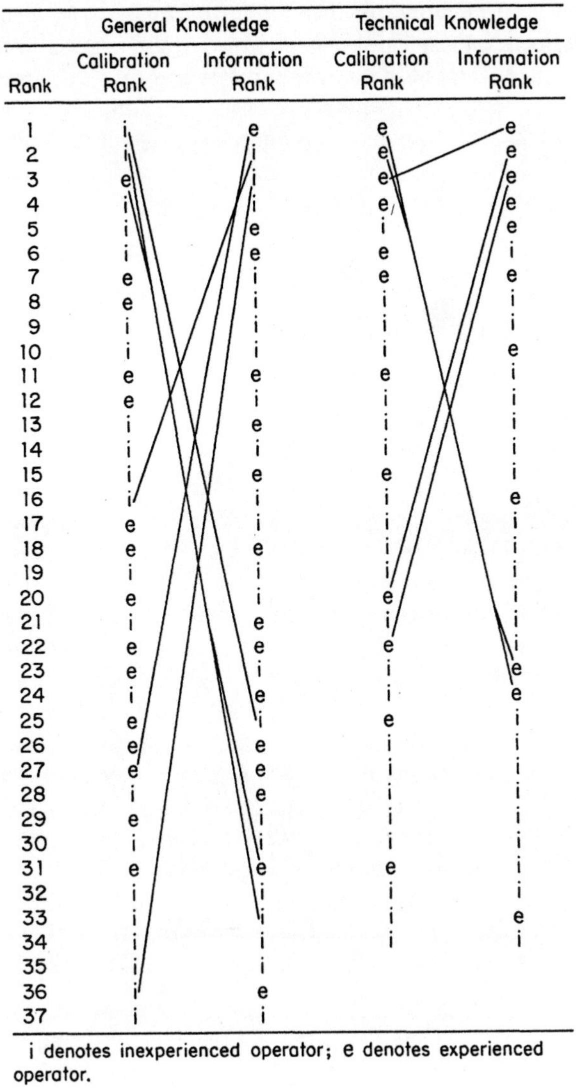

Figure 10.1 Graphical representation of the correlations between calibration and entropy ranks for the top three ranked subjects on the general knowledge and technical tests.

calibration score. A priori, one would expect a negative correlation between goodness in entropy and calibration, and this indeed is found.

There is no a priori reason to expect the group of experienced assessors to perform better in their field of expertise than the inexperienced group in both scores. This requires an explanation. The evident explanation, of course, is that experience teaches assessors to be better experts in the sense of being better calibrated and more informative.

"Goodness" of calibration and entropy seems not to generalize outside the area of expertise.

If calibration and entropy are meaningful parameters for evaluating expert probability assessments, there can be no doubt that the experts in this study have ample room for improvement. This raises the interesting question whether it is possible to upgrade assessment quality through training. After all, that which "nature" teaches inefficiently is what training programs should teach efficiently. To date very little work has been done on this question. In any case, training in probability assessment is only possible once we can define and measure "goodness."

THE MANAGERS EXPERIMENT

Delft Hydraulics is an independent institute for consultancy and research in the field of hydrodynamics, hydrology, hydraulics, and water resources management. The firm carries out contract research for clients, and contracts are awarded on the basis of competitive bids. In planning future manpower and resource needs, the firm's management is confronted with many project proposals, many of which are highly uncertain. One large department of this institute has been using a system for quantifying uncertainty as subjective probability since 1986. The system was largely compatiable with the discrete test described in Chapter 8, and hence it was possible to measure calibration and entropy, and to study performance. The data described below were obtained after an analysis of department records. An analysis of this data was presented in Bhola et al. (1991)[1].

The department in question operates with an annual turnover of about 8 million U.S. dollars and contracts about 100 projects yearly. Each project proposal is accompanied by the project leader's subjective assessment of the probability that the project proposal will be accepted. For each project, there is one such estimate. The use of subjective probability enables management to gain insight in the quality of assessing uncertainty by the various project leaders. In addition, insight was gained into the properties of project leaders and projects that influence the quality of assessments. In the initial phase, two questions were deemed of interest:

1. Do more experienced project leaders produce better estimates?
2. Is the quality of assessment affected by the type of project?

The Subjects

There were 14 project leaders for whom previous assessments and data regarding eventual realizations were available. Table 10.2 shows the rank ordering of the experts in terms of age and years experience in the firm.

The Results

The subjective probability assessments were discretized into 10%, 20%, ..., 90% probability bins. Response and sample entropy were computed for an expert

[1]We are grateful to D. Roeleven for catching some errors in the codes used in a previous version of this analysis.

Table 10.2 Ranking of Ages and Years of Employment of Project Leaders

Expert No.	Age Rank	Years Experience Rank
1	4.5	6
2	9.5	9
3	14	14
4	11	11
5	9.5	12
6	12	7
7	1	2
8	7	10
9	6	1
10	13	13
11	3	4
12	2	3
13	8	8
14	4.5	5

assessing n items as $H(P)/n$ and $H(S)/n$, respectively, where $H(P)$ and $H(S)$ are defined in Formulae (8.6) and (8.7), respectively. If RI_e denotes the statistic defined in (8.5) for expert e, then the calibration score $C(e)$ of e is defined as

$$C(e) = \text{Prob}\{RI > RI_e | e \text{ is perfectly calibrated and all items are independent}\}$$

In addition, the base rate resolution index defined in Chapter 9 [see formula (9.9)] is computed for each expert. This index measures the degree to which the sample distributions in each bin, per expert, differ from the overall sample distribution, per expert. Unnormalized weights are computed by dividing the calibration score by the response entropy [compare formula (9.12)]. The results for each expert are shown in Table 10.3. For all events whose assessments are shown in Table 10.3, the fate of the project proposal is known. Table 10.4 shows the allocations and realizations for each bin.

Several points regarding Tables 10.3 and 10.4 require comment. First of all, most of the experts are quite well calibrated. For only two experts is the probability of a relative information score [(8.5)] exceeding the observed value less than 5%. The chi square approximation is unreliable for the number of items shown in Table 10.4, so the calibration scores were determined by direct calculation. Also notable is the fact that the sample entropy is generally *lower* than the response entropy. In part, this is explained by the fact that many bins contained a small number of items. Hence, the sample distribution for true and untrue events was often (0, 100%), or (100%, 0), and these extreme distributions, of course, have zero entropy. Nevertheless, inspecting group performance for the 20% and 80% bins indicates that these experts may display some *under*confidence. Unlike the experiment with mechanical engineers, there is no significant rank correlation between response entropy and calibration scores. Neither is there significant rank correlation between the calibration scores and the base rate resolution index.

Table 10.5 shows the aggregate calibration scores, with projects grouped

Table 10.3 Calibration, Response, and Sample Entropy Base Rate Resolution Scores and Unnormalized Weights for the 14 Experts

Expert	Calibration Score	Response Entropy	Sample Entropy	Weight	BSR Index
1	0.541	0.508	0.250	1.064	0.716
2	0.099	0.523	0.201	0.190	0.995
3	0.049	0.545	0.452	0.000	0.942
4	0.105	0.495	0.231	0.212	0.699
5	0.307	0.493	0.337	0.622	0.834
6	0.034	0.434	0.174	0.000	0.749
7	0.771	0.535	0.414	1.442	0.743
8	0.424	0.503	0.351	0.843	0.945
9	0.465	0.521	0.503	0.892	0.870
10	0.371	0.379	0.000	0.980	1.000
11	0.722	0.508	0.453	1.421	0.900
12	0.901	0.480	0.139	1.878	0.907
13	0.347	0.557	0.212	0.622	0.217
14	0.304	0.419	0.225	0.726	0.938

Table 10.4 Binwise Allocations of the 14 experts

Expert	Number of Events and True Events in Each Bin									Number of Events	True Events
	(10%)	(20%)	(30%)	(40%)	(50%)	(60%)	(70%)	(80%)	(90%)		
1	1 0				4 3			1 1	3 3	9	7
2	5 0	1 0	3 3		6 2			3 3	1 1	19	9
3					9 2	1 0		3 3	6 4	19	9
4	2 1			1 0	2 2			1 1		6	4
5	5 0		2 0	1 0	6 3			1 0	3 2	18	5
6	3 2			1 0	2 2			1 1	4 4	11	9
7	1 0			1 1	7 4			4 3	4 4	17	12
8	5 1		3 0	1 1	6 4			1 1	2 2	18	9
9	4 1				8 3				3 3	15	7
10	11 0		1 0		2 2				3 3	17	5
11	12 0				9 5			2 2	7 6	20	13
12			1 0		1 0	2 1	1 1		5 5	10	7
13			1 1		3 2	1 1	3 3	1 1		9	8
14	4 1			2 0	1 0				3 3	10	4

Table 10.5 Results for Projects, per Contractor

	Contractor				
	A	B	C	D	E
Calibration	0.04	0.65	0.58	0.60	0.13

Contractors A, B, and C are Dutch; D and E are foreign.

Table 10.6 Results for Projects Grouped by Turnover

Turnover (1000 Dfl)	Calibration	Response Entropy
0–25	0.45	0.49
25–100	0.26	0.51
100–250	0.50	0.50
>250	0.95	0.50

according to potential contractor. Curiously, the score was worse for contractor A, which is the contractor with which Delft Hydraulics has the most experience.

There is a correlation between the percentage projects in the Netherlands of each expert and his calibration score. The Spearman rank correlation is 0.58 (both ranked from high to low), which is significant at the 5% level. In other words, project leaders with more projects in the Netherlands tended to have higher calibration scores.

Table 10.6 shows the aggregated scores for projects broken down into four categories of monetary turnover. It is seen that the very large projects are assessed better than others.

There is a negative correlation between calibration and age. The Spearman rank correlation coefficient, −0.82, is significant at the 5% level. Figure 10.2 graphs the calibration score against ranked ages. This negative correlation may indicate that the younger project leaders are better probability assessors. However, the younger leaders have a different mix of projects among the different contractors than the older leaders. It is possible that younger leaders have proportionally more projects in the Netherlands, and that this explains their better performance.

To decide whether age or percentage projects in the Netherlands,

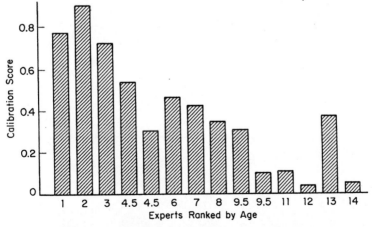

Figure 10.2 Calibration and age rank of the 14 experts.

%(A + B + C), best explains the data, Kendall's tau coefficient for partial rank correlation is computed. This coefficient measures the rank correlation between two variables, when the influence of a third variable is eliminated by keeping it constant. The results are

Rankcorr. (cal. score %(A + B + C)|age constant) = 0.102

Rankcorr. (cal. score, age|%(A + B + C) constant) = 0.577

Only the latter rank correlation is significant, hence the variable "age" better explains the differences in calibration than the variable "percentage projects in the Netherlands."

CONCLUSION

Significant differences between the designs of these two experiments make generalizations difficult. In the first experiment all subjects assessed the same items. Since the subjects were required to assess 1% and 99% quantiles, they had to judge smaller probabilities than the subjects in the second experiment. This may well explain the overall differences in calibration between the two experiments. The first experiment shows a positive correlation between experience (hence age) and calibration, while the second experiment presents us with a negative correlation. Taken together, these two experiments provide a useful warning against superficial conclusions.

Both experiments show that calibration and entropy can be meaningfully used to analyze expert probability assessments. Systematic use of these techniques open the possibility of training experts to give better assessments. Training programs could focus on the specific problems revealed by the data, and the impact of training could be monitored by subsequent measurements. One conclusion emerging from both experiments is this: Quantifying expert uncertainty as subjective probability can be a useful tool in rational decision making.

III

COMBINING EXPERT OPINIONS

11

Combining Expert Opinions; Review of the Literature

This chapter reviews various models for combining expert opinion found in the literature, and motivates some of the choices underlying the models in the following three chapters. We assume that the combination results in a probability distribution for the "decision maker." Three broad classes of models are discussed. The Bayesian models discussed in the second section all take their point of departure from Bayes' theorem and require the decision maker to supply a prior probability. The first section is devoted to weighted combinations without prior distributions. It is useful to recall from Chapter 6 that subjective probabilities exist for rational *individuals*. Most decision-making bodies are not individuals but groups. There is no reason why groups should have a preference structure like that of rational individuals, and no reason why groups, for example, the scientific community, should have prior distributions. The third section is devoted to psychological scaling models. These models are rather unlike the models in the first two sections. They have received little attention from Bayesians studying the "expert problem," but have demonstrated their appeal among people involved with applications.

WEIGHTED COMBINATIONS OF PROBABILITIES

Mathematical Background

Much has been learned in recent years about the mathematical properties of various rules for combining probabilities. Excellent summaries can be found in French (1985) and Genest and Zidek (1986). In this section we restrict our attention to the class of combination rules derived from "elementary weighted means." This class is wide enough to include the most interesting possibilities, and its mathematical properties are well understood. Within this class we can derive many important results very easily, using standard results in analysis. The *locus classicus* for these results is Hardy, Littlewood, and Polya (1983), to which we refer for proofs.

To get the discussion started, we may recall the expert estimates of the probability per section-hour of pipe failure taken from the Rasmussen Report (see Table 2.5). The 13 estimates ranged from 5E-6 to E-10. The analyst performing the risk analysis of a nuclear power plant must choose some estimate for this quantity. How is he to choose? Should he take the largest or the smallest value? Should he take the arithmetical average of the expert estimates (4,7E-7), or perhaps the geometrical average (7.5E-9)? Should he perform some weighted combination of the expert assessments? It is clear that the choice of combination rule will have a great impact on his final result. Is there any principle to guide his choice?

In fact there are general principles to which the analyst can appeal in choosing a rule of combination. Suppose we have experts $1, \ldots, E$, and that each expert i gives a probability vector $p_{i1}, \ldots, p_{in}$ for the elements $A_1, \ldots, A_n$ of some partition of the set S of possible worlds. Further, let $w_1, \ldots, w_E$ be nonnegative weights that sum to unity. For any real number r we define the

Elementary r-norm

weighted mean: $M_r(j) = \left(\sum_{i=1}^{E} w_i p_{ij}^r \right)^{1/r}$

r-norm probability: $P_r(j) = \dfrac{M_r(j)}{\sum_{k=1}^{n} M_r(k)}$

These expressions generalize for continuously many alternatives; probabilities are replaced by probability densities, summations are replaced by integrals, and the qualification "elementary" is dropped. $P_r(j)$ is a probability, the denominator assures that the terms $P_r(j)$ sum to unity. The interpretation of M_r may be gleaned from the following facts (see Hardy, Littlewood, and Polya, 1983, chap. 2):

1. $M_r(j) \to \prod p_{ij}^{w_i}$ as $r \to 0$ [this defines $M_0(j)$].
2. If $r < s$ then $M_r(j) < M_s(j)$.
3. If $r \to \infty$ then $M_r(j) \to \max_{i=1,\ldots,E}\{p_{ij}\}$.
4. If $r \to -\infty$ then $M_r(j) \to \min'_{i=1,\ldots,E}\{p_{ij}\}$.
5. Define $M_r(j+k) = [\Sigma_i w_i(p_{ij} + p_{ik})^r]^{1/r}$, and assume p_{ij}/p_{ik} is not constant in i. Then the following (in)equalities hold:

    If $r > 1$ then $M_r(j+k) < M_r(j) + M_r(k)$.
    If $r = 1$ then $M_r(j+k) = M_r(j) + M_r(k)$.
    If $r < 1$ then $M_r(j+k) > M_r(j) + M_r(k)$.

    If p_{ij}/p_{ik} is the same for all i, then equality holds in all three of the above cases.

6. Under the same conditions as (5), if

$r > 1$	then	$\Sigma_j M_r(j) < 1$
$r = 1$	then	$\Sigma_j M_r(j) = 1$
$r < 1$	then	$\Sigma_j M_r(j) > 1$

$M_1(j)$ is the *weighted arithmetic mean*, $M_0(j)$ is the *weighted geometric mean*, and $M_{-1}(j)$ is the *weighted harmonic mean* of the experts' assessments for alternative j.

For a fixed set of weights, choosing r larger gives more influence to the larger assessments for each alternative. In the limit for $r \to \infty$, the largest assessment for each j is chosen and these are normalized to determine the probability P_∞. Similar remarks apply for choosing smaller r. All the probabilities P_r possess what is called the *zero preservation property*; that is, if $p_{ij} = 0$ for all i, then $P_r(j) = 0$. P_0 possesses this property in the extreme; if any expert assigns alternative j probability 0, then so will the decision maker who uses P_0.

Marginalization

From (6) we see that $P_1(j) = M_1(j)$; and for $r \neq 1$ the probability $P_r(j)$ is affected by the terms $M_r(k), k \neq j$. This can lead to curious effects. Suppose $r > 1$, and suppose after deriving P_r, the decision maker subdivides the alternative A_n into two disjunct subalternatives A_{n1} and A_{n2}. Let P_r' denote the probability derived after hearing the experts probabilities for these subalternatives. From (5) we see that the denominator of P_r will be less than that of P_r', hence for $j \neq n$, $P_r(j) > P_r'(j)$.

If the combination rule is such that the probabilities are unaffected by refinements of the partition of alternatives $A_1, \ldots, A_n$, then the rule is said to possess the *marginalization property*. From the above it is clear that P_1 is the only probability derived from elementary weighted means that possesses the marginalization property. Marginalization in this context is equivalent to the so-called *strong setwise function property* (McConway, 1981): the decision maker's probability for any event A depends only on the experts' assessments for event A.[1]

A combination rule that does not possess the marginalization property is downright queer. Consider a simple example. Two experts whom I esteem equally are consulted about the probability that my flashlight, which I forgot to unpack after last year's vacation, still works. Both experts give the flashlight a probability of 0.8 of not working. I decide to combine their opinions via a normalized geometric mean (i.e, P_0). It is easy to check that my probability for the flashlight not working is also 0.8. A discussion ensues whether failure is most likely due to a dead battery or corroded contacts (only these possible failure modes are considered, and the eventuality that both occur is excluded). The first expert assigns probabilities 0.7 and 0.1 to these events, whereas the second assigns probabilities 0.1 and 0.7, respectively. On receiving this information my probability that the flashlight won't work drops to 0.73. Further disagreement about *which* contacts might be corroded causes my probability to drop again. If the first expert decides that a dead battery has probability 0.8 and corroded contacts probability 0, my probability drops to 0.51. If the second expert then changes his probability for corrosion and dead battery to 0.8 and 0, respectively, then my probability that the flashlight fails drops to 0. All the while, the experts agree that the flashlight has a probability 0.8 of not working.

---

[1] In a more general setting marginalization and the strong setwise function property are not equivalent; however, marginalization and the zero preservation property imply the strong setwise function property. McConway (1981) and Wagner (1982) prove that P_1 is the only combination rule satisfying the strong setwise function property when the number of alternatives is at least 3.

Independence Preservation

The combination rule P_0 has one property that some people consider important (Laddaga, 1977). Let A and B be unions of the alternatives $A_1, \ldots, A_n$. Suppose we elicit only the experts' probabilities for A and B, and suppose that all experts regard these events as independent, that is, $p_i(A \cap B) = p_i(A)p_i(B)$. Writing

$$M_0(A) = \prod_{i=1}^{n} p_i(A)^{w_i}$$

and noting that

$M_0(A \cap B) = M_0(A)M_0(B)$

$M_0(A \cap B) + M_0(A \cap B') + M_0(A' \cap B) + M_0(A' \cap B')$

$$= [M_0(A) + M_0(A')][M_0(B) + M_0(B')];$$

$$\frac{M_0(A \cap B) + M_0(A \cap B')}{[M_0(A) + M_0(A')][M_0(B) + M_0(B')]} = \frac{M_0(A)}{M_0(A) + M_0(A')}$$

it is easy to verify that $P_0(A \cap B) = P_0(A)P_0(B)$. Moreover, the value $P_0(A)$ is the same as would be found if the experts were asked only for their probabilities for A, and not for B. Hence, if the experts are asked *only* about events that they all regard as independent, then the rule P_0 has the marginalization property and preserves independence.

To some people, the above fact might appear to be a powerful argument in favor of the combination rule derived from the weighted geometric mean. However, in the context of combining *subjective* probabilities, the preservation of independence is not so desirable as it might appear at first sight. We recall from Chapter 7, if an experts' subjective probabilities for a sequence of events are independent then he is intransigent in the following sense: No matter what relative frequencies are observed, he is unwilling to alter his probabilities for the unobserved events. Independence preservation means that when all experts are intransigent, the decision maker should be intransigent as well, even if the experts disagree.

Some authors point to another advantage of the rule P_0, not shared by P_r, $r \neq 0$. P_0 has the property that the result of first combining, and then processing, the results of new observations via Bayes' theorem is the same as first letting the experts process the results of the new observations and then combining their updated probabilities. A rule having this property is called *externally Bayesian* (see Genest and Zidek, 1986, for references, and French, 1985, for criticism). Of course, this rule is only meaningful if the decision maker has a prior distribution. When the decision maker does not have a prior distribution, there is only one serious contender, namely P_1, the weighted arithmetical average.

Determining Weights

Authors of the Bayesian persuasion point out that no substantive proposal has been made for determining the weights to be used in weighted combination. French (1985) summarizes the obstacles to any operational definition of the weights. Such a definition must overcome the problems of

> Correlation between experts
> Uneven calibration of experts
> Possible dishonesty of experts

In the following chapter a theory of weights will be presented that aspires to overcome the last two problems, and there are some encouraging results with regard to correlation. In this section we review the literature on determining weights.

It is not quite true that no substantive proposals have been made, though the effort put into *determining* weights palls in comparison with the effort spent in studying what to do with the weights once we have them.

Winkler (1968) lists four ways of determining weights:

1. Assign all experts equal weights.
2. Rank the experts in preference, lower rank corresponding to lower preference, and assign weights proportional to ranks.
3. Let the experts weigh themselves.
4. Use proper scoring rules.

The fourth proposal is not worked out, and from Chapter 9 we can anticipate the problems encountered when the traditional notion of a scoring rule for individual events is maintained.

An elegant suggestion was made by De Groot (1974) and independently by Lehrer and Wagner (1981). We return to the matrix $\mathbf{p} = [p_{ij}]$ of assessments from expert i of alternative j. Each expert i is asked to report the weight w_{ij} that he would assign to the opinion of expert j (i also assigns a weight to himself). After learning the assessments from all the other experts, expert i should change his probability, on this view, to

$$\mathbf{p}'_{ij} = \sum_k w_{ik} p_{kj}; \qquad \mathbf{p}' = \mathbf{w}\mathbf{p}$$

where the second expression is the matrix notation for the first. There being no reason to stop at $\mathbf{p}'$, the experts should change again to $\mathbf{p}'' = \mathbf{w}(\mathbf{w}\mathbf{p}) = \mathbf{w}^2\mathbf{p}$, and so on. Under certain general conditions which we shall not describe here, this process converges to a matrix $\mathbf{w}^\infty \mathbf{p}$, where the rows of w^∞ are all the same. In other words, by iteratively revising their own opinions in the above manner, the experts all converge toward the *same* probability vector. Practical methods exist for approximating the matrix w^∞.

The "De Groot weights" mentioned in Chapter 5 are the weights given by any row of $\mathbf{w}^\infty$. Despite the evident mathematical appeal of this procedure there are three serious drawbacks:

1. The method requires that the same experts assess all alternatives.
2. The notion of "honesty" in assigning weights to the opinions of one's colleagues needs defining. There is no guarantee that honesty, whatever that means, is encouraged by this method.

3. In practice, using De Groot weights leads to loss of accountability; people simply will not publish the results of such self-weighing, and since $\mathbf{w}^\infty$ can be derived from the outcome $\mathbf{w}^\infty \mathbf{p}$ and the original assessments $\mathbf{p}$, the original matrix $\mathbf{p}$ will not be published either.

BAYESIAN COMBINATIONS

Roughly speaking, Bayesian models treat expert assessments as "observations" in the sense of Chapter 6 and use Bayes' theorem to update the decision maker's prior distribution on the basis of these observations. Many Bayesian models have been proposed in the literature, those discussed below may be regarded as prototypes. We adhere to the convention that a capital letter indicates a random variable, and a lower-case letter, a realization. Hence, "$X = x$" means that random variable X takes the value x.

Apostolakis and Mosleh

Mosleh and Apostolakis (1982, 1986) present Bayesian models, which they have applied in risk analysis. Letting $P(x)$ denote the probability density of the decision maker for some unknown quantity X, we suppose that expert i, $i = 1, \ldots, E$ gives estimates X_i of X. Let $\mathbf{X}$ denote the vector $X_1, \ldots, X_E$. Bayes' theorem reads

$$P(x \mid \mathbf{X}) = kP(\mathbf{X} \mid x)P(x)$$

where k is a normalization constant and the likelihood $P(\mathbf{X}|x)$ is the decision maker's probability of receiving assessment $\mathbf{X}$ given that the true value is x. A model is specified by specifying $P(x)$ and $P(\mathbf{X}|x)$. Assuming that the experts' assessments are independent,

$$P(\mathbf{X} \mid x) = \prod P(X_i \mid x)$$

the latter problem reduces to determining the terms under the product on the right-hand side. Apostolakis and Mosleh suggest two models for accomplishing this.

1. On the *additive error model*, expert i's estimate X_i is treated as the sum of two terms:

$$X_i = x + e_i$$

where x denotes the true value and e_i and additive error term. The model assumes that the errors e_i are normally distributed with mean m_i and standard deviation σ_i. The choice of m_i and σ_i reflects the decision maker's appraisal of expert i's bias and accuracy.

Under these assumptions the likelihood $P(X_i|x)$ of observing estimate X_i given that the true value is x, is simply the value of the normal density with mean $x + m_i$ and standard deviation σ_i.

It is interesting to study the decision maker's posterior expectation for x, given the advice $\mathbf{X}$, under the above assumptions. As in all Bayesian models, the decision maker must first provide his prior $P(x)$. Let us assume that $P(x)$ is the normal density with mean μ and standard deviation σ. A standard calculation[2] shows

$$E(x\mid \mathbf{X}) = \int xP(x\mid \mathbf{X})\,dx$$

$$= \sum_{i=1}^{n+1} w_i(X_i - m_i)$$

where

$$w_i = \frac{\sigma_i^{-2}}{\sum_{j=1}^{n+1} \sigma_j^{-2}} \qquad X_{n+1} = \mu \qquad m_{n+1} = 0 \qquad \sigma_{n+1} = \sigma$$

Hence, the decision maker's updated expectation is just a weighted sum of the experts' expectations, corrected to remove the appraised biases, and the decision maker's prior expectation. The weights are determined by the assessed accuracies σ_i, including the decision maker's assessment of this own accuracy, $\sigma = \sigma_{n+1}$. The decision maker thus treats himself as the $(n + 1)$th expert.

2. On the *multiplicative error model* expert i's assessment is treated as the product

$$X_i = xe_i$$

of the true value x and an error term e_i. Taking logarithms on both sides of the above equation, we reduce this case to the additive error model for the observation $\ln X_i$. The assumption of normality in that model entails that X_i is lognormally distributed. Assuming independence as above, the decision maker's posterior expectation in this case is

$$E(x\mid \mathbf{X}) = \prod_{i=1}^{n+1} \frac{X_i^{w_i}}{e^{w_i m_i}}$$

If the experts' assessments are not independent, then the likelihood function cannot be written as a product of the likelihoods for the single-expert assessments. However, under the normality assumption, we can specify a joint normal likelihood function. Dependences between the experts are accounted for by specifying the correlation coefficients between the experts' assessments.

The additive error model was introduced by Winkler (1981). Lindley and Singpurwalla (1986) describe a more general model in which the experts give

---

[2]This result is mentioned in Lindley (1983). One shows

$$P(x\mid \mathbf{X}) = C \exp -\frac{1}{2}\left[\sum_{i=1}^{n+1} (x/\sigma_i - a_i)^2\right]$$

$$= C' \exp -\tfrac{1}{2}(Kx - L)^2$$

where

$$K^2 = \sum_{i=1}^{n+1} 1/\sigma_i^2$$

$$KL = \Sigma(X_i - m_i)/\sigma_i^2$$

The constant C' serves to normalize the density $P(x\mid X)$. The mean of $P(x\mid X)$ is $E(x\mid X) = L/K$.

variances in addition to the point estimates X_i. Procedures for subjectively estimating correlation coefficients are described in Kadane et al. (1980) and Gokhale and Press (1982).

In a variation of this model (Mosleh and Apostolakis, 1986), the experts' assessments X_i consist of vectors of percentiles from an unknown distribution with parameters $\alpha = (\alpha_1, \ldots, \alpha_k)$. A prior distribution $P(\alpha)$ is updated given the experts' advice $\mathbf{X}$ by solving

$$P(\alpha|\mathbf{X}) = kP(\mathbf{X}|\alpha)p(\alpha)$$

The distribution for X is found by solving

$$P(X|\mathbf{X}) = \int P(X|\alpha)p(\alpha|\mathbf{X})\,d\alpha$$

Another variation has been developed for the case that experts assess a probability of an uncertain event. If p_i is the assessed probability from expert i that the event occurs, then $X_i = \ln[p_i/(1 - p_i)]$ is the "log odds" for this assessment, and can be treated as in the additive error model (Lindley, 1985). Clemen and Winkler (1987) have applied this model to probabilistic weather forecasts with the prior distribution derived from the base rate frequencies and found relatively poor performance.

This model is attractive, conceptually simple, and has demonstrated its worth in applications. However, there are two principle drawbacks:

1. These models place a heavy assessment burden on the decision maker. Not only must he specify his prior distribution, he must also specify two parameters for each expert plus a correlation coefficient for each pair of experts. For the 13 experts assessing pipe failure probabilities that comes to 104 assessments, in addition to his prior. The model gives no guidance how this is to be done, and there is no provision for updating the decision maker's estimates of the experts' biases and accuracies on the basis of past performance.
2. For more general classes of distributions, the correlation coefficients do not determine the joint distribution, and this approach would not work.

Winkler's Natural Conjugate Theory

Winkler (1968) developed a rather different theory for incorporating expert probability assessments in decision making, called the *natural conjugate method*. We begin with Bayes' theorem:

$$P(x|D) = \frac{P(D|x)P(x)}{P(D)} \tag{11.1}$$

where P is the decision maker's probability density for some unknown quantity x, and D represents some observational data relevant to x. $P(D|x)$ and $P(x)$ are called *natural conjugates* if $P(x)$ and $P(x|D)$ belong to the same class of distributions, that is, if they can be described by the same parameters. Natural conjugates are very convenient in Bayesian inference, since the updating can be expressed as an updating of the parameters of the prior distribution.

We illustrate Winkler's approach with a simple example. Let x denote the probability of heads with a bent coin; $0 \leqslant x \leqslant 1$. The sort of data we can gather about x is of the form "r heads in n tosses," assuming that the tosses form a Bernoulli sequence. We put $D = (r, n)$. Under these assumptions,

$$P(D|x) = \binom{n}{r} x^r (1 - x)^{n-r} \qquad (11.2)$$

From the definition of conditional probability we know

$$P(D) = \int P(D | x) P(x) \, dx \qquad (11.3)$$

Suppose that $P(x)$ is a *beta density* with parameters $(r', n' - r')$, $0 \leqslant r' \leqslant n'$ [see formula (7.9)]:

$$P(x) = \frac{x^{r'-1}(1 - x)^{n'-r'-1}}{B(r', n' - r')} \qquad (11.4)$$

Inserting (11.4), (11.3), and (11.2) in (11.1) and using (7.9), we see that $P(x|D)$ is a beta density with parameters $(r + r', n - r + n' - r')$. Hence, knowing D and knowing the parameters of the prior distribution allows us to write the posterior immediately.

We see that the binomial likelihood function (11.2) and the beta prior (11.4) are natural conjugates, since the posterior is also a beta density. As described in the supplement to Chapter 7, the prior distribution admits a simple interpretation in terms of *equivalent observations*. We saw that observing r heads in n tosses induced a change in the parameters from $(r', n' - r')$ to $(r + r', n + n' - r - r')$. Hence, it is natural to consider the original parameters r' and n' as equivalent to having already observed $(r' - 1)$ heads in $(n' - 1)$ tosses, with the uniform prior distribution which is a beta density with parameters $(1, 1)$. The expectation of (11.4) is r'/n' [Eq. (7.11)]. This is therefore the value we should use to estimate x if we have no new data. After observing D our belief state is equivalent to starting with the uniform prior and observing $(r + r' - 1)$ heads in $(n + n' - 1)$ tosses. The value n' is associated with the variance of (11.4) and indicates how confident we are in the prior assessment. Keeping r'/n' fixed, the variance decreases as n' gets larger, reflecting greater confidence in the assessment r'/n' [Eq. (7.12)].

Winkler's idea is simply this. When an expert gives his probability density function for a variable of interest, we interpret this as an equivalent observation. Of course, this is only possible if his density has the required form. Suppose experts $1, \ldots, E$ give beta densities with parameters $(r_i, n_i - r_i)$, $i = 1, \ldots, E$ for x. We consider first the case where the experts are not weighted. The decision maker starts with the prior of the first expert in (11.1). Each successive expert is then treated as an observation $D_i = $ "r_i heads in n_i tosses," and fed into (11.1). The result is a beta density with parameters $(r, n - r)$ where $r = \Sigma r_i$, $n = \Sigma n_i$.

Notice that n and r can be regarded as weighted sums of the expert's parameters, with weights $w_i = 1$. If we wish to let the advice of some experts weigh more heavily than that of others, we simply choose different (nonnegative) weights and adopt a beta posterior with parameters $r = \Sigma w_i r_i$, $n = \Sigma w_i n_i$.

The w_i's do not sum to 1. Winkler proposes to use $w = \Sigma w_i$ as an index of the degree of dependence between the experts' assessments. If the experts are independent, then we should think of the equivalent observations (r_i, n_i) as coming from disjunct subpopulations. Hence, taken together, the experts correspond to an equivalent observation from a population of size Σn_i, and $w_i = 1, i = 1, \ldots, E$. It is natural to choose E as an upper bound for w. If the experts are maximally dependent, then the equivalent observations should all correspond to the same subpopulation, hence the n_i's and r_i's should all be the same. Of course, this will seldom occur, but Winkler says that $w = 1$ is an appropriate lower bound for w, reflecting maximal dependence.

It should be noted that Winkler's theory is not restricted to beta priors with binomial likelihood functions. Natural conjugates exist for a wide variety of distribution families. The most complete discussion of natural conjugate distributions is found in Raiffa and Schlaifer (1961).

Winkler's natural conjugate theory is attractive both mathematically and conceptually. The notion of an equivalent observation is a handy way of encoding experts' density assessments and discussing possible dependence. The restriction to natural conjugate families is not as severe as it might appear at first sight, as such families offer ample freedom in modeling.

The notion of equivalent observations should give us pause, however. Considering the coin with unknown probability x of heads, an expert who gives a (r, n) beta prior is in the belief state corresponding to perfect ignorance, followed by observing $r - 1$ heads in $n - 1$ tosses. This belief state is then used by the decision maker as if $r - 1$ heads had *actually been observed* in $n - 1$ tosses. The expert's opinion about the data is just as good as the data itself. From the examples in Part I, we may characterize this feature of the theory as foolhardy. However, it would be easy enough to build in additional conservatism in this respect, by suitably reducing the total weight w, perhaps even allowing $w < 1$. It would also be interesting to see how the experts' weights would be affected by real observations. The theory offers resources for generalization in these directions, but such has not been done to date.

An interesting variation on the idea of equivalent observations is Huseby's (1987) idea of "imaginary observations." Huseby's theory uses overlapping imaginary data sets, determined subjectively by the decision maker, to express dependence. The experts' assessments are not constrained to a natural conjugate class. Instead they give quantiles of their distributions, and the quantiles are fitted to distributions from an appropriate natural conjugate class. The theory assumes the experts are well calibrated, and has no facility at present for dealing with possible miscalibration.

Morris' Theory

Morris (1974, 1977) developed a Bayesian theory of expert use that is conceptually important, even if the assumptions underlying the theory are prohibitively strong. The theory attempts to account for past performance and attempts to show how a decision maker can use Bayes' theorem to recalibrate an expert. The theory is of interest because it raises interesting problems, problems which are circumvented in the Bayesian theory presented in Chapter 12. The theory is directed to probability densities. A later article (Morris, 1983) addresses discrete probability functions. *Management Science*, vol. 32, no. 3, 1986, contains a discussion of Morris' theory, particularly the 1983 theory.

Single Expert

We have one expert who gives a density function $f(x)$, with cumulative distribution $F(x)$, for a continuous random variable X. The updating technique requires that f be determined by its mean and variance. Morris assumes that f is a normal density. We can therefore represent the data from this expert as $D = (m, v)$, where m and v are the mean and variance, respectively, of the expert's density. The problem is to determine the decision maker's posterior density $P(x|m, v)$. Once again, we begin with Bayes' theorem,

$$P(x|m, v) = \frac{P(m, v|x)P(x)}{P(m, v)}$$

As $P(m, v)$ does not depend on x, we may absorb this term into a normalization constant. The crucial term to be determined is $P(m, v|x)$. To do this Morris makes two assumptions. The first is *scale invariance*:

$$P(m, v|x) = P(m|v, x)P(v|x) = P(m|v, x)P(v)$$

(The first equality follows from the definition of conditional probability.) Since $P(v)$ does not contain x, we can absorb this term into the proportionality constant as well. Scale invariance says the decision maker's probability that the expert gives variance v is independent of the unknown value of x. This assumption is quite gratuitous. In assessing log failure probabilities, for example, it is prima facie plausible that higher probabilities will be better known and hence have smaller variance Cooke (1986), Martz (1986) confirmed this conjecture.

We determine $P(m|v, x)$ by introducing a change of variable. Instead of asking for the probability density for m, given x and v, we might just as well ask for the probability that the true value x corresponds to the rth quantile of the expert's distribution, given that x is the true value, and given that the expert has chosen variance v. Consider v fixed and introduce the so-called performance indicator function ϕ:

$$\phi(m, X) = \text{quantile of expert's distribution realized by } X$$

$$= F(X|m, v)$$

$$= \int_{-\infty}^{x} f(z|m, v)\, dz \tag{11.5}$$

Note that ϕ is a function of the random variable X. As the mean gets larger with $X = x$ fixed, the quantile corresponding to x gets smaller. In fact,

$$\int_{-\infty}^{m_0} P(m|x, v)\, dm = \int_{\phi_0}^{1} P(\phi|x, v)\, d\phi; \qquad \phi_0 = \phi(m_0, x) \tag{11.6}$$

We now take derivatives of both sides of (11.6) with respect m_0. On the left-hand side we simply get $P(m_0|x, v)$, the term we want to determine. The right-hand side yields

$$-\frac{P(\phi_0|x, v)\, d\phi_0}{dm_0} \tag{11.7}$$

Since f is normal, $df(z|m)/dz = -df(z|m)/dm$ and:[3]

$$\frac{d\phi}{dm} = \frac{d}{dm}\int_{-\infty}^{x} f(z|m, v)\, dz = -\int_{-\infty}^{x} df(z|m, v) = -f(x|m, v).$$

Substituting this into (11.7), and the result into the derivative of (11.6) with respect to $m = m_0$, yields

$$P(m_0|x, v) = P(\phi_0|x, v)f(x|m_0, v) \tag{11.8}$$

Morris now introduces his second assumption, *shift invariance:*

$$P(\phi|x, v) = P(\phi|v)$$

This means roughly the following: for any $r \in [0, 1]$, the probability density of seeing the true value of X correspond to the rth quantile of the expert's distribution is independent of X.

This is a very strong assumption, as a simple example makes clear. Suppose the decision maker is the director of a company, who asks an adviser what the competitor's price will be for the next year. The adviser states his variance v, and retires to consider his mean value m. The decision maker thinks the price will be about \$20, but he isn't very sure. He is confident in his adviser, and thinks there is a probability of $\frac{1}{2}$ that the true price will fall between the 25% and 75% quantiles of whatever distribution the adviser gives. At this moment an industrial spy delivers a memo recently purloined from the competitor in which the competitor's price is revealed to be \$5. The competitor is making a surprise move and starting a price war. Now, if the decision maker's opinion of his adviser satisfies shift invariance, he would still believe that \$5 will fall between the adviser's 25% and 75% quantiles with probability $\frac{1}{2}$. However, if hearing the price \$5 leads him to think that the adviser's 25% quantile will probably be greater than \$5, then shift invariance is violated.

The probability density $P(\phi = r|v)$ is the decision maker's probability density that the true value will realize the rth quantile of the expert's distribution. Morris calls this probability density the *performance function* Φ:

$$\Phi(r): = P(\phi = r|v)$$

Φ is a function of X and the mean m of F. It reflects the decision maker's subjective opinion of how well calibrated his adviser is. If the decision maker believes the expert to be well calibrated, then $\Phi(F(x)) = dF(x)$. Using this notation and shift invariance, (11.8) becomes

$$P(m|x, v) = \Phi(F(x))f(x|m, v) \tag{11.9}$$

---

[3]Note that this step is not possible for other two-parameter families of densities, for example, the beta or the lognormal families.

Hence, the decision maker's posterior after receiving the expert's advice is

$$P(x\,|\,m,\,v) = k\Phi(F(x))f(x\,|\,m,\,v)p(x) \qquad (11.10)$$

where k is a normalization constant. Φ recalibrates the expert's density according to the decision maker's opinion of the expert.

Although (11.9) reflects the decision maker's subjective probability, we know from Chapter 7 that subjective probabilities sometimes track relative frequencies. Hence, it may be possible to find data for determining Φ. Let us write $X = X_n$, and suppose we have variables $X_1, \ldots, X_{n-1}$ and assessments $F_1, \ldots, F_n$ from the expert. Morris considers the supposition that for all $r\in[0, 1]$, the events

$$\{F_i(X_i) \leqslant r\} \qquad i = 1, \ldots, n$$

are exchangeable. This way of speaking is treacherously ambiguous. We must specify exchangeability with respect to a distribution of the decision maker. In the present context, the above events must be considered exchangeable with respect to the decision maker's distribution *conditional* on x_n and v_n, and perhaps on other information $(x_i, m_i, v_i?)$. We recall from Chapter 7 that the decision maker's probability for $\{F_n(X_n) \leqslant r\}$, given the outcomes of the previous events, will approach the relative frequency of occurrence of these events. Knowing this probability, we may determine Φ as

$$\Phi(r) \approx \frac{d}{dr} P(F_n(X_n) \leqslant r\,|\,\text{outcomes of previous events})$$

Multiple Experts

Morris' theory for multiple experts proceeds along the same lines as for the single expert, and we shall not give the derivations. Suppose for convenience that there are two experts giving assessments F and G, characterized by mean values $m = (m_1, m_2)$ and variances $v = (v_1, v_2)$. If the experts are judged to be independent by the decision maker then the generalization of (11.10) is straightforward:

$$P(x\,|\,m,\,v) = k\,\Phi(F(x))\,\Phi(G(x))\,f(x\,|\,m_1v_1)\,g(x\,|\,m_2v_2)\,P(x) \qquad (11.11)$$

If they are judged to be dependent, then the dependence is absorbed into the joint performance function $\Phi(F(x), G(x))$:

$$P(x\,|\,m,\,v) = k\,\Phi(F(x),\,G(x))\,f(x\,|\,m_1v_1)\,g(x\,|\,m_2v_2)\,P(x) \qquad (11.12)$$

Estimating the joint performance function is acknowledged to be a formidable task.

Exchangeability for Multiple Experts

The exchangeability assumption causes problems when more than one expert is involved. We discuss this issue briefly, without invoking the specific assumptions in Morris' model. Suppose X and Y are continuous variables, F_X, F_Y are the cumulative distributions of the first expert, and G_X, G_Y the cumulative distributions of a second expert, for X and Y. Assume that all these distributions, as well as the decision maker's priors, are invertible. Assume that the decision maker knows the experts' distributions, but does not know the true values of X and Y. For all $r\in[0, 1]$, assume that $\{F_X(X) \leqslant r\}$ and $\{F_Y(Y) \leqslant r\}$ are exchangeable with respect to the decision maker's distribution, and similarly for G_X and G_Y. Since exchangeable events have the same probability of occurrence, and since F_X and F_Y are invertible, for all $r \in [0, 1]$,

$$P(F_Y(Y) \leqslant r) = P(F_X(X) \leqslant r) = P(X \leqslant F_X^{-1}r)$$

and a similar equation holds for the G's. Hence:

$$P(F_X(X) \leqslant r) = P(G_X(X) \leqslant G_X F_X^{-1}r) = P(G_Y(Y) \leqslant G_X F_X^{-1}r)$$

Since these hold for all $r \in [0, 1]$,

$$F_Y = F_X G_X^{-1} G_Y$$

In other words, the first expert's distribution for Y is completely determined by his distribution for X and by the second expert's distributions. If the decision maker assumes exchangeability, the first expert need not be consulted for Y! In practice, of course, the last equation will never be satisfied, unless the experts conspire. Similar problems arise if exchangeability is considered conditional on other information. This feature is a serious problem for Bayesian models that attempt to update the decision maker's opinion of his experts on the basis of their past performance.

PSYCHOLOGICAL SCALING

We conclude this chapter with a brief discussion of psychological scaling models. These models are described more fully in Chapter 14. The discussion here serves to set these models off against the weighted combination and Bayesian models, and to review the background literature.

These models can be traced back to Weber's law and Fechner's law for comparing intensities of physical stimuli. They became popular after the pioneering work of Thurstone (1927). Bradley (1953) developed a variant that has found wide application in consumer research. A good modern treatment of Thurstone's models is found in Torgerson (1958), and a thorough mathematical treatment of the Thurstone and Bradley models is found in David (1963). Psychological scaling models require a large number of experts, on the order of 10, and appear attractive in cases where the experts are unfamiliar with numerical assessment.

The models are designed for estimating relative intensities of psychological stimuli on the basis of pairwise comparisons. The stimuli may be anything from beauty to taste to intensities of physical stimuli. The idea of applying this method to the estimation of subjective probabilities seems to have originated with Pontecorvo (1965), and Blanchard, Mitchell, and Smith (1966) (see Humphreys [1988] for additional references). Hunns and Daniels (1980) (see also Hunns, 1982) revived interest by conducting an impressive exercise estimating human error probabilities. Seaver and Stillwell (1983) and Comer et al. (1984) pursued this line. The *Handbook of Human Reliability* (Swain and Guttmann, 1983) endorses the method as potentially fruitful. A good summary of these applications, and comparisons with other methods, is found in Kirwan (1987). Applications to reliability of mechanical components are described in Chapter 14.

The idea behind these models is best explained via a simple example. We deal here only with the simplest model of Thurstone. Suppose a large number of experts are available to assess the ages of four celebrities, Ron, Don, Lon, and John. We assume that each expert has some internal value for each age, and that the experts are unable to verbalize these numbers reliably. However, they are able to determine whether, in their opinion, any given celebrity is older than any other.

We assume that the internal values are normally distributed over the population of experts, with the mean value for each celebrity equal to his true age. We assume these distributions are independent, and all have variance σ^2. Hence, when a decision maker chooses an expert and asks him/her whether Ron is older than John, the decision maker effectively samples from these two independent distributions. If the expert answers "yes" then value drawn for Ron is larger than the value for John. Each expert is queried about each pair of celebrities.

Let R, D, L, and J be independent normal variables distributed as the internal values of Ron, Don, Lon, and John's ages in the expert population, with means $r, d, l,$ and j. Under these assumptions the distributions of scale values may be represented as in Figure 11.1.

The probability that the decision maker "draws" values for D and L such that $D > L$ is determined by the relative spacing of d and l. For the distributions in Figure 11.1, this probability is less than $\frac{1}{2}$, but is substantially greater than the probability of drawing an R and J such that $R > J$. By using the experts' responses to estimate these probabilities, the model yields estimates for the relative spacing of the means $r, d, l,$ and j.

We see here the most conspicuous difference between the psychological scaling models, and the models presented in the previous sections. Psychological scaling models do not lead to numerical estimates, but to scales with one or more degrees of freedom. The assumption of normality in the above model leads to a scale with two degrees of freedom. In Chapter 14 a model is presented for estimating failure probabilities based on the exponential distribution. This leads to a scale with one degree of freedom.

The experts' pairwise comparisons can be examined for consistency (i.e., absence of intransitivities) and concordance. Measures for these quantities will be described in Chapter 14. Measures of goodness of fit have been defined for both the normal and the exponential model. Via simulation it is also possible to generate confidence bounds.

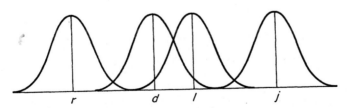

Figure 11.1 Distributions for scale values in population of experts for ages of Ron, Don, Lon, and John, with mean values r, d, l, and j.

The psychological scaling models have two very attractive characteristics:

1. They require only "qualitative" input from the experts. Experience with these models indicates that experts feel comfortable with this type of elicitation and can answer up to 200 paired comparison questions in under an hour.
2. They lead automatically to a sort of "consensus" estimate, with confidence bounds if simulation is used.

Nothing in life is free, and the drawbacks are equally evident:

1. A large number of experts are required.
2. The models make very strong assumptions regarding the experts' psychological assessment mechanisms.
3. The models lead to scales with one or more degrees of freedom, and empirical values are required to derive numerical estimates.

There have been attempts to validate the results of the Thurstone model, as applied to the estimation of (log)human error probabilities. Hunns and Daniels (1980) report excellent agreement with empirical probabilities in a train collision study. Comer et al. (1984) found good agreement between the results of this model and direct estimation. Embrey and Kirwan (1983) and Williams (1983) report substantial product moment correlation between the results of the model and the log probabilities for human errors on simple tasks (the probabilities were determined by experiments on simulators). For more complicated tasks, the correlation was positive, but low.

CONCLUSION

Of the models described in this chapter, the psychological scaling models, in particular the Thurstone model, have seen the most application. The results may be described as guardedly encouraging.

Of the Bayesian models, only the simplest models have actually been applied to real problems, and these applications assume that the experts are independent. None of the models successfully uses the theory of exchangeability to update the decision maker's assessment of the experts. In this sense, the requirement of empirical control has yet to be satisfied. The Bayesian model described in Chapter 13 does satisfy this principle, as well as the other principles discussed in Chapter 5.

The weighted mean models have undoubtedly found many ad hoc applications. Systematic applications are hard to find. In Part I, applications using the arithmetic means with De Groot weights (Hofer, Javeri, and Laffler, 1985) and self-weights (Bernreuter et al., 1984) were mentioned. Such weights raise serious problems with respect to accountability, neutrality, and empirical control. The (unnormalized) geometric mean with equal weights was used in the *IEE Guide to the Collection and Presentation of Electrical, Electronic, and Sensing Component Reliability Data for Nuclear Power Generating Stations* (1977). Given the simplicity and intuitive appeal of the arithmetic weighted mean model, it would be desirable to have a theory of weights which satisfied the principles of Part I.

12

The Classical Model

The classical model constructs a weighted combination of expert probability assessments. The weights are based on the theory of proper scoring rules as described in Chapter 9, and reward good calibration and low entropy (or equivalently, high information) with calibration playing a dominant role. The designation "classical" derives from a close relation between calibration scoring and hypothesis testing in classical statistics. The variables assessed may be uncertain quantities with a discrete or continuous range of possible values. For continuous variables, the model is based on eliciting quantiles from the experts' distributions. Although this chapter draws on results from previous chapters, it is written to stand alone. Hence definitions of many concepts familiar from previous chapters are repeated here.

The model is designed to satisfy the principles of reproducibility, accountability, empirical control, neutrality, and fairness, as set forth in Chapter 5. The results obtained in Chapter 9 show that the weights encourage honesty in the experts, at least in the long run, as required by the principle of neutrality. The use of a calibration score builds in the element of empirical control. The other principles are satisfied simply by making the model and the data explicit. The analyst is not required to judge experts, or to judge distributions of variables. He has only to follow heuristic guidelines. The model also provides a mechanism for evaluating performance, either of the experts, of the model itself, or of other combination models. On the basis of experience to date as discussed in Chapter 15, the model outperforms the "best expert" and outperforms the simple arithmetic average of experts' assessments. Initial development was sponsored by the Dutch Ministry for environment, and later phases have been supported by the European Space Agency and the European community. The model lends itself easily for computer implementation and can handle large expert problems.

The basic model is described both for uncertain events and continuous variables. Subsequently, variations and possible enhancements are discussed. The procedure for optimizing the decision maker's distribution is explained, and the problem of correlated experts is discussed. A final section draws conclusions.

NOTATION AND DEFINITIONS

By a "probability distribution" we shall mean either a probability mass function (in the case of discrete variables) or a probability density function (in the case of continuous variables). A probability distribution is not to be confused with a cumulative distribution function. "ln" denotes the natural logarithm.

Let $p = p_1, \ldots, p_n$ be a probability distribution over alternatives $\{1, \ldots, n\}$. The *entropy* or *negative information* $H(p)$ of p is defined as

$$H(p): = -\sum_{i=1}^{n} p_i \ln p_i \tag{12.1}$$

$H(p)$ assumes its minimal value 0 if $p_i = 1$ for some i, and assumes its maximal value $\ln n$ if $p_i = 1/n$, $i = 1, \ldots, n$. $H(p)$ is commonly regarded as an index of the lack of information in the distribution p, high values of H indicate low informativeness.

Let $q = q_1, \ldots, q_m$ be another distribution, and let pq denote the product distribution. pq is the joint distribution of two independent variables having marginal distributions p and $\dot{q}$. It is easily verified that

$$H(pq) = H(p) + H(q) \tag{12.2}$$

Equation (12.2) generalizes for products of finitely many independent variables.

Let $s = s_1, \ldots, s_n$ be a probability distribution, and assume $p_i > 0$, $i = 1, \ldots, n$; then the *relative information of s with respect to p* $I(s, p)$ is defined as

$$I(s, p) := \sum_{i=1}^{n} s_i \ln \frac{s_i}{p_i} \tag{12.3}$$

$I(s, p)$ is always nonnegative and $I(s, p) = 0$ if and only if $s = p$. $I(s, p)$ is commonly taken as an index of the information learned if one initially believes that p is correct and subsequently learns that s is correct. It is easy to verify that

$$H(p) = \ln n - I(p, u) \tag{12.4}$$

where u is the uniform distribution over $1, \ldots, n$; that is, $u_i = 1/n$, $i = 1, \ldots, n$.

Let s denote a sample distribution generated by N independent samples from the distribution p. Let χ_d^2 denote the cumulative distribution function a chi square variable with d degrees of freedom. Then

$$\text{Prob}\{2NI(s, p) \leqslant x) \to \chi_{n-1}^2(x) \qquad \text{as } N \to \infty \tag{12.5}$$

(Equation 12.5) says that the statistic $2NI(s, p)$ is asymptotically χ^2-distributed as the number N of independent samples from p goes to infinity. The sum of independent χ^2 variables is again a χ^2 variable and the number of degrees of freedom of the sum is the sum of the numbers of degrees of freedom.

BASIC MODEL; UNCERTAIN EVENTS

We imagine that experts assess the probabilities of M uncertain events with indicator functions $1_1, \ldots, 1_N$. The events must be chosen such that they are independent in the opinion of the user. Whenever possible, the events should be defined such that within a reasonable time it is unambiguously clear whether the events have occurred. We assume for purposes of exposition that this is the case for all events.

A set of experts $e = 1, \ldots, E$ assesses the probability of each uncertain event by assigning the corresponding indicator functions to one of B probability bins. Each probability bin is associated with a distribution over the possible outcomes "occurred" and "not-occurred." The bins will be characterized by the probability p_b of occurrence, $1 > p_b > 0$, $b = 1, \ldots, B$. When no confusion can arise, p_b is also used to denote the distribution associated with the bth bin. $p_e(m)$ denotes the probability associated with the bin to which expert e assigns variable 1_m.

On the basis of the observed values of $1_1, \ldots, 1_N$ and the experts' assignments, weights w_e will be determined for each expert $e = 1, \ldots, E$, satisfying

$$w_e \geqslant 0 \qquad \Sigma w_e = 1 \qquad \text{if } w_e > 0 \text{ for some } e \tag{12.6}$$

These weights will then be used to determine the decision maker's probability P for subsequent, as yet unobserved uncertain events 1_{N+k}

$$P\{1_{N+k} = 1\} = \sum_{e=1}^{E} w_e p_e\{1_{N+k} = 1\} \tag{12.7}$$

assuming that w_e is positive for some e. Modeling parameters can always be chosen to assure that this is the case. These weights are *global* in the sense that the same weights apply to all "unobserved variables" 1_{N+k}, $k = 1, \ldots$. They are also dynamic in the sense that with each new observation the weights can recomputed.

To define the weights w_e we restrict attention for the moment to one expert e. The following notation refers to the assignments of expert e for the variables $1_1, \ldots, 1_N$, and the dependence on e and N is suppressed in the notation. Let

$$n_b = \text{number of variables assigned to bin } b; \ b = 1, \ldots, B \tag{12.8}$$

$$n = (n_1, \ldots, n_B) \qquad N = \Sigma n_b$$

$$s_b = \text{sample distribution of variables in bin } b; \ b = 1, \ldots, B$$

$$s = (s_1, \ldots, s_B)$$

$$S = \frac{1}{N} \Sigma n_b s_b; \text{ total sample distribution.}$$

$$C(e) = 1 - \chi_B^2[\Sigma \, 2n_b I(s_b, p_b)] = \text{calibration score} \tag{12.9}$$

$$H_e(n) = \frac{1}{N} \Sigma \, n_b H(p_b) := \text{average response entropy} \tag{12.10}$$

$$1_\alpha(x) = 1 \text{ if } x \geqslant \alpha, \text{ and } = 0 \text{ if } x < \alpha, \ \alpha \in (0, 1) \tag{12.11}$$

$$w'_{e\alpha} = \frac{1_\alpha(C(e)) \times C(e)}{H_e(n)} \tag{12.12}$$

$$w = \Sigma \, w'_{e\alpha}$$

$$w_{e\alpha} = \frac{w'_{e\alpha}}{w}$$

The quantity $C(e)$ takes values close to 1 if e's calibration is very good and takes values close to 0 if e's calibration is very poor. The weights (12.12) reward good calibration and low response entropy, and the calibration score dominates the entropy score, as noted in Chapter 9. For all $\alpha \in (0, 1)$ the weights $w'_{e\alpha}$ are weakly asymptotically strictly proper, as discussed in Chapter 9 [see (9.12)]. This is the basic weight recommended for use in the classical model with uncertain events. It should be emphasized that the choice of the calibration term in (12.12) is largely determined by the requirement that the weights should have desirable asymptotic properties. The choice of the information term is determined by intuitions regarding the representation of high informativeness (see Chapters 8 and 9).

Note that an expert can receive zero weight. This occurs if the probability of seeing an overall deviation between the sample distributions and bin probabilities greater than $I(s_1, p_1), \ldots, I(s_B, p_B)$ is less than α, under the assumption that the items are independent and distributed according to the probabilities of the bins to which they are assigned. This is equivalent to saying that the user regards each expert as a statistical hypothesis (namely that the variables are independent and that their marginal probabilities correspond to the expert's assessments). The close relation to classical hypothesis testing accounts for the designation "classical model." It is essential for the propriety of the weights (12.12) that $\alpha > 0$, that is, that zero weight is a real possibility.

This having been said, it must be emphasized that weighing experts differs from testing statistical hypotheses in two respects, as discussed in Chapter 9. First, the decision maker is not only interested in the calibration of his experts, but also in their informativeness. Second, the significance level α is not chosen in the same way as in hypothesis testing. This aspect is explained under "Optimization; Virtual Weights" below.

BASIC MODEL; CONTINUOUS VARIABLES

In many applications it may be more natural to work with (practically) continuous variables. Experts are asked to give fixed quantiles for continuous variables. Calibration and information scores will be used to define weights, which, in turn, will be used to define global, dynamic weights for combining expert assessments in a manner similar to (12.7) above. We adhere to the convention of denoting random variables with capital letters, and their possible realizations with lowercase letters.

$X_1, \ldots, X_N$ practically continuous variables $\hspace{4em}$ (12.13)

$f_1, \ldots, f_R; 0 \leqslant f_1 < \cdots < f_R \leqslant 1$; quantiles elicited from experts' distributions; we adopt the convention that $f_0 = 0$, $f_{R+1} = 1$.

Q_{ie}; $i = 1, \ldots, N$; $e = 1, \ldots, E$; cumulative minimal information distribution for X_i satisfying the constraint that the quantiles $f_1, \ldots, f_R$ agree with expert e's assessment for variable i (strictly speaking, this definition is meaningful only if the 0th and 100% quantiles are finite). If x_{ire} is the f_rth quantile of expert e for X_i, then $Q_{ie}(x_{ire}) = f_r$.

$[x_{i0}, x_{iR+1}]$; intrinsic range for variable X_i; x_{i0} and x_{iR+1} are called the lower and upper cutoff points for X_i, respectively, and they must satisfy $x_{i0} < x_{ire} < x_{iR+1}$, for all $r = 1, \ldots, R$, and all $e = 1, \ldots, E$. We adopt the notation: $x_{i0e} = x_{i0}$; $x_{iR+1e} = x_{iR+1}$.

$p_r = f_r - f_{r-1}$, $r = 1, \ldots, R+1$; "theoretical probability" associated with the event $Q_{ie}(X_i) \in (f_{r-1}, f_r]$; the event $Q_{ie}(X_i) \in (f_{r-1}, f_r]$ is termed "probability outcome r."

$p = (p_1, \ldots, p_{R+1})$

χ_R^2 = cumulative χ^2 distribution function with R degrees of freedom

It will be observed that each variable must be supplied with an "intrinsic range," containing all the quantiles elicited from the experts. In some cases the choice of the cutoff points might be motivated, for example, if X_i were a relative frequency or a percentage, then 0 and 1 might provide suitable cutoff points. However, in many cases the choice of cutoff points must be made ad hoc. This is one point at which the analyst must simply make a decision. The choice affects only the measure of information, and model performance is quite robust with respect to this choice. In the computer implementation used to analyze the data in Chapter 15, a simple "10% overshoot" above and below the interval generated by the set $\{x_{ire} | e = 1, \ldots, E; r = 1, \ldots, R\}$ is used. Because the intrinsic range depends on the assessments of all experts, the information score of a given expert may change slightly as experts are added or removed.

When the value x_i is observed, then for each expert exactly one probability outcome is "hit." In this way, observations of x_i; $i = 1, \ldots, N$, generate a sample distribution over the probability outcomes for each expert. The sample distribution depends on N, but we suppress this in the notation as N is considered fixed. For the notation in (Equation 12.14) we restrict attention to a single expert e.

$s = (s_1, \ldots, s_{R+1})$ sample distribution over probability outcomes r,

$r = 1, \ldots, R+1$ (12.14)

$C(e) = 1 - \chi_R^2[2NI(s, p)]$

$$I(e) = \frac{1}{N} \sum_{i=1}^{N} \left[\ln(x_{iR+1} - x_{i0}) + \sum_{r=1}^{R+1} p_r \ln \frac{p_r}{x_{ire} - x_{ir-1e}} \right]$$

$w_e' = C(e) \times I(e) \times 1_\alpha(C(e))$ (12.15)

$W = \Sigma w_e'$

$w_e = \dfrac{w_e'}{W}$ assuming $W > 0$

$I(e)$ is the average over $i = 1, \ldots, N$ of the relative information in the densities Q_{ie} with respect to the uniform distribution over the intrinsic range for variable X_i (see the discussion in Chapter 8 following footnote 2). The discussion of the weights (9.12) applies to (12.15) above. $I(e)$ must be bounded and bounded away for 0 to be weakly asymptotical strictly proper in the sense of Proposition 9.6.

VARIATIONS AND ENHANCEMENTS

The principles underlying the basic model sketched above have been discussed in the foregoing chapters. They include the principles for expert opinion in science, Savage's axioms for rational preference, the theory of proper scoring rules, and the properties underlying the choice of the weighted arithmetic average combining rule discussed in the previous chapter. However, we cannot derive a model for combining expert opinions exclusively from "first principles." At several points in the preceding chapters alternative modeling approaches were indicated. The present section is not devoted to exploring all such possible alternatives; rather three variations and/or enhancements are mentioned, which seem particularly important and which can be evaluated in light of the experiences in Chapter 15.

Item Weights

The weights (12.12) and (12.15) above are global. It would be possible to replace the average entropy or relative information with respect to the uniform distribution U_i by terms which measure the (lack of) information for each variable X_i separately. This is most natural with regard to continuous variables. In this case the term $I(e)$ in (12.15) would be replaced by

$$I(Q_{ie}, U_i) = \ln(x_{iR+1} - x_{i0}) + \sum_{r=1}^{R+1} p_r \ln \frac{p_r}{x_{ire} - x_{ir-1e}}$$

yielding the unnormalized weight

$$w'_{ie\alpha} = C(e) \times I(Q_{ie}, U_i) \times 1_\alpha(C(e)) \tag{12.16}$$

for each X_i, $i = 1, \ldots$. Normalization is performed per item in the obvious way. This allows an expert to downweight or upweight himself on individual items, according as his quantiles are further apart or, respectively, closer together.

It is hardly a foregone conclusion that the use of item weights would result in better assessments for the decision maker. In the engineers experiment discussed in Chapter 10 a negative correlation between calibration and information was found. It may well be that experts are less well calibrated on just those items for which they are more informative. If so, the decision maker's calibration could be degraded by the use of item weights. The results presented in Chapter 15 fail to provide convincing evidence in favor of item weights.

Uniform Versus Logarithmic Scale

The basic model for continuous variables appeals to the uniform distribution as a "background measure" in two places. First, the distributions Q_{ie} are chosen to be minimally informative with respect to the uniform distribution, consistent with e's quantile assessments for X_i. Second, the information scores $I(e)$ or $I(Q_{ie}, U_i)$ are computed with respect to the uniform distribution. Any other measure could in principle serve as a background measure. If X_i takes values very close to 0 or very large, it may be more natural to consider the possible values of X_i on a logarithmic scale. That is, we could replace the variable X_i by its log transform $\ln X_i$. This is equivalent to using the log uniform distribution as a background measure. Of course, the same scale must be used for all experts.

Such a choice is not without consequences and should not be made lightly. For example, suppose unit probability mass is to be distributed over the interval of percentages (1%, 100%). If the distribution is uniform, then the intervals (1%, 25%) and (75%, 100%) each receive a quarter of the mass. If the distribution is log uniform, then the intervals (1%, 3.16%) and (31.6%, 100%) each receive a quarter of the mass. The log uniform distribution concentrates mass at lower values than the uniform distribution. In the absence of further information the following is suggested: Choose a uniform scale if the realization is expected to lie between 0.001 and 1,000; and choose otherwise a logarithmic scale.

Numerical Accuracy/Discounting Calibration

It seems strange at first sight to couple these two notions. However, they are very much related, as numerical accuracy limits our ability to distinguish calibration scores. Any numerical routine has a limited accuracy. In implementing the basic model for continuous variables, the chi square distribution has been computed to four significant places. This entails that calibration scores down to 10^{-4} can be distinguished, and that 10^4 is the greatest possible ratio of calibration scores.

As the number of realizations increases, the calibration scores tend to go down. If we think of the calibration score as a significance test, then increasing the number of realizations increases the power of the test. Suppose we test the hypothesis that the probability of heads equals $\frac{1}{2}$, by tossing a coin whose probability of heads is really $\frac{2}{3}$. The hypothesis may look reasonable after 15 tosses, but would probably look very unreasonable after 500 tosses. Hence, by increasing the number of realizations without limit, sooner or later every expert who is not *perfectly* calibrated will receive the lowest computable calibration score (in the present implementation, 10^{-4}).[1] At this point, of course, calibration scores are no longer distinguished, and the weights can depend only on information scores.

If we wish to restore the dominance of calibration over information, then there are two mathematical techniques at our disposal for this purpose. First, and most evident, we could extend the accuracy of the numerical routines, enabling lower calibration scores to be distinguished. Second, we could reduce the power of the test, replacing $2NI(s, p)$ in (12.14) by $2N'I(s, p)$, for $N' < N$. The ratio N'/N will be called *the power of the calibration test*, and N' is the *effective number of realizations at power level N'/N*.

Examples in Chapter 15 illustrate that these two techniques are roughly equivalent mathematically. However, the second is preferable. Not only is it easier to implement, but it also draws attention that the *degree to which calibration scores are distinguished is a modeling parameter, whose value must be determined by a decision of the analyst*. Failure to realize this may lead to poor model performance. For example, suppose that calibration scores are distinguished down to 10^{-12}, and

---

[1]We may have to wait a long time. In the Dutch meteorological experiment described in Chapter 15, many calibration scores were greater than 10^{-3} after some 2500 realizations. In quantile tests this limit seems to be reached more quickly.

that on a modest set of realizations one expert achieves calibration score 10^{-6} while all others receive the minimal score. The model would effectively assign all weight to the best-calibrated expert, and would produce a poorly calibrated decision maker. The point is, it may be imprudent to let a very poorly calibrated expert dominate other experts who are even worse. In fixing the numerical accuracy of the routines we effectively limit the ratio of calibration scores. For example, in the present implementation the accuracy is 10^{-4}; hence a calibration score of 10^{-3} can never dominate by a factor greater than 10, etc. If low scores are being caused by a large number of realizations, then we might *improve* model performance by reducing the effective power of the test; hence we should have to choose an optimal power level. Examples will be discussed in Chapter 15.

ISSUES

We discuss three issues that may arise in the application of the basic model.

Seed Variables

It will often arise that the decision maker needs assessments for events, none of which will be observed within a required time frame. This typically occurs in risk analysis, where probabilities for unlikely and nonrepetitive events must be assessed. In this case the model must be "seeded" with other events, whose outcomes are known, or become known within a short time. These seed variables must be drawn from the experts' area of expertise, but need not pertain to the problem at hand. Weights are then determined on the basis of seed variables and used to define the decision maker's distributions for the variables of interest. The choice of meaningful seed variables is difficult, and critical.

The number of seed variables required depends on the number of bins and on the bin probabilities p_b. According to standard statistical practice, the chi square approximation for $2n_b I(s_b, p_b)$ is acceptable if

$$n_b p_b \geqslant 4, \; n_b(1 - p_b) \geqslant 4$$

The decision maker is not testing hypotheses, but combining assessments. His concern is that $C(e)$ distinguish well- from less-well-calibrated experts, and that w'_e be asymptotically proper. The accuracy of the chi square approximation is therefore less crucial than in hypothesis testing. Of course, one could forego the chi square approximation and calculate the distribution of $2n_b I(s, p_b)$ explicitly. Even when this is done, however, the number of seed variables required with uncertain events compares unfavorably with that for continuous variables when three quantiles are elicited. In this case, 8 to 10 seed items seems sufficient to observe substantial differences in calibration. Of course, the fewer the number of realizations, the less robust the calibration scores are likely to be.

In the more recent applications of the classical model, considerable effort has been put into the identification of meaningful seed variables. This effort is felt to pay off; not only can such seed variables be found, they greatly enhance confidence in expert judgment generally, and in the classical model in particular.

Disaggregation

If the number of uncertain items is large, it may become possible to disaggregate the variables into distinct sets and compute weights for each such set. This makes sense whenever there are enough realizations, and when the scores substantially differ between sets of variables. The Dutch meteorological experiment discussed in Chapter 15 affords examples where disaggregation was worthwhile.

Experts Assessing Different Seed Items

In practice it often arises that experts have assessed variables in the past for which realizations are now available. Such variables can be used as seed variables. However, it will frequently happen that the seed variables are different for different experts. The classical model can function in this situation, by taking the following points into account:

- The effective power of the calibration tests should be set equal to the smallest number of items which any expert has assessed.
- For uncertain events, if the base rates over the seed variables are different for different experts, $1/H(e(n))$ in (12.12) should be replaced by $1/N \sum n_b I(p_b, S)$ (summation over $b = 1, \ldots, B$, where S depends on the expert e.), where S is the overall sampling distribution.
- For continuous variables, the intrinsic range must be determined by the analyst; if this cannot be done meaningfully, then either item weights must be used, or information must not be considered at all.

MEASURING AND OPTIMIZING PERFORMANCE; VIRTUAL WEIGHTS

For each choice of α in (12.12) or (12.15), a distribution P_α of the decision maker is defined. This distribution can also be scored with respect to calibration and entropy. Hence, an unnormalized "virtual weight" $w_{dm}(\alpha)$ for the decision maker is defined as a function of α. This is the weight that a "virtual expert" would receive when he gives the decision maker's distribution, and is scored along with the actual experts. The virtual weight is always global, even when the experts are combined via item weights.

Suppose we add the virtual expert to the set of experts and calculate the decision maker's probability anew. Let P_{dm} be the decision maker's probability based on the original experts and P' the new probability. Where P_e denotes the distribution of expert e,

$$P_{dm} = \frac{\sum w_e P_e}{\sum w_e}$$

$$P' = \frac{\sum w_e P_e + w_{dm} P_{dm}}{w_{dm} + \sum w_e}$$

Substituting the first equation above into the second, we see that $P' = P_{dm}$. Hence adding the virtual expert to the set of experts would not result in a new distribution for the decision maker.

Moreover, since the decision maker's virtual weight depends on the significance level α, we can choose α so as to maximize the decision maker's virtual weight,[2] thereby defining a unique set of weights $w_e = w_{e\alpha}$, where

$$\alpha' = \underset{\alpha \in (0,\,1)}{\mathrm{argmax}}\, w_{dm}(\alpha)$$

The optimal virtual weight of the decision maker can be compared with the global weights of the experts, or with the virtual weight of other decision makers generated by other combinations of the experts' opinions. For example, this allows us to compare the performance of the basic model with the use of item weights.

The use of virtual weight optimization to determine α is a significant feature of the classical model. It underscores the difference between forming weighted combinations of expert opinions and classical hypothesis testing. Of course, we could also simply choose weights for experts that optimize the calibration and information scores of the decision maker. However, this optimization problem is mathematically untractable on all but very small sets of experts. It is very nonrobust, and experts' weights would, in general, have no relation to the quality of their assessments. In the language of optimization, the restriction to weights that are asymptotically strictly proper scoring rules introduces a constraint that renders the optimization tractable.

In all applications performed to date, the optimized decision maker under the basic classical model has consistently outperformed the best expert, and also outperformed the decision maker gotten by simple arithmetic averaging of the expert's distributions. This, however, is a fact of experience and not a mathematical theorem, as the following section illustrates.

CORRELATION

In the foregoing chapters attention has been drawn to the fact that expert assessments are frequently correlated. It is therefore appropriate to address this issue.

Expert assessments (either 'within' or between experts) will generally be correlated, as they are based on common sources of information. Such correlation is usually benign, and always unavoidable. Another sort of correlation arises when experts conspire to influence the decision maker by giving assessments at variance with their true opinions.

For example, if experts conspire to give the same distribution for all variables, then they will obviously receive the same score, and the weight assigned to their assessments will be unfairly multiplied by the number of experts in the conspiracy. The weights encourage honesty, but if, in spite of this encouragement, the experts give dishonest assessments, the model is powerless to redress this. However, the

---

[2]For this idea I am indebted to Simon French.

optimization feature in choosing the significance level will assign these assessments weight 0 if, indeed, the decision maker is better off without them on the seed variables. In discussing correlation we therefore assume that the experts all respond honestly.

A peculiar sort of correlation can actually degrade model performance, such that the decision maker's virtual weight is lower than the experts' weights. This phenomenon has never been observed in practice, but it is of theoretical interest.

To illustrate, assume there are two experts and four uncertain events. The events with indicators x_1 and x_2 occur, and the events with indicators y_1 and y_2 do not occur. The experts assign these events to 40% and 60% probability bins, as shown in Table 12.1. Obviously, the experts' calibration and entropy scores will coincide, so they will each receive weight $\frac{1}{2}$. The decisions maker's assignment can be easily derived and is also shown in Table 12.1. It is clear that the decision maker's distribution is more entropic, and it is also clear that he will be less well calibrated than either expert.

Table 12.1 An Example in which the Decision Maker is Less Well Calibrated and More Entropic than Either Expert.

	Probability Bin		
	40%	50%	60%
Expert 1	x_1, y_1		x_2, y_2
Expert 2	x_1, y_2		x_2, y_1
Decision maker	x_1	y_1, y_2	x_2

The x's occur and the y's do not occur. Each expert receives weight $\frac{1}{2}$.

There are two striking features of this example: (1) The experts are equally well calibrated (so that the optimal choice of α cannot lead to one of them being rejected), and (2) their assessments are positively correlated for the events that occur and negatively correlated for those which do not occur.

CONCLUSIONS

The classical model is not a mathematically closed theory. One could hardly expect to derive a theory of weights from first principles. It is a practical tool and must be judged as such. Once the principles for applying expert opinion in science have been satisfied, the only important question is "does it work?" Applications are discussed in Chapter 15. We may conclude this chapter with a few general remarks.

A fundamental assumption of the classical (as well as the Bayesian) model is that the future performance of experts can be judged on the basis of past performance. The success of any implementation depends to a large measure on defining relevant variables whose true values become known in a reasonable time frame. This requires resourcefulness on the part of the analyst as well as the sympathetic cooperation of the experts themselves. It is essential that the experts understand the model and generally appreciate its potential usefulness.

Experts may have biases. Their biases may be expected to fall into two general categories, probabilistic biases and domain biases. Probabilistic biases, such as the base rate fallacy, anchoring, overconfidence, representativeness, involve the misperception of probabilities. Domain biases are connected with individuals' preferences relating to their specific fields. An expert may be "sold" on a particular design (e.g., his own), and may have a visceral distrust of other designs. In principle, the model can deal with both types, though the latter is much more difficult to identify and verify. Identification will generally require knowledge of the individuals involved, and verification will require substantial specific data. Once identified, a domain bias may be neutralized by a judicious disaggregation of the data set.

The whole question of domain biases should be approached gingerly. In a sense, domain biases are the essence of expert opinion. One uses more than one expert in the hope that such biases will interfere destructively. The user must not embark on a crusade to eliminate domain biases. If he suspects a significant domain bias but lacks the data to verify it, he must under no circumstances attempt to neutralize the bias via a post hoc choice of model parameters. This would compromise his objectivity. The user is not himself an expert and must not choose sides in professional squabbles.

Finally, we emphasize that the classical (as well as the Bayesian) model requires experts with some implicit insight into probability theory and some facility in estimating numerical values. The assessment task may involve considerable "cognitive depth." If the experts have little training or feeling for this type of task, the psychological scaling models may be more appropriate.

13

The Bayesian Model

Bayesian models require that the user supply prior probability distributions and process expert assessments by updating these distributions via Bayes' theorem. The model of Mendel and Sheridan (1986, 1989; see also Mendel, 1989) uses Bayes' theorem to recalibrate and combine expert assessments of continuous variables via Bayes' theorem.

The only other model (Morris, 1977) which putatively accomplishes the same has serious computational and philosophical drawbacks (see Chap. 11). The Mendel-Sheridan model compares favorably with other Bayesian models in the following respects:

1. Default egalitarian prior distributions are given by the model itself.
2. The expert assessments are not restricted to a given class of distributions, but rather the experts give quantiles of their distributions for continuous variables (which makes this model compatible with the classical model).
3. Bayes' theorem automatically accounts for correlations in the expert assessments—correlation coefficients need not be assessed by the decision maker.
4. The computational algorithm has been streamlined for ease of implementation.
5. The model has demonstrated its worth in experiment.

The Bayesian model is compatible with the continuous variable version of the classical model.

Because of the features mentioned above, the Mendel-Sheridan model fully conforms to the principles formulated in Chapter 5. There are features of this model that restrict its practical application at this moment, and these will become apparent in the following.

In addition to the model of Mendel and Sheridan, a variant involving the notion of partial exchangeability will be presented below. The variant repairs a theoretical shortcoming in the Mendel-Sheridan model, but is still under development at present.

This chapter is written jointly with M. Mendel. Comments of Simon French on a previous draft of this chapter are gratefully acknowledged.

BASIC THEORY

The Two-Expert Problem: Motivation

In order to motivate the concepts used in the Mendel-Sheridan model we first discuss the two-expert assessment problem. Suppose that two experts each assess the distribution of a continuous uncertain quantity and report their 5%, 50%, and 95% quantiles. Each expert's quantiles divide the real line into four "probability outcomes." An expert's probability of the true value falling beneath his 5% quantile, and thus hitting the first probability outcome, is 5%, etc.

With only one expert, the Bayesian model presented below will recalibrate the expert such that the decision maker's probability for each probability outcome will approach the observed relative frequency with which the outcome is hit.

When a second expert comes into play the situation for the decision maker becomes much more complex. We assume that the experts' quantiles never coincide. Then, instead of three distinguished points on the real line, the decision maker now has six. These give rise to seven "probability outcomes"; and the decision maker's choice of prior and posterior probabilities for these outcomes is quite subtle. In order for the decision maker to extract all information from the assessments, he must distinguish the relative placements of the experts' quantiles. For example, if we let "|" denote the quantiles of the first expert, and "|" denote those of the second, then one possible arrangement of quantiles is shown in Figure 13.1. The "x" indicates the true value. The pair of numbers above each of the seven outcomes indicates the numbers of the outcomes of the first and second expert, respectively, which would be hit if the true value should fall in that outcome. The true value shown below falls into the third outcome of the first expert, and into the second outcome of the second expert.

The decision maker will have to distinguish the various ways in which the experts' quantiles could be arranged. An easy way to do this is to consider a 4 by 4 matrix, as shown in Figure 13.2, in which the possible probability outcomes of the first expert are set out against those of the second. A possible arrangement of quantiles corresponds to a path through this matrix that starts in the upper left corner [the (1, 1) cell], ends in the lower right corner [the (4, 4) cell], and that leaves each cell via an edge touching the lower right corner of the cell. The cells on a given path correspond to the possible combinations of outcomes of the two experts. The path corresponding to the arrangement of quantiles in Figure 13.1 is shown as " ▨ " in Figure 13.2. The true value is shown as hitting the (3, 2) cell.

There are 20 distinct paths in total, and all of them pass through the upper left and the lower right cells. Only one path passes through cell (4, 1) and only one passes through cell (1, 4). In Figure 13.3, the number in each cell corresponds to the number of paths passing through that cell.

The decision maker is going to use these experts to assess a number of

(1, 1) | (2,1) | (3,1) | (3,2) x | (3,3) | (4,3) | (4, 4)

Figure 13.1

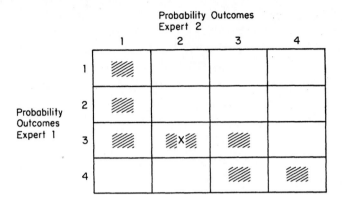

Figure 13.2

variables and is going to learn from experience. The question is, experience with respect to what? Roughly speaking, the Mendel-Sheridan model considers the above matrix as a random variable whose possible values are the cells that might be hit. Before obtaining the experts' advice on the current variable, the decision maker uses the theory of exchangeability to update his probability distribution for this random variable. The variant proposed below introduces exchangeability assumptions only *after* the experts' assessments for the current decision variable are known, that is, only after a path has been chosen. In the following the notation and definitions for treating the general case of E experts and R quantiles is introduced. It will be convenient to illustrate the concepts with reference to the above figures.

NOTATIONS AND DEFINITIONS

The notation used here agrees with that in the classical model, and agrees as much as possible with the notation in Mendel and Sheridan. It is convenient to introduce an underlying sample space Ω with generic element ω. The σ-field of measurable

	Probability Outcomes Expert 2			
	1	2	3	4
1	20	10	4	1
2	10	12	9	4
3	4	9	12	10
4	1	4	10	20

Probability Outcomes Expert 1

Figure 13.3 Number of paths passing through each cell.

events will not be specified, and it is assumed that all variables are measurable. The sample point ω also determines the advice of the experts, as, indeed, the user is uncertain what this advice will be before hearing it. The functional dependence of advice on the sample point will generally be suppressed in the notation.

$\{1, \ldots, E\}$ set of experts;   (13.1)

$\Omega = $ underlying sample space $\omega \in \Omega$, ω a sample point;

$X_i : \Omega \to R$; $i = 1, \ldots, N$ (practically) continuous variables;

$X_0 : \Omega \to R$; (practically continuous) current decision variable;

Q_{ie}; $e = 1, \ldots, E$, $i = 0, \ldots, N$; expert e's cumulative distribution function for X_i;

$f_1, \ldots, f_R$; quantiles elicited from each expert for each variable;

$0 = f_0 < f_1 < \cdots < f_R < f_{R+1} = 1$;

$p_r = f_r - f_{r-1}$; theoretical probability for the "probability outcome" $Q_{ie}(X_i) \in (f_{r-1}, f_r)$, $r = 1, \ldots, R+1$.

$a_{ie} = x_{i1e}, \ldots, x_{iRe}$; where $Q_{ie}(x_{ire}) = f_r, r = 1, \ldots, R$; a_{ie} is expert e's advice for X_i;   (13.2)

$a_i = a_{i1}, \ldots, a_{iE} = $ advice from experts $1, \ldots, E$ for X_i;

$\Phi = \{(r_1, \ldots, r_E) \mid r_e \in \{1, \ldots, R+1\}, e = 1, \ldots, E\}$; Φ corresponds to the set of cells in the matrix of Figure 13.2.   (13.3)

$Z_i : \Omega \to \Phi$; $i = 0, \ldots, N$, where $Z_i(\omega) = (r_1, \ldots, r_E) \leftrightarrow X_i(\omega)$ hits the r_eth probability outcome of expert e; $e = 1, \ldots, E$.

Let the elements of Φ be lexicographically[1] ordered by $j = 1, \ldots, (R+1)^E$. We shall write $Z_i(\omega) = j$ to mean $Z_i(\omega) = $ the jth element of Φ in the lexicographical ordering.

$Y_j(\omega) = |\{Z_i(\omega) = j \mid i = 1, \ldots, N\}|$; Y_j counts the number of times that j has been observed for the variables preceeding the current decision variable   (13.4)

$Y = Y_1, \ldots, Y_{(R+1)}E$

$P = $ the decision maker's probability.

THE MENDEL-SHERIDAN MODEL; ASSUMPTIONS

The Mendel-Sheridan model makes a number of modeling assumptions relating the quantities defined above.

$$Z_0, \ldots, Z_N \text{ are exchangeable with respect to } P. \qquad (13.5)$$

This implies, among other things, that:

$$P(Z_i) = P(Z_j) \qquad i, j = 0, \ldots, N \qquad (13.5a)$$

$$P(Z_0 \mid Z_1, \ldots, Z_M) = P(Z_0 \mid Y). $$

$P(Z_0)$ is shorthand for $P(Z_0 = j)$ for j arbitrary, etc. It must be borne in mind that

---

[1] Ordered pairs of reals are lexicographically ordered when $(x, y) \geq (u, v)$ if $x > u$ or $x = u$ and $y > v$, otherwise $(x, y) < (u, v)$. The generalization for E-tuples of reals is straightforward.

the user's probability in these expressions is not conditioned on the expert's advice a_0 for the current decision variable. These expressions reflect the user's probability of seeing various outcomes "hit" without yet knowing what the experts' quantiles are.

$$P(Z_0 | a_0, \ldots, a_M, Z_1, \ldots, Z_M) = P(Z_0 | a_0, \ldots, a_M, Y) = P(Z_0 | a_0, Y) \quad (13.6)$$

Put somewhat simply, (13.6) says that the user is only interested in the experts' advice for the current decision variable, and the frequency with which the various values of j have been hit on the previous variables.

$$P(Z_0 = r_1, \ldots, r_E) = \prod_{i=1}^{E} p_{r_i} \quad (13.7)$$

Assumption (13.7) states that before receiving any advice and before performing observations, the user is prepared to take the expert's advice at face value, and to regard the experts as independent.

$P(Z_0 | Y)$ and $P(Z_0 | a_0, Y)$ are minimally informative, subject to the constraints expressed in the foregoing assumptions. $\qquad (13.8)$

The Mendel-Sheridan Posterior Model

The user is ultimately interested in $P(X_0 | a_0, Y)$. Bayes' theorem will be used to determine $P(Z_0 | a_0, Y)$, and the former distribution will be determined by minimizing information relative to the latter. The distribution $P(Z_0 | a_0, Y)$ will be determined by a two-stage application of Bayes' theorem, starting from $P(Z_0)$. The entire inference trajectory is

$$P(Z_0) \to P(Z_0 | Y) \to P(Z_0 | a_0, Y) \to P(X_0 | a_0, Y)$$

Determining P(Z₀|Y)

Let $\pi = (\pi_1, \ldots, \pi_{(R+1)^E})$, $\pi_j \geq 0$, $\Sigma \pi_j = 1$. Assumption (13.5) and De Finetti's representation theorem entail that [see Eq. (7.13)]:

$$P(Z_0 = j | y) = \int \pi_j \sum_{k=1}^{(R+1)E} \pi_k^{Y_k} \, dF(\pi) \quad (13.9)$$

for some suitable prior probability measure dF over the possible values of π. If dF can be determined from the constraints (13.5) to (13.8), then the first step in the inference will in principle be solved. Mendel and Sheridan give the density $f = dF$, which minimizes the information with respect to the uniform distribution subject to the constraint (13.7):

$$\int \ln \frac{f(\pi) \, d\pi}{f_{\text{uniform}}(\pi) f(\pi)} \quad (13.10)$$

The solution is

$$f(\pi) = \frac{\beta e^{\beta \pi}}{1 - e^{\beta \pi}} \quad (13.11)$$

where $\beta \pi = \Sigma \beta_i \pi_i$, and the vector β is determined from (13.7). When (13.11) is substituted into (13.9), the result cannot be expressed in closed form.

Mendel and Sheridan tackle this problem by choosing a Dirichlet distribution f_{dir} that has minimal information relative to the density in (13.11). Referring to the discussion in the supplement to Chapter 7, the Dirichlets are characterized by $(R + 1)^E$ parameters α_j, $j = 1,\ldots,(R + 1)^E$. The parameter $T = \Sigma\alpha_j$ is also introduced. The parameters of the Dirichlet distribution may be interpreted in terms of "equivalent observations"; $\alpha_j - 1$ corresponds to the number of "equivalent observations" of the events "$Z_i = j$," $i = 1,\ldots,$ N. In Bayesian multinomial inference, the Dirichlet is a conjugate prior for the multinomial likelihood function. By choosing the parameters of the Dirichlet, the user expresses his prior confidence in the outcome j by saying, in effect, "I am as confident in j as I would be if I had no previous information and then observed $\alpha_j - 1$ occurrences of j out of T trials." The posterior is then also a Dirichlet, and [see Eq. (7.17)]:

$$P(Z_0 = j \mid Y) = \frac{Y_j + \alpha_j}{N + T} \tag{13.12}$$

When $N = 0$,

$$P(Z_0 = j) = \frac{\alpha_j}{T} \tag{13.13}$$

the left-hand side of which is determined by (13.7).

Mendel and Sheridan state on the basis of numerical results that (13.12) is adequately approximated by

$$P(Z_0 = j \mid Y) \approx \frac{Y_j + 1}{N + P(Z_0 = j)^{-1}} \tag{13.14}$$

where $P(Z_0 = j)$ is given by (13.7). For a discussion of the quality of these approximations, we refer to Mendel and Sheridan (1986).

Determining $P(Z_0 \mid a_0, Y)$

When the decision maker learns the experts' advice a_0 on the current decision variable, he must condition (13.14) on this advice. Learning the experts' advice on the decision variable has the effect of making some values of j in (13.14) impossible. Indeed, learning this advice is equivalent to selecting a path in the matrix of Figure 13.2. Once a path is selected, only cells on this path can be hit.

Mendel and Sheridan regard the reception of the advice a_0 as *equivalent* to learning that certain values of j have probability 0. We shall see that this is not strictly true. However, on this assumption, the conditionalization of (13.14) on a_0 is quite simple. For those values of j which are consistent with a_0, one simply uses the values given in (13.14), normalized such that their sum is unity. For other values of j, the probability is set equal to 0. More precisely,

$$\Gamma = \sum_{i; i \text{ consistent with } a_0} p(Z_0 = i \mid Y) \tag{13.15}$$

$$P(Z_0 = j \mid a_0, Y) = \frac{p(Z_0 = j \mid Y)}{\Gamma} \qquad \text{if } j \text{ is consistent with } a_0;$$

$$= 0 \qquad \text{otherwise.}$$

Mendel and Sheridan note that the order of conditionalization on Y and a_0 is irrelevant.

Determining $P(X_0 | a_0, Y)$

The problem of determining $P(X_0 | a_0, Y)$ is trivial if the range of X_0 is bounded. In this case a minimal information solution subject to (13.15) yields a density function that is a step function. The probability mass for j in (13.15) is uniformly distributed over the corresponding interval in the range of X_0.

If the range of X_0 is not bounded the density in the "tails" cannot be determined by a minimal information argument, and the analyst must choose the bounds. It must be noted that the choice of cutoff points is more critical in the Bayesian model than in the classical model. Consider the single-expert case, and suppose that 20% of the realizations are observed to fall beneath the expert's 5% quantiles. After a sufficient number of observations, the expert will be recalibrated such that the assessed 5% quantile will be treated as the 20% quantile. Beneath this quantile, the probability mass will be distributed uniformly. Hence the new 5% quantile will be placed at one-fourth the distance between the 20% quantile and the lower cutoff point. The placement of the recalibrated 5% quantile is linearly dependent on the choice of the lower cutoff point.

It must also be noted that the prior distribution determined by (13.7) tends to be more concentrated than either of the experts' distributions. Referring to the assessments shown in Figure 13.1, interpreted as 5%, 50% and 95% quantiles, the decision maker's prior is pictured in Figure 13.4. More than 92% of the mass is concentrated in cells $(3, 2)$ and $(3, 3)$:

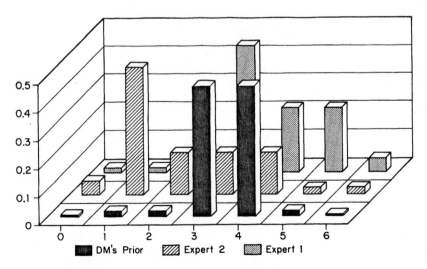

Figure 13.4 Decision maker's prior corresponding to 5%, 50%, and 95% assessments in Figure 13.1.

EVALUATION OF MODEL PERFORMANCE VIA BILINEAR LOSS FUNCTIONS

The model's performance has been evaluated by computing a loss as a function of an estimate and a realization. This is done for each variable, and the losses are added to determine a cumulative loss. In the figures below a bilinear loss is used. If Θ is the true value, and x the estimate for Θ, then the expected bilinear loss has the form

$$E(L(x, \Theta)) = \int_{-\infty}^{x} (x - \Theta)K_1 \, p(\Theta) \, d\Theta + \int_{x}^{\infty} (\Theta - x)K_2 \, p(\Theta) \, d\Theta$$

where $K_1 > 0$, $K_2 > 0$. The Bayesian estimate under a loss function is that value x' of x for which the expected loss is minimum. It is not difficult to show that under the above loss function the Bayesian estimate corresponds to the $K_2/(K_1 + K_2)$th quantile of the distribution for Θ.

Mendel and Sheridan (1986) have computed the cumulative losses for the bilinear loss functions $L(x, \Theta)$ shown in Figure 13.5. Two experts, A and B (graduate students at MIT), assessed 53 items from the Guinness book of records. Figure 13.6 (*a* and *b*) shows the cumulative losses for experts A and B, compared to that of the recalibrated experts derived by applying the Bayesian model to these experts singly. Figure 13.6(*c*) shows the loss when the prior distribution for both experts is used, compared with the loss when this prior distribution is updated. Figure 13.7 shows the losses of the model applied to the experts singly, compared to the losses when the experts are combined.

Model performance may be called encouraging when the number of observations is fairly large. Note that the improvement wrought by the model is especially marked for the loss function L_3, which punishes underestimation. People do tend to underestimate record items. In Chapter 15 the Mendel-Sheridan model is applied in a case study involving only 13 observations, and model performance is less satisfactory. Finally, note that loss has the dimension of the variable being

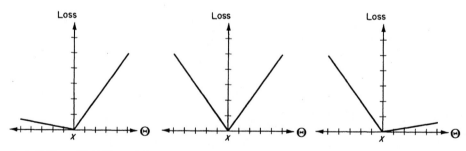

L_1: Left Slope : Right Slope = 1:8 L_2: Left Slope : Right Slope = 1:1 L_3: Left Slope : Right Slope = 8:1

Figure 13.5 Bilinear loss functions. L_1 punishes primarily overestimation, L_2 is symmetric and L_3 punishes primarily underestimation.

estimated. A change of unit, say from kilograms to grams, will cause the computed loss to increase by a factor of 1000. Hence, the meaning of cumulative loss for variables measured in different units (kilograms, dollars, etc.) must be sought in the analyst's own utility function.

A PARTIALLY EXCHANGEABLE BAYESIAN MODEL

The following variant on the Mendel-Sheridan model is motivated by the subtle but very important distinction between *conditioning on the experts' advice* and conditioning on a subset of the possible values of Z_0. Roughly speaking, in the Mendel-Sheridan model the user updates his distribution over the possible values of Z_0, so as to approach the observed frequency distribution generated by $Z_1, \ldots, Z_N$. Upon hearing a_0, he conditionilizes his Z_0 distribution on the subset corresponding to the cells on the path defined by a_0.

To make things easier, let us assume that the decision maker simply adopts the frequency distribution (this is true in the limit). Then, upon hearing a_0, the decision maker conditions the frequency distribution on the cells corresponding to a_0. Now, the frequency with which a cell is hit is the product of the frequency with which the cell is offered, and the frequency with which it is hit, *given that it is offered*. Referring to Figure 13.3, we see that the frequencies with which cells are offered can differ dramatically. The cell $(1, 1)$ is on every path, and hence is offered with each variable $X_1, \ldots, X_N$. On the other hand, cell $(4, 1)$ is offered on only 1 of the 20 possible paths. Hence, one would anticipate that the frequency of $(4, 1)$ would be quite low relative to the frequency of $(1, 1)$, reflecting the fact that $(1, 1)$ is offered much more often than $(4, 1)$. In spite of this, cell $(4, 1)$ may be hit each time it is offered. In other words, the frequency of $(4, 1)$, *given that it is offered*, may be quite different than the frequency of $(4, 1)$ divided by the sum of the frequencies of $(1, 1)$, $(2, 1)$, $(3, 1)$, $(4, 1)$, $(4, 2)$, $(4, 3)$, $(4, 4)$ [which is the frequency of hitting $(4, 1)$ conditional on the single path through $(4, 1)$].

This all goes to show that conditioning on the experts' advice is not equivalent with conditioning on the path determined by that advice. Indeed, the event "a_0" is not in the field generated by the variable Z_0. Step (13.15) in the Mendel-Sheridan model must be regarded as an estimate. This estimate may be acceptable for the most frequently offered paths, as borne out by experimental results. However, for infrequent paths it may be poor, and may get worse as more data is collected.

Theoretically, the situation could be redressed by considering Z_0 as exchangeable with those previous variables giving rise to the same path as a_0. In other words, the variables $Z_0, \ldots, Z_N$ are considered as *partially exchangeable* given $a_0, \ldots, a_N$; and Z_i, Z_j are exchangeable if a_i and a_j generate the same path, $0 \leqslant i < j \leqslant N$. Formula (13.13) would then apply when "Y_j" is interpreted to refer to the number of hits in the subsequence of variables exchangeable with Z_0.

The problem with this theoretical solution is the following. The number of possible paths is large and grows very fast as the number of quantiles and the number of experts increase. In the case considered in Figure 13.3, the number of paths is 20. The general case has been calculated. For three experts estimating two

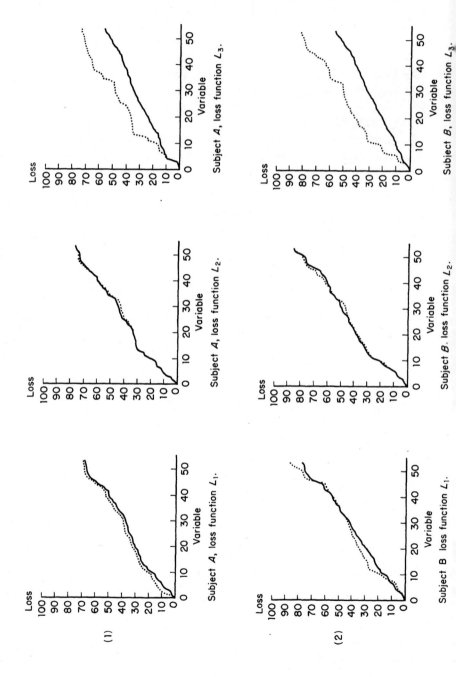

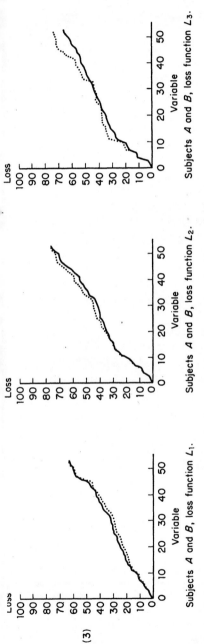

Figure 13.6 Cumulative losses for calibrated observer (solid) and uncalibrated observer (dashed) for the loss functions in figure 13.5.

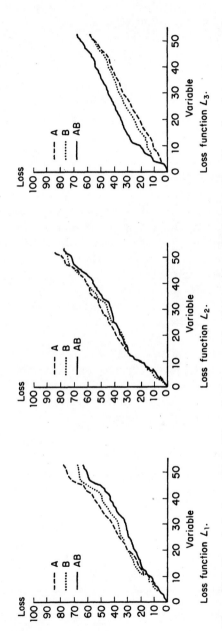

Figure 13.7 Cumulative losses for (calibrated) observer with subject *A* alone (dashed), *B* alone (dotted) and *A* and *B* together for the loss functions in figure 13.5.

quantiles, the number of possible paths is 90^2. For three experts estimating three quantiles, this number is 1680, for five experts estimating two quantiles, it is 113,400. Hence, we should need astronomical amounts of data to build up frequency distributions as the number of experts or the number of quantiles increases. If the experts assess an infinite number of quantiles (e.g., by giving entire distributions) the model breaks down.

Undoubtedly, additional theoretical work could lead to more practical models. For example, one could introduce the *intensity* of a cell as the number of times the cell is hit divided by the number of times the cell is offered. Upon receiving advice a_0, the decision maker could adopt the renormalized intensities of the cells on the path corresponding to a_0 as his posterior for Z_0. However, a satisfactory Bayesian analysis for such a procedure remains to be given.[3]

CONCLUSIONS

As with the classical model, a successful implementation stands or falls with the ability to define variables that are relevant and whose true values become unambiguously known within a reasonable time frame. This requires resourcefulness on the part of the analyst and sympathetic cooperation from the experts themselves. It is therefore essential that the experts understand and appreciate the model, and that the analyst take the utmost care in defining the variables.

The Bayesian model is less flexible than the classical model. In its present form it admits only quantile assessments and requires that all experts assess the same variables. Moreover, because of its minimally informative default priors, the Bayesian model will surely require more variables to warm up. Evidence of this is presented in Chapter 15. Hence this model will be more dependent on "seed variables" than the classical model. On the other hand, the Bayesian model is unquestionably more powerful, and under ideal conditions may be expected to yield better results.

---

[2] Consider 6 points on the real line representing the 6 quantile assessments. The 2 corresponding to the first expert's quantiles can be chosen in 15 ways. Of the remaining 4 points, 2 can be assigned to the second expert in 6 ways, and the third expert's choice is fixed. For M experts with K quantiles the general formula is: # arrangements $= \binom{MK}{K} \cdot \binom{(M-1)K}{K} \cdot \ldots$.

[3] Let O_j denote the event that cell j is offered, and let h_j denote the event that cell j is hit. Let a_0 generate a path containing cell j. If we assume that $P(h_j \text{ and } a_0 \mid h_j \text{ and } O_j) = P(a_0 \mid O_j)$, then an easy calculation shows that $P(h_j \mid a_0) = P(h_j \mid O_j)$. If this assumption holds strictly, the expected intensities along a path will automatically add to 1. If this assumption holds only approximately, renormalization will be required.

14

Psychological Scaling Models: Paired Comparisons

We now turn to the psychological scaling models. For background and general discussion, refer to Chapter 11. This chapter provides the tools for applying three psychological scaling models, the Thurstone model, the Bradley-Terry model and the NEL (negative exponential lifetime) model. The latter two models are computationally identical, but differ in interpretation. The NEL model is specifically designed for estimating constant failure rates. Derivations and proofs for these models are readily available in the literature and will not be given here. In a concluding section the relation to the principles for expert opinion in science is discussed. Before turning to the models we first discuss generic issues.

GENERIC ISSUES

We adopt the following notation throughout.

n Number of experts;

$A(1), \ldots, A(t)$ Objects to be compared;

$V(1), \ldots, V(t)$ True values of the objects;

$V(i, e)$ Internal value of expert e for object i;

When applied to expert probability assessment, the objects will generally be events, and the values will be the "true" probabilities of the events. We assume that each expert judges each pair of objects and expresses a preference for one of the two objects in each pair, by comparing his internal values. Thus "$A(i)$ is preferred to $A(j)$ by e" is understood to mean that $A(i)$ is judged more probable then $A(j)$ by e, that is, $V(i, e) > V(j, e)$. We also write $A(i) > A(j)$ to say that $A(i)$ is preferred to $A(j)$, when the expert is identified by the context.

In practical applications, despite instructions from the analyst, some experts

---

This chapter was written jointly with M. Stobbelaar

will sometimes be unwilling or unable to judge some pairs of objects. Hence the data will contain *void comparisons*. In most cases, the formulas given below generalize easily to the case where some experts are unable or unwilling to express preferences for some pairs, and these generalizations will be indicated.

In analyzing paired comparison data, the first question to be addressed, before choosing a model, is "Is there a significant difference in the objects with respect to preference?" This question may be posed with regard to an individual expert, or with regard to the set of experts as a whole.

SIGNIFICANT PREFERENCE FOR ONE EXPERT

Suppose for expert e, the values $V(i, e)$ are barely distinguishable. For each pair, he expresses some preference, but the preferences are "unstable." How might we detect this from the data from this expert?

The following procedure suggests itself. Suppose $A(1)$, $A(2)$, and $A(3)$ are barely distinguishable in preference. If the expert says $A(1) > A(2)$ and $A(2) > A(3)$, then $A(1)$ and $A(3)$ will still be very close in preference, and the expert might well say $A(3) > A(1)$. These three objects would then constitute a *circular triad* in the expert's preference structure. In the language of Chapter 6, the expert's preferences are intransitive. All intransitivities in the preferences of the objects can be reduced to circular triads.

When an expert assesses a large number of pairs, we should not be surprised if a few circular triads arise. This poses no particular problem, as each object is compared many times with other objects, and the models will extract an underlying trend, if there is one.

As the number of circular triads increases, we would begin to doubt whether the expert has sharply defined underlying preferences. David (1963) provides a procedure for testing the hypothesis that each preference is determined at random.

Let $a(i, e)$ be the number of times that expert e prefers $A(i)$ to some other object. David shows that the number $C(e)$ of circular triads in e's preferences is given by

$$C(e) = \frac{t(t^2 - 1)}{24} - \frac{1}{2}\sum_{i=1}^{t} [a(i, e) - \frac{1}{2}(t - 1)]^2 \qquad (14.1)$$

Kendell (1962) has calculated the probabilities that various values of $C(e)$ would be exceeded, with 2 to 10 objects, under the hypothesis that the preferences are randomly determined. These are shown in Table B.4, Appendix B. For more than seven objects he shows that

$$C^{\sim}(e) = v + \left(\frac{8}{t - 4}\right)\left[\left(\frac{1}{4}\right)\binom{t}{3} - C(e) + \frac{1}{2}\right] \qquad (14.2)$$

with

$$v = \frac{t(t - 1)(t - 2)}{(t - 4)^2}$$

is approximately chi square distributed with v degrees of freedom.

If the random preference hypothesis cannot be rejected at the 5% level on the basis of the preference data, then the analyst should consider dropping this expert from the set of experts.

SIGNIFICANT PREFERENCES FOR THE SET OF EXPERTS

Assume that the preferences of each expert are significant, in the sense that the random preference hypothesis is rejected for each expert. It may still be the case that no significant preference emerges from the group as a whole. This will arise if the experts disagree sufficiently in their preferences. Hence, we detect significant preferences in the group of experts by considering the degree of agreement in their preferences. There are two tests which can be used for this purpose, a test involving the "coefficient of agreement u" and a test involving the "coefficient of concordance W."

Coefficient of Agreement u

Let $a(ij)$ denote the number of times that some expert prefers object $A(i)$ to $A(j)$. Then $a(ji) = n - a(ij)$. If the experts agree completely, then half of the $a(ij)$s will be equal to n, and the other half equal to 0. (Note that complete agreement does not preclude circular triads.) Define

$$\Sigma^\sim = \sum_{i \neq j} \binom{a(ij)}{2}$$

where the summation runs over $t(t - 1)$ terms. Kendall (1962) defines the *coefficient of agreement u* as

$$u = \frac{2\Sigma^\sim}{\binom{n}{2}\binom{t}{2}} - 1. \tag{14.3}$$

u attains its maximum, 1, when there is complete agreement. Under the hypothesis that all agreements of the experts are due to chance, the distribution of u can be determined. Kendall (1962) tabulates the distribution of $\Sigma^\sim$ for small values of n and t (Table B.5, Appendix B). Furthermore, for large values of n and t,

$$u' = \frac{4\left[\Sigma^\sim - \binom{n}{2}\binom{t}{2}(n-3)/2(n-2)\right]}{n-2}$$

is approximately chi square distributed with

$$\frac{\binom{t}{2}n(n-1)}{(n-2)^2}$$

degrees of freedom. The hypothesis that all agreements are due to chance should be rejected at the 5% level.

Coefficient of Concordance *W*

The sum of the ranks $R(i)$ is calculated by

$$R(i) = \sum_e R(i, e)$$

where $R(i, e)$ is the rank of $A(i)$ obtained from the responses of expert e; the value of $R(i, e)$ ranges from 1 to t. Siegel (1956) defines

$$W = \frac{S}{\frac{1}{12}n^2(t^3 - t)}$$

where S is the sum of squares of the observed deviations from the mean of $R(i)$:

$$S = \sum_i [R(i) - t^{-1}\sum_j R(j)]^2$$

In case of complete agreement W equals 1, and gets smaller as agreement diminishes.

For the null hypothesis that the preferences are at random, Siegel (1956) presents a table of critical values of S, for t between 3 and 7 and n between 3 and 20. This table is included as Table B.6, Appendix B. For t larger than 7,

$$W' = \frac{S}{\frac{1}{12}nt(t + 1)}$$

is approximately chi square distributed with $t - 1$ degrees of freedom (Siegel, 1956).

THE THURSTONE MODEL

Thurstone (1927) developed an elegant mathematical model for analyzing paired comparison data. The best modern reference is Torgerson (1958). The least squares method of solution was proposed by Mosteller (1951a, b, and c) whose three articles on the subject are still worth reading. In some applications the model is used with one judge who expresses his preference for each pair several times, and a probabilistic mechanism for determining preference is assumed. Expert opinion applications involve several experts, and each expert judges each pair once. No probabilistic model is made for the individual preference judgments, but assumptions are made about the distribution of responses in the population of experts.

Assumptions

For $i, j = 1, \ldots, t$, it is assumed that

The values $V(i, e)$ are normally distributed over the population of experts, with mean $\mu(i)$ and standard deviation $\sigma(i)$.   (14.4)

$\mu(i) = V(i);$ (14.5)

$\sigma(i) = \sigma,$ that is, $\sigma(i)$ does not depend on i;   (14.6)

$V(i, e) - V(j, e)$ is normally distributed over the population of experts with mean $\mu(ij) = \mu(i) - \mu(j)$ [this follows from (14.4)] and standard deviation $\sigma(ij) = (\sigma^2 + \sigma^2)^{1/2}$.   (14.7)

In general we have:

$$\sigma(ij) = [\sigma(i)^2 + \sigma(j)^2 - 2\rho(ij)\sigma(i)\sigma(j)]^{1/2} \qquad (14.8)$$

where $\rho(ij)$ is the correlation coefficient of $V(i, e)$ and $V(j, e)$. (14.7) adds that the distributions $V(i, e)$ and $V(j, e)$ are uncorrelated. Slightly weaker assumptions are mathematically tractable; in particular, (14.7) can be replaced by the assumption that $\rho(ij)$ does not depend on i and j (Mosteller, 1951a). In the context of estimating small probabilities, several authors (Comer et al. 1984, Hunns, 1982, Kirwan et al. 1987) recommend replacing (14.4) with

> The values ln $V(i, e)$ are normally distributed over the population of experts, with mean $\mu(i)$ and standard deviation $\sigma(i)$.   (14.4a)

It is plausible that (14.4) or some variant would reasonably describe the distributions $V(i, e)$.[1] Assumptions (14.5), (14.6), and (14.7) are very strong.

Solution of the Thurstone Model

Suppose expert e is drawn randomly from the population of experts. The probability that e prefers $A(i)$ to $A(j)$ is given by

$$p(V(i, e) - V(j, e) > 0) = p\left(\frac{V(i, e) - V(j, e) - \mu(ij)}{\sigma(ij)} > \frac{-\mu(ij)}{\sigma(ij)}\right) = p\left(X > \frac{-\mu(ij)}{\sigma(ij)}\right)$$

where X is a standard normal variable. This probability is estimated by the percentage $\%(ij)$ of experts who prefer $A(i)$ to $A(j)$. Since the density of X is symmetric about zero,

$$p\left(X > \frac{-\mu(ij)}{\sigma(ij)}\right) = p\left(X < \frac{\mu(ij)}{\sigma(ij)}\right) = \Phi\left(\frac{\mu(ij)}{\sigma(ij)}\right)$$

where Φ is the cumulative distribution of the standard normal variable; hence,

$$\Phi^{-1}(\%(ij)) \approx \frac{\mu(ij)}{\sigma(ij)}$$

Set $x(ij) = \Phi^{-1}(\%(ij))$. Since $\sigma(ij)$ does not depend on i and j, by choosing a suitable scaling constant we may arrange that $\sigma(ij) = 1$, and hence write

$$\mu(i) - \mu(j) \approx x(ij) \qquad (14.9)$$

There are $\binom{t}{2}$ equations of the form (14.9), one for each pair of objects, and t unknowns $\mu(1), \ldots, \mu(t)$. If the estimation $\%(ij)$ was perfect, and if the modeling assumptions hold, then all equations of the form (14.9) would be satisfied, and could be solved for $\mu(1), \ldots, \mu(t)$. However, the estimates will not be perfect, and the

[1] It should be noted however, that the least squares solution algorithm yields an optimum (unbiased and minimum variance) estimate for the means of the normal distributions (see, e.g., Hanushek and Jackson, 1977, chap. 2). If (14.4a) is used, these estimates are optimal for log $\mu(i)$, but not for $\mu(i)$.

system (14.9) will be overdetermined for $t \geqslant 4$. If (14.7) were replaced by (14.8), the correlation coefficients $\rho(ij)$ would also be unknowns, and the resulting system would be underdetermined.

We can solve the overdetermined system by finding estimates $\mu'(i)$ of $\mu(i)$ such that the sum of the squared errors

$$\sum_{i=1}^{t} [\mu'(i) - \mu'(j) - x(ij)]^2 \tag{14.10}$$

is minimal. Adding a constant to $\mu(i)$, $i = 1, \ldots, t$, would not affect the value of the above sum; hence we may add a constant such that $\Sigma \mu'(i) = 0$. The solution is then

$$\mu'(i) = \frac{1}{t} \sum_{j=1}^{t} x(ij) \tag{14.11}$$

Goodness of Fit

The solution (14.11) minimizes the squared error (14.10), but the error may still be large. How large is too large? One way to answer this question is the following. Assume that assumptions (14.4), (14.6), and (14.7) hold with $\mu'(i)$ replacing $\mu(i)$, that is, assume that our solution is really correct. We now ask, what is the probability of seeing a set of preferences $\%(ij)$ that lead to an error at least as great as (14.10)? If this probability is too small, then we should conclude that some of our modeling assumptions are false. Mosteller (1951c) gives the following test. Define

$$y(ij) = \mu'(i) - \mu'(j)$$
$$\Theta(ij) = \arcsin [\%(ij)]^{1/2}$$
$$\phi(ij) = \arcsin [\Phi(y(ij)]^{1/2}$$

where Φ is the cumulative standard normal distribution and arcsin is measured in degrees. Mosteller shows that

$$D = \frac{\Sigma_{i<j} n(\Theta(ij) - \phi(ij))^2}{821} \tag{14.12}$$

is approximately chi square distributed with $\binom{t}{2} - t + 1$ degrees of freedom. Hence, it is possible to estimate the probability of seeing a deviation at least as great as D. If this probability is less than 5%, the modeling assumptions (14.4), (14.6), and (14.7) should not be maintained. Mosteller (1951c) remarks that this test may not be effective in detecting violations of (14.6). It is important to note that assumption (14.5), perhaps the most important, is not tested in this way.

Transformation

If our data survive all the foregoing tests, then the solution (14.10) can be used. However, this solution involves an arbitrary positive scaling constant, used to arrange that $\sigma(ij) = 1$, and an arbitrary shift parameter, used to arrange that $\Sigma \mu'(i) = 0$. Hence, if $\mu'(i)$ is a solution then so is

$$\mu''(i) = a\mu'(i) + b \qquad i = 1, \ldots, t \tag{14.13}$$

for arbitrary constants $a > 0$ and b. If we know two values $V(i)$ and $V(j)$, for some i

and j, then we can solve (14.13) for a and b, and determine the remaining true values via (14.13).

Confidence Bounds

Using assumptions (14.4) to (14.7) it is possible to determine confidence bounds for the solution (14.10) via simulation. The procedure is as follows:

> Solve the initial model for $\mu'(i)$, $i = 1, \ldots, t$.
>
> Sample values $V'(i, e)$, $e = 1, \ldots, n$ from the normal distribution with mean $\mu'(i)$ and $\sigma = 1$.
>
> Define $\%(ij)'$ as the number of e such that $V'(i, e) > V'(j, e)$, divided by n.
>
> Solve the model again using $\%(ij)'$ instead of $\%(ij)$, and find solutions $m(i)$, $i = 1, \ldots, t$.

If we repeat the above say 1000 times, always keeping the $\mu'(i)$ fixed, we get 1000 values $m(i)$ for each i. Excluding the 50 smallest and 50 largest, the range spanned by the remaining values defines a 90% confidence interval.

These confidence bands are derived from the modeling assumptions (14.4) to (14.7). They do not reflect uncertainty in these assumptions, but rather reflect sampling fluctuations in the choice of experts. As the number n of experts increases, the bounds determined in this way shrink; modeling uncertainty does not.

Void Comparisons

The solution (14.10), the goodness of fit test (14.12), and the procedure for determining confidence bounds can all be applied in the case of void comparisons. The solution requires only the proportion of experts preferring one alternative to another. The goodness of fit test can be applied by replacing n by the number $n(ij)$ of experts comparing $A(i)$ to $A(j)$.

THE BRADLEY-TERRY MODEL

The Bradley-Terry model was introduced by Bradley and Terry (1952) and further developed in Bradley (1953). Ford (1957) contributed an important convergence result, and a good mathematical exposition is found in David (1963).

Assumptions

The model assumes that each object $A(i)$ is associated with a true scale value $V(i)$, and that the probability $r(ij)$ that $A(i)$ should be preferred over $A(j)$ is given by

$$r(ij) = \frac{V(i)}{V(i) + V(j)} \qquad (14.14)$$

The $V(i)$ are determined only up to a constant scale factor, hence we may assume that $\Sigma V(i) = 1$.

When several experts judge each pair of objects once, it is assumed that $r(ij)$ is the same for all experts, and that the judgments are independent. In other words, the judgments can be regarded as independent coin-tossing experiments with probability of heads $= r(ij)$.[2]

Solution

The proportion $\%(ij)$ is taken as an estimate of $r(ij)$, and the overdetermined system that results by substituting $\%(ij)$ for $r(ij)$ in (14.14) must be solved. The method of least squares is not tractable, owing to the form of (14.14), and the modeling assumptions suggest another strategy. One seeks values $V(1),\ldots,V(t)$ such that, under the modeling assumptions, the probability of seeing the outcomes $\%(ij)$ is as large as possible.[3] Finding such $V(i)$ is equivalent to solving the following system (see David, 1963):

$$V(i) = \frac{a(i)}{\Sigma' \, n[V(i) + V(j)]^{-1}} \tag{14.15}$$

where $a(i) =$ the number of times $A(i)$ is preferred by some expert to some other object, and Σ' denote summation over j, with $j \neq i$.

The solution to (14.15) can be found by iteration. Begin with initial values $V^0(1),\ldots,V^0(t)$ and define the first iteration by

$$V^1(i) = \frac{a(i)}{\Sigma' \, n[V(i) + V^k(j)]^{-1}} ; \tag{14.16}$$

where $k = 1$ if $j < i$, $k = 0$ if $j > i$.

This process converges under general conditions to the unique solution (see footnote 2), and is easily programmed on a personal computer.

Goodness of Fit

The quantity

$$F = 2\left\{ \sum_{\substack{i,j=1 \\ i \neq j}}^{t} a(ij)\,\ln r(ij) - \sum_{i=1}^{t} a(i)\ln V(i) + \sum_{i<j} n\ln[V(i) + V(j)] \right\} \tag{14.17}$$

is asymptotically chi square distributed with $(t-1)(t-2)/2$ degrees of freedom (Bradley, 1953).

Void Comparisons

When not all experts compare all pairs, the iteration process is slightly altered and

---

[2] With this interpretation, circular triads can arise from the probabilistic mechanism according to which preference is determined. Nonetheless, the coefficient of agreement should be used to test and hopefully reject the null hypothesis: All probabilities of preference are equal to $\frac{1}{2}$.

[3] This is known as the maximum likelihood solution. Ford (1957) has shown that the solution is unique and that the iterative process converges to this solution if the following condition is met: it is not possible to divide the set of objects into two nonempty subsets, such that no object in one subset is preferred by any judge above some object in the second subset.

n is replaced in (14.16) by the number $n(ij)$ of experts who compare object i with object j. Similarly, in testing for goodness of fit, $n(ij)$ replaces n in (14.17).

Confidence Intervals

Confidence intervals can be determined by simulation in a manner analogous to the Thurstone model. Once a solution is found for the scale values $V(1), \ldots, V(t)$, these are used to define preference probabilities via (14.14). Simulated preference judgments are then performed by the computer for the n experts and the $t(t-1)/2$ pairs of objects. From this simulated data a simulation solution is determined via (14.16). After, say, 1000 simulation solutions have been found, always keeping the original solution fixed, 90% confidence intervals may be determined as in the Thurstone model.

THE NEGATIVE EXPONENTIAL LIFETIME (NEL) MODEL

The NEL model was developed to assess failure rates in mechanical components.

Assumptions

The objects are mechanical components. For each pair of objects the experts are asked "Which of these two components, operating independently, will fail first?" It is assumed that all components are as good as new at time $T = 0$, and "$A(i) < T$" means that $A(i)$ fails before time T. The following assumptions are made:

$$P(A(i) < T) = \int_0^T r(i)e^{-r(i)t}\, dt \qquad (14.18)$$

In other words, $A(i)$ is assumed to have an exponential life distribution with failure rate $r(i)$. Further, it is assumed that the probability that an expert says "$A(i)$ fails before $A(j)$" is given by

$$P(A(i) < A(j))$$

under the assumption that $A(i)$ and $A(j)$ are independent. An elementary calculation[4] shows that

$$P(A(i) < A(j)) = \frac{r(i)}{r(i) + r(j)} \qquad (14.19)$$

In other words, the model assumes that the expert answers the question put to him by performing a mental experiment, letting $A(i)$ and $A(j)$ operate independently with life probability (14.18), and observing which fails first.

---

[4] This can be seen by writing

$$P(A(i) < A(j)) = \int_0^\infty \int_x^\infty r(i)r(j)e^{-r(i)y - r(j)x}\, dy\, dx$$

and performing the integration.

Solution, Goodness of Fit, Confidence Intervals, Void Comparisons

It will be observed that (14.19) has the same form as (14.14). This means that the solution procedure described in the Bradley-Terry model will apply here, as well as the goodness of fit test, the procedures for generating confidence intervals, and the procedure for handling void comparisons.

Transformation

The solution (14.16) determines the values $V(1), \ldots, V(t)$ only up to a scaling constant. If for component i the true failure rate $r(i)$ is known or can be estimated, then the failure rates for the other components can be found by setting

$$r(j) = \frac{r(i)V(j)}{V(i)}$$

RELATION TO PRINCIPLES FOR EXPERT OPINION IN SCIENCE

We conclude this chapter by noting that the psychological scaling models stand in a rather different relation to the principles set forth in Part I for using expert opinion in science than the models described in the previous two chapters. Reproducibility, accountability, neutrality, and fairness pose no particular problem. Empirical control is another matter.

The tests of goodness of fit do not constitute empirical control. These tests determine whether the paired comparison data are sufficiently consistent with the modeling assumptions, but do not involve a comparison of model output with empirical values.

If the confidence intervals are interpreted as subjective confidence intervals for the decision maker, then empirical control is possible. If true values become observable at some later point, then one should expect that 90% of the true values fall within the 90% confidence bounds. The hypothesis that the model output is well calibrated may then be tested in the manner described in Chapter 9.

However, this interpretation of the confidence intervals may not be warranted. The confidence intervals reflect uncertainty due to the choice of experts, and do not reflect uncertainty regarding the modeling assumptions themselves. As the number of experts increases, the confidence intervals shrink, whereas the uncertainty in the modeling assumptions need not decrease.

If the model output is in the form of probabilities for uncertain events, then these can in principle be treated as input into the classical model and subjected to the empirical control inherent in the classical model.

In either case, empirical control is in principle available for the model output as a whole. Empirical control for the individual assessments of the experts is not possible.

15

Applications

This chapter discusses applications of the models presented in Part III. Three of these applications ("ESTEC-1," "DSM-1," and DSM paired comparisons) were developed in the course of research supported by the Dutch Ministry of Housing, Physical Planning and Environment, and are described in Cooke et al. (1989).[1] The studies "DSM-2" and DSM-3 were supported by the Dutch State Mines (Akkermans, 1989; Claessens 1990), and ESTEC-2, and -3 were supported by the European Space Agency (Meima and Cooke, 1989; Offerman, 1990a). A study involving weather forecasting data was supported by Delft Hydraulics (Roeleven, 1989).

The first section of this chapter reviews applications of the classical model, and discusses the variations and enhancements proposed in Chapter 12. The second section discusses Bayesian analyses of the data sets in the first section. This is possible because the Bayesian model in Chapter 13 is compatible with the continuous version of the classical model. The third section discusses a joint application of the psychological scaling model and the classical model. The supplement to this chapter contains extensive data from these applications.

APPLICATIONS OF THE CLASSICAL MODEL

Seven applications of the classical model described in Chapter 12 have been undertaken to date. Table 15.1 provides a summary description of these studies. In the first six applications experts assessed their 5%, 50%, and 95% quantiles for quantities with continuous ranges. ESTEC-1, -2, -3, and DSM-2 developed data for use in quantitative risk analyses; DSM-1 was intended to provide reference values for a paired comparison study. DSM-3 used expert judgment to quantify uncertainties in key parameters of a ground water transport model. (The results of

Many (ex)students and colleagues have helped in gathering and analyzing the data recorded in this chapter, in particular, D. Roeleven, M. Kok, D. Akkermans, M. Stobbelaar, D. Solomatine, J. van Steen, C. Preyssl, B. Meima, F. Vogt, and M. Claessens.

[1] The numerical results given in this report, also in Preyssl and Cooke (1989) differ slightly from those reported here due to refinements in the numerical routines.

Table 15.1 Applications of the Classical Model

Study	Problem	No. of Experts	No. of Items/Seed Items
ESTEC-1	Propulsison system	4	48/13
DSM-1	Flange connections	10	14/8
DSM-2	Crane failure	8	39/12
DSM-3	Ground water transport	7	38/10
ESTEC-2	Space debris risk	7	58/26
ESTEC-3	Composite materials	6	20/12
RDMI	Meteor'l forecasting	4–5	≥2000

ESTEC = European Space Technical Centre, DSM = Dutch State Mines, RDMI = Royal Dutch Meteorological Institute.

this study came too late for complete treatment here, and only summary data is presented.) The RDMI study concerned uncertain events. Data supplied by the Royal Dutch Meteorological Institute was analyzed to determine whether a combination of expert opinions according to the classical model would yield better forecasts. The first six studies required seed variables; in the RDMI study realizations were retrievable for every assessment.

Summary Descriptions

The study DSM-1 is described in the last section of this chapter, and the RDMI study is described in the following section. Summary remarks must suffice for the others. The ESTEC-1 study grew out of discussions between design and reliability engineers over a proposed design of a propulsion system. Four experts participated in the study, two senior design engineers (experts 1 and 2), a junior design engineer (expert 3) and a reliability engineer (expert 4). The elicitation was conducted with each expert individually by a member of the analysis team. Nine of the seed items concerned empirical failure frequencies of spaceflight systems;[2] four concerned "general reliability items." The results and expert scores were fed back informally to the experts. The results and method were accepted by the project supervision, and the European Space Agency is currently implementing an extended expert judgment program building on these methods (Preyssl and Cooke, 1989; Cooke, French, and van Steen, 1990).

The study ESTEC-2 applied expert judgment to assess critical variables used in calculating the risks to manned spaceflight due to collision with space debris. The current design base models of the debris environment stem from 1984, and in some circles are felt to underestimate the risk. New data on the debris environment are to become available in 1991, affording the opportunity to compare the predictions of this study with realizations. The 26 seed variables concerned the number of radar-tracked objects injected into orbit in the years 1961 to 1985 (for future years this is a variable of interest). The manner of elicitation was somewhat idiosyncratic and utilized graphic methods. Six of the seven experts participating in

---

[2] The seed items are included in supplement A, but the names of the systems have been removed.

this study were specialists in space environment modeling and shielding. One expert was a reliability specialist. The results were fed back to the experts and met with general assent.

The ESTEC-3 study fed into a study of the effectiveness of various new composite materials in reducing the risks of debris and meteorite collision in manned space flight. Six experts participated in the study. Fourteen seed variables were originally defined, but two were subsequently excluded when the values of the realizations were thrown in doubt. Feedback to the experts led to a follow-up study of on-line training to improve expert calibration (Offerman, 1990b).

The DSM-2 study was initiated by the chemical process concern DSM after receiving indications that crane activities in the neighborhood of process facilities could constitute a significant risk. Eight experts, including crane drivers, signallers, a supervisor, a job planner, a technical inspector, and a department head participated in the study. These experts had little or no knowledge of probability and statistics. Elicitation was accomplished via personal interview, asking first for a best estimate, then for an interval for which the expert was 90% sure that it contained the true value. The results of this study confirmed the apprehensions regarding crane risks.

In DSM-3, expert judgment was used to quantify uncertainty in important parameters of a ground water transport model used to predict future contamination with hazardous materials. The seven experts were all geohydraulogists. Data from permeameter- and pump measurements were used to define seed variables. Experts were asked to assess transmissivity of soil in locations where transmissivity had already been measured. They were given information about the soil obtained from drilling. The assessment tasks for seed variables correspond closely to the assessment tasks for the variables of interest.

Overall Evaluation

The classical model yields probabilistic assessments for an optimized decision maker, as described in Chapter 12. These assessments can be evaluated by considering their calibration and information scores for those variables whose realizations are known, and comparing these with scores of other assessments. Table 15.2 presents results comparing the basic model assessments with the optimized decision maker gotten using "item weights" (see Chap. 12), the decision maker gotten by assigning the experts equal weight and taking the arithmetic average of their distributions, and the best expert, that is, the expert receiving the greatest weight in the basic classical model.

The RDMI results presented in Table 15.2 represent aggregations over some 6000 assessments plus realizations. These data are analyzed according to the "uncertain event" version of the classical model, and hence (lack of) information is measured as mean relative entropy [see (12.10)]. In some cases the calibration score $C(e)$ [(12.9)] is very small, in the order of 10^{-12}; hence the values of the scoring variable $\Sigma 2n_b I(s_b, p_b)$ are given. The RDMI case is described in greater detail below. A full description is available in Dutch (Roeleven, 1989) and a description in English is in preparation.

Table 15.2 Comparison of the Optimized Decision Maker Under the Basic Classical Model with the Decision Maker Generated by Item Weights (See Chap. 12) by Assigning Experts Equal Weight, and with the "Best Expert."

Optimized DM	Performance Comparisons		
	Item Weight DM	Equal Weight DM	Best Expert

Continuous Variables

	Calibration	Mean Seed Inform.	Unnormalized (Virtual) Weight	Normalized Weight
ESTEC-1 Op. DM	0.43	1.72	0.74	0.49
Item wgt. DM	0.15	2.21	0.33	0.30
Eq. wgt. DM	0.43	1.42	0.61	0.44
Best exprt	0.14	2.95	0.41	0.27
DSM-1 Op. DM	0.66	1.37	0.91	0.45
Item wgt. DM	0.66	1.40	0.92	0.46
Eq. wgt. DM	0.53	0.81	0.43	0.27
Best exprt	0.54	1.53	0.84	0.42
DSM-2 Op. DM	0.84	1.37	1.13	0.98
Item wgt. DM	0.70	1.39	0.97	0.97
Eq. wgt. DM	0.50	0.69	0.34	0.93
Best exprt	0.005	2.46	0.01	0.01
DSM-3 Op. DM	0.70	3.01	2.11	0.55
Item wgt. DM	0.40	3.97	1.59	0.41
Eq. wgt. DM	0.05	3.16	0.16	0.04
Best exprt	0.40	3.97	1.59	0.38
ESTEC-2 Op. DM*	0.78	0.32	0.25	0.996
Item wgt. DM	0.78	0.41	0.32	0.97
Eq. wgt. DM	0.90	0.16	0.14	0.99
Best exprt	0.0001	2.29	0.0002	0.001
ESTEC-3 Op. DM	0.27	1.44	0.39	0.95
Item wgt. DM	0.28	1.66	0.47	0.96
Eq. wgt. DM	0.12	0.93	0.11	0.85
Best exprt	0.005	2.53	0.01	0.03

Uncertain Events

	Calibration Index†	Mean Entropy‡		Calibration Index†	Mean Entropy‡
RDMI (2, 4)§			RDMI (3, 5)		
Op. DM	20.1	0.42	Op. DM	103.9	0.43
Eq. wgt. DM	56.3	0.43	Eq. wgt. DM	242.1	0.45
Best exprt	133.3	0.372	Best exprt	263.7	0.40

"Calibration" and "mean seed information" correspond to the quantities $C(e)$ and $I(e)$ of formula (12.14). For the measurement of relative information, seed variables for which $0.001 < \text{realization} < 1000$ are referred to a uniform background measure, other seed variables are referred to a loguniform background measure.

*The optimal DM was obtained by scoring calibration at the 70% power level; the calibration scores shown refer to the 100% power level (see Chap. 12 for explanation).

†Due to the very large number of items the chi square variable (with 10 degrees of freedom) $\Sigma 2n_b \, I(s_b, p_b)$ is shown; lower values correspond to better calibration [(12.9)].

‡Lower values indicate greater information [see (12.10)].
§These results are aggregated over experts' predictions for periods 2 and 4, respectively; for periods 3 and 5, see text. Items weights were not used in analyzing this data.

The following points emerge from Table 15.2:

- The optimal DM is always better calibrated and less informative than the best expert; however, his unnormalized virtual weight dominates that of the best expert.
- The optimal DM is better calibrated *and* more informative than the equal weight DM, except in the ESTEC-2 and DSM-3 studies; in unnormalized weight, the optimal DM always dominates the equal weight DM. The equal weight DM is inferior to the best expert in one of the studies.
- The use of item weights yields a marginal improvement in two studies (DSM-1, ESTEC-3) and significantly degrades performance in ESTEC-1.

Supplement A to this chapter contains the expert and optimal DM scores, the DM's assessments for the seed variables, and the realizations, for the continuous-version applications (with the exception of DSM-3, which was finished too late for inclusion). Supplement B contains the experts' assessments for the seed variables for these applications.

In accordance with the discussion in Chapter 12, a uniform background measure was used for all seed variables whose realization fell between 0.001 and 1000; otherwise a loguniform background measure was used. To get an impression of the importance of the background measure, Table 15.3 gives the results for the optimal DM in the five continuous-version studies after changing the background measure. "U" and "L" indicate that all variables have been referred to the uniform or loguniform background measure, respectively. In one study, ESTEC-1, this made a substantial difference in the optimal DM. In the other studies the effect was marginal.

The ESTEC-2 study illustrates the complications that can arise as the number of seed variables becomes large. Table 15.4 shows the scores under the power levels 1.0, 0.70 and 0.60. At power 1.0 all experts receive the minimal calibration score of 10^{-4}. At power 0.60 expert 6 is a factor 10 better calibrated than the other experts. The optimal DM is obtained at power level 0.70, at which expert 6's calibration dominates by a factor 5. When the accuracy of the chi square distribution is

Table 15.3 Results of Changing the Background Measure

Study	Cal (DM)	Mean Seed Info. (DM)	Unnormalized Weight (DM)	Normalized Weight (DM)
ESTEC-1L	0.15	1.07	0.51	0.43
ESTEC-1U	0.20	1.97	0.40	0.31
DSM-1L	0.66	2.58	1.70	0.47
DSM-1U	0.66	0.77	0.51	0.43
DSM-2L	0.70	1.94	1.35	0.98
ESTEC-2L	0.57	1.74	0.99	0.997
ESTEC-2U	0.78	0.32	0.25	0.996
ESTEC-3L	0.65	0.82	0.54	0.996
ESTEC-3U	0.27	1.92	0.52	0.95

L denotes log scale for all variables. U denotes uniform scale for all variables.

Table 15.4. Expert and Optimal DM Scores for ESTEC-2 at Power Levels 1.0, 0.70, and 0.60

Case name: ESTEC-2   16.7.89   CLASS system

Results of Scoring Experts

Weights: global   DM optimization: yes
Significance level: 0   Calibration power: 1.0

Expert Name	Calibr.	Mean Rel. Infor.		Number Realiz.	Unnorm. Weight	Normalized Weight	
		Total	Realiz.			No DM	With DM
1	0.00010	1.655	1.848	26	0.00018	0.23115	0.00599
2	0.00010	2.269	1.337	26	0.00013	0.16729	0.00434
3	0.00010	1.747	1.084	26	0.00011	0.13567	0.00352
4	0.00010	2.901	2.285	26	0.00023	0.28594	0.00741
5	0.00010	1.465	0.831	26	0.00008	0.10399	0.00270
6	0.00010	1.793	0.299	26	0.00003	0.03744	0.00097
7	0.00010	1.452	0.308	26	0.00003	0.03851	0.00100
DM	0.07000	0.795	0.429	26	0.03004		0.97408

Case name: ESTEC-2   16.7.89   CLASS system

Results of Scoring Experts

Weights: global   DM optimization: yes
Significance level:0   Calibration power: 0.7

Expert Name	Calibr.	Mean Rel. Infor.		Number Realiz.	Unnorm. Weight	Normalized Weight	
		Total	Realiz.			No DM	With DM
1	0.00010	1.655	1.848	26	0.00018	0.20105	0.00075
2	0.00010	2.269	1.337	26	0.00013	0.14550	0.00054
3	0.00010	1.747	1.084	26	0.00011	0.11800	0.00044
4	0.00010	2.901	2.285	26	0.00023	0.24870	0.00093
5	0.00010	1.465	0.831	26	0.00008	0.09045	0.00034
6	0.00050	1.793	0.299	26	0.00015	0.16280	0.00061
7	0.00010	1.452	0.308	26	0.00003	0.03350	0.00012
DM	0.78000	0.764	0.316	26	0.24610		0.99628

Case name: ESTEC-2   16.7.89   CLASS system

Results of Scoring Experts

Weights: global   DM optimization: yes
Significance level: 0   Calibration power: 0.6

Expert Name	Calibr.	Mean Rel. Infor.		Number Realiz.	Unnorm. Weight	Normalized Weight	
		Total	Realiz.			No DM	With DM
1	0.00010	1.655	1.848	26	0.00018	0.17290	0.00082
2	0.00010	2.269	1.337	26	0.00013	0.12513	0.00060
3	0.00010	1.747	1.084	26	0.00011	0.10148	0.00048
4	0.00010	2.901	2.285	26	0.00023	0.21388	0.00102
5	0.00010	1.465	0.831	26	0.00008	0.07778	0.00037
6	0.00100	1.793	0.299	26	0.00030	0.28001	0.00133
7	0.00010	1.452	0.308	26	0.00003	0.02881	0.00014
DM	0.92000	0.757	0.243	26	0.22318		0.99524

extended to six significant decimals, expert 6's calibration dominates by a factor 10, and the DM in this case is identical to the DM with power 0.6.

THE ROYAL DUTCH METEOROLOGICAL STUDY

This study is of particular interest, as it involves a very large number of realizations. The data emerge from an experiment conducted by the Royal Dutch Meteorological Institute at Zierikzee in The Netherlands over the period 1980 to 1986. The data were made available for a research project carried out jointly by the T.U. Delft and the Delft Hydraulics Laboratory, and summary results are described here.

The predictions concerned three magnitudes: precipitation, visibility, and wind speed. Wind speed is the primary variable of interest at Zierikzee, as it impacts most directly on the integrity of the sea dikes. The resident experts have the most experience with this variable and are assisted by computer models yielding categorical (i.e., not probabilistic) predictions. Visibility has become a variable of interest in recent years as the station at Zierikzee became involved in supporting navigational safety in neighboring sea lanes. The experts at Zierikzee had no experience in predicting precipitation, and this variable was included for purposes of comparison with other studies. Results from the first 2 years of this experiment were published by Daan and Murphy (1982) and Murphy and Daan (1984).

Four (later five) experts assessed the probabilities that these magnitudes would exceed certain fixed threshold values, and the assessments were discretized to take values in the intervals (0%, 10%), [10%, 20%),.... Hence, there were 10 probability bins, corresponding to the average probabilities 5%, 15%,...,95%. The data structure is somewhat complicated, as only one expert was on duty at one time, and each threshold exceedence event was assessed for periods of 6, 12, 18, 24, and 30 hours into the future. The assessments were staggered as the experts relieved one another at 12-hour intervals, and there was always one expert on duty. Hence, the 24-hour assessments of the "morning expert" concerned the same events as the 12-hour assessments of the "evening expert." The data structure is shown in Figure 15.1. All predictions and realizations have been retrieved.

Because of the choice of threshold values, and because of the inherent

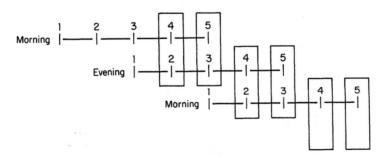

Figure 15.1 Data structure for combining assessments of morning and evening experts. Vertically aligned assessments concern the same exceedence event. Assessments within one block were combined.

Table 15.5. Unnormalized Weights for Annual, Cumulative DMs, Equal Weight DM, Log odds DM, and 'Aggregate expert' (see text).

Unnormalized Response Weight for the Combination of Periods 2 and 4

| Year | Threshold | DM | | | | Experts |
		Annual	Cumulative	Equal	Log-Odd	Period 2
1982	FF22	2.50083*		1.80919	1.49413	0.76766
	VV4	0.48998		0.18199	0.50863*	0.10254
	VV10	0.02953		1.83412*	1.02889	0.00262
1984	FF22	0.24761	0.05743	0.00702	0.26244*	0.00402
	VV4	0.13880	0.36060*	0.11166	0.07916	0.00075
	VV10	0.23615	0.19918	0.26211	1.62923*	0.00523
1985	FF22	2.67633*	2.63388	1.23094	0.34779	0.24495
	VV4	0.15451	0.01652	0.03438	0.45423*	0.00079
	VV10	1.72290*	1.41895	0.84495	1.01982	0.00946
1986	FF22	1.17939	0.17496	0.91024	2.50876*	0.05787
	VV4	0.00688*	0.00041	0.00010	0.00516	9 E-8
	VV10	0.11262*	0.01354	0.09181	0.02264	<0.5 E-9
Calindex		20.07		56.27	40.99	133.28
Res. ent.		0.419		0.434	0.400	0.371

Unnormalized Response Weight for the Combination of Periods 3 and 5

| Year | Threshold | DM | | | | Experts |
		Annual	Cumulative	Equal	Log-Odd	Period 3
1982	FF22	0.24236		0.38277	2.25414*	1.73017
	VV4	1.79570*		1.11183	0.74028	0.09367
	VV10	0.00024		0.28832*	0.04349	0.00032
1984	FF22	0.54751*	0.19399	0.08050	0.14416	0.00505
	VV4	0.00005	0.00725*	0.00128	0.11022*	0.00004
	VV10	0.00006	0.00036*	0.00059	0.00002	1 E-8
1985	FF22	0.92357*	0.00594	0.56786	0.12644	0.34340
	VV4	0.05219	0.01434	0.00625	0.12779*	0.00081
	VV10	0.36606	0.29209	0.00054	0.47969*	0.15394
1986	FF22	0.92357*	0.45599	0.10284	0.35244	0.20517
	VV4	0.05219*	0.00000	0.00000	0.00036	<0.5 E-9
	VV10	0.36606*	0.00000	0.00169	0.00038	<0.5 E-9
Calindex		102.22		103.90	102.62	242.09
Res. ent.		0.433		0.456	0.416	0.399

Asterisk indicates best performance per event-year.

predictability of some of the exceedence events, some of the exceedence events were excluded from the analysis. These were events that were very predictable, and the expert assessments were strongly clustered about the extreme probability values. For such events combination of expert opinions is irrelevant, as the assessments agree in large measure. All assessments 6 hours into the future were excluded, as these strongly clustered into the extreme probability bins.[3] Preliminary analysis identified three exceedence events that were sufficiently unpredictable to make combination of expert assessments meaningful. These were the events "wind speed exceeds 22 knots" (coded FF22), "visibility is less than 4 km" (coded VV4), and "visibility is less than 10 km" (coded VV10). For these three exceedence events the assessments of periods 4 and 5 of the "morning expert" could be combined with the assessments of periods 2 and 3 of the "evening expert." In Figure 15.1, vertically aligned assessments concerned the same events, and assessments within one block were combined.

For each individual expert and each separate period, weights were computed according to the basic classical model. Since each event was predicted by only two experts, and since a given pair of experts occurred relatively infrequently in the data set, no optimization was applied. Rather the expert–period weights were simply normalized to form the assessments of the decision maker. The weights were calculated in two ways. The "annual DM" is computed using weights derived from the experts' performance in the preceeding year, and the "cumulative DM" is computed using weights derived from all previous years. These weights determine the DMs' assessments in the current year. Because of changes of personnel, new experts entered the data set in 1983, hence there is no DM for 1983. Moreover, because of sickness, leaves of absence, etc., the numbers of assessments of the individual experts differed substantially, and equalization of calibration power was applied.

Table 15.5 shows the unnormalized weight of the annual and cumulative DM, per exceedence event, per year. The weights for the "aggregate expert" corresponding to the predictions less far into the future are also shown. In combining the assessments for period 3 and period 5, this is the "period 3 expert," and in combining the assessments for period 2 and period 4, this is the "period 2 expert." Of course, on different days the period 3 expert is actually a different individual. The expert weights shown in Table 15.5 are derived from the aggregate period performance, and are not the weights used in combining, as the latter depend on the individual experts. The assessments of the period 3 aggregate expert are consistently better than those of the period 5 aggregate expert (and similarly for periods 2 and 4).

Clemen and Winkler (1987) found that averaging the log-odds of the experts' assessments to form the log-odds of the decision maker yielded the best performance in a study combining precipitation forecasts. This is equivalent to taking the

---

[3] In most cases the realization was defined as an average over the 6-hour interval, making it very easy to predict the variables for the first period.

normalized geometric mean of the experts probabilities.[4] The second best performance was gotten by taking the arithmetic average of the experts' probabilities, that is, assigning the experts equal weight. The log-odds combination gives relatively more weight to assessments near zero and one. If low entropy correlates with good calibration, then the log-odds rule gives good performance. This was indeed the case in the RDMI data set. Table 15.5 also shows the results of the "log-odds DM" and the "equal DM." An asterisk indicates the best performance per event-year. The total Calindex and Response Entropy scores are also shown (see also Table 15.2).

From Table 15.5 it is evident that the annual DM outperforms the other DMs and outperforms the best expert. The annual DM dominates all others in 11 of the 24 cases. The log-odds DMs dominate in 9 cases, the equal weight DM dominates three times, the cumulative DM once, and the best expert is never dominant. Without a structured combination of expert assessments, the DM would perforce adopt the assessments of the period 3 or, the period 2 expert depending on the time at which the assessment was needed. Any of the DMs would have yielded better performance, and the classical model computed on an annual basis yields the best performance.

In total, the cumulative DM made 2000 assessments per variable, the yearly DM made 2400 assessments per variable, and the "best expert" made 2560 assessments per variable. Interestingly, the yearly DM generally outperforms the cumulative DM, indicating that fluctuations in expert performance are significant.

In Chapter 12 it was noted that correlation in expert assessments could degrade model performance if the correlation on the set of events that actually occurred differed greatly from the correlation on the set of events that did not in fact occur. In the RDMI data set the correlation was about 0.5 on both sets.

The Royal Dutch Meteorological Institute, of course, did not have access to the classical model for combining expert opinions while the experiment was underway. This analysis is after the fact. However, had the model been on-line, it would have improved the experts' probabilistic forecasts.

BAYESIAN ANALYSIS

As indicated in Chapter 13, the Bayesian model of Mendel and Sheridan is compatible with the classical model for continuous variables. Hence, it is possible to feed the expert data from the above studies into the Bayesian model and evaluate the model's performance. A full analysis is presented for the ESTEC-1 case, and

---

[4] Let p_1 and p_2 denote the probabilities of experts 1 and 2, and let q be the probability of the decision maker. Then q solves

$$\ln \frac{q}{1-q} = \frac{1}{2}\left(\ln \frac{p_1}{1-p_1} + \ln \frac{p_2}{1-p_2}\right)$$

so that

$$q = \frac{(p_1 p_2)^{1/2}}{(p_1 p_2)^{1/2} + [(1-p_1)(1-p_2)]^{1/2}}$$

summary results for the DSM-1, -2, ESTEC-2, and -3 are given in Tables 15.7 and 15.8.

From a Bayesian viewpoint, it is most natural to evaluate performance by calculating cumulative loss for the seed variables. This involves choosing a loss function for each seed item and computing the loss as a function of the realization and the Bayesian estimate for that item. As emphasized in Chapter 13, adding losses from different items presupposes that the items are measured in units of equal utility. For example, if one item is measured in kilograms and another in meters, then the disutility of one kilogram error on the first item must be equal to the disutility of one meter error (in the same sense) for the second. Such judgments, of course, are highly subjective and problem dependent, and do not normally form part of a quantitative analysis.

In this analysis we simply assume that the units in which the items are measured are units of equal utility, and we compute loss under three bilinear loss functions for which the Bayesian estimate corresponds to the 5%, 50%, and 95% quantiles (see Chap. 13). If loss is proportional to r times the distance between the estimate and the true value when overestimating and s times this distance when underestimating, then his minimum expected loss is obtained when his estimate is the $s/(r + s)$th quantile of his distribution.

The Bayesian analysis is first performed for each of the four experts individually. Figure 15.2 shows the cumulative losses for expert 1 and for the "Bayesian updated expert 1" (the variables all concern relative frequencies, and the loss is displayed on a log scale). The Bayesian updated expert 1 is the result of applying the Bayesian model to expert 1 and updating his distribution after each true value is learned. Three loss functions are used, for which the optimal estimates are the 5%, 50%, and 95% quantiles, respectively. Figure 15.3 to 15.5 provide the same information for experts 2 to 4. Figure 15.6 shows the cumulative losses of the

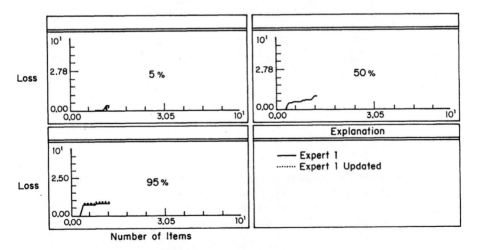

Figure 15.2 Bilinear cumulative loss for Bayesian estimates at 5%, 50%, and 95% quantiles for expert 1, compared with expert 1 updated.

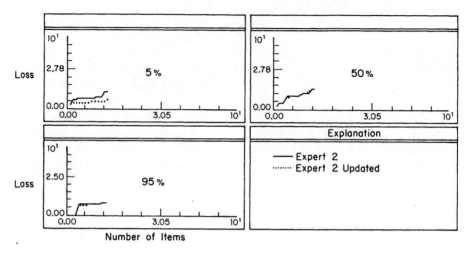

Figure 15.3 Bilinear cumulative loss for Bayesian estimates at 5%, 50%, and 95% quantiles, for expert 2, compared with Bayesian updated expert 2.

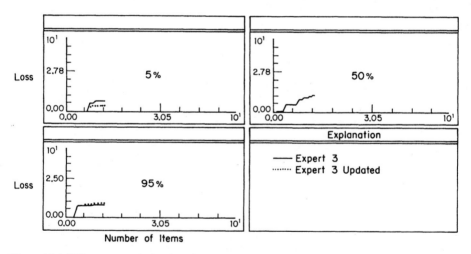

Figure 15.4 Bilinear cumulative loss for Bayesian estimates at 5%, 50%, and 95% quantiles for expert 3, compared with Bayesian updated expert 3.

Mendel and Sheridan combination of experts 3 and 4 (these are the "best" experts in the classical model), and compares the losses with experts 3 and 4 individually. The graphs in Figure 15.6 have been blown up, and we see that the Mendel and Sheridan combination yields a cumulative loss that is between that of experts 3 and 4.

Table 15.6 shows the results of feeding the Bayesian updated experts back into the classical model. For each variable, we extract from the Bayesian updated

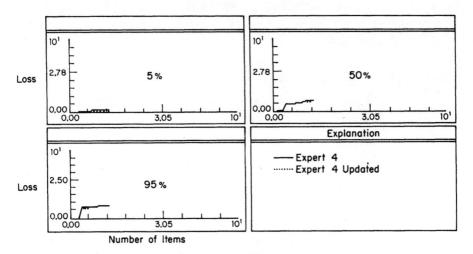

Figure 15.5 Bilinear cumulative loss for Bayesian estimates at 5%, 50%, and 95% quantiles for expert 4, compared with Bayesian updated expert 4.

distribution the 5%, 50%, and 95% quantiles, and determine the calibration and information scores corresponding to these assessments. In each case we see that the Bayesian updating improves calibration at the expense of information. Indeed, on overconfident assessors the updating has the effect of driving the outer quantiles further apart, making the updated assessor less overconfident.

Also shown are the optimized decision maker and the Mendel-Sheridan combination of experts 3 and 4. It is remarkable that the Mendel-Sheridan combination is quite poorly calibrated. This illustrates the features discussed at length in Chapter 13, namely, that this combination involves a prior that is more informative than the experts' distributions, and requires a large number of variables to move significantly away from the prior. The information scores for the original experts differ slightly from those in Table 15.3 because the additional experts influence slightly the intrinsic ranges for each variable.

The Bayesian model seems to perform well on this data, when applied to individual experts. Whether analyzed in terms of loss functions or in terms of unnormalized weight, the updated experts perform well, the unnormalized weight of updated experts 3 and 4 dominates that of the classical DM. One imporotant caveat must be made here. As mentioned in Chapter 13, the Bayesian model is very sensitive to the choice of cutoff points for each variable. In the classical model applied to the original experts, the cutoff points for each variable were determined as "a little larger/smaller" than the highest/lowest quantile for each variable. If we had chosen the cutoff points much more spread out, this would affect all the original experts' information scores in more or less the same way, and would not influence the placement of their quantiles. However, in performing the Bayesian updating the cutoff points had to be more spread out, to allow the model "room to recalibrate." The cutoff points for all variables were set at 1E-13, $1 - 1\text{E-}13$ (the estimates all concerned relative frequencies). This choice directly influences the

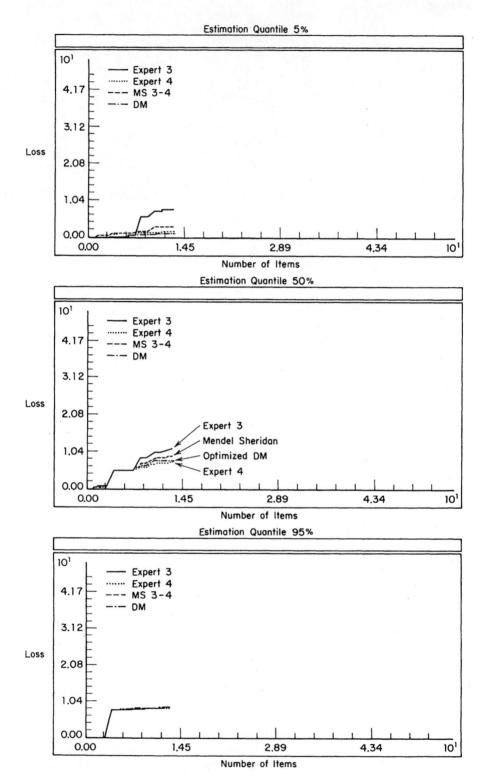

Figure 15.6 Bilinear cumulative loss for Bayesian estimates at 5%, 50%, and 95% quantiles, for optimized decision maker and updated optimized decision maker.

placement of the 5% and 95% updated quantiles. A different choice, say 1E-99, $1 - E$-99, would give the original experts much higher information relative to the updated experts (computed on a logarithmic scale). In applying the Bayesian model, even in the single-expert case, the user must be acutely aware of the affects of choosing cutoff points.

Table 15.6 underscores the reservations in Chapter 13 regarding the Mendel-Sheridan Bayesian combination of expert assessments. Practical guidelines for the applicability of this model, or of possible variants, cannot be given at present.

Table 15.7 shows the cumulative bilinear losses for the experts and for the classical DM on all five continuous-version applications. Interestingly, the total losses do not correspond particularly well with the classical weights. In three of the five studies the classical DM has the lowest total loss.

Given the problems with a Bayesian combination of experts, it is natural to explore the possibility of updating the individual experts using Bayes' theorem, and combining the updated experts according to the classical model. Such a method is philosophically inconsistent, as the classical model is based on the theory of proper scoring rules, whereas updating before combination destroys the propriety of the weights. Nonetheless, curiosity cannot be suppressed, and the results are shown in Table 15.8. Comparing the results with Table 15.2, in all five cases the DM's unnormalized and normalized weight is worse in Table 15.8. The data give little encouragement for this type of hybrid model.

Table 15-6. Results of Feeding the Bayesian Updated Experts and the Mendel-Sheridan Combination of Experts 3 and 4 into the Classical Model and Comparing the Calibration, Information, and Weights with the Four Original Experts and with the Optimized Decision Maker Class DM. Changes in the intrinsic ranges (see text) cause the information scores to differ slightly from those in supplement A for the non-updated experts.

Case name: ESTECBTH		9.7.89					CLASS system

Results of Scoring Experts

Weights: global   DM optimization: no
Significance level: 0   Calibration power: 1.0

			Mean Rel. Infor.		Number	Unnorm.	Normalized Weight	
Expert Name		Calibr.	Total	Realiz.	Realiz.	Weight	No DM	With DM
1	Exp. 1	0.04000	2.210	2.210	13	0.08842	0.01533	0.01419
2	Exp. 2	0.00100	4.592	4.592	13	0.00459	0.00080	0.00074
3	Exp. 3	0.14000	3.037	3.037	13	0.42522	0.07373	0.06826
4	Exp. 4	0.14000	2.049	2.049	13	0.28687	0.04974	0.04605
5	Update-1	0.45000	2.213	2.213	13	0.99564	0.17265	0.15983
6	Update-2	0.14000	3.683	3.683	13	0.51564	0.08941	0.08277
7	Update-3	0.45000	2.528	2.528	13	1.13741	0.19723	0.18259
8	Update-4	0.92000	1.539	1.539	13	1.41610	0.24556	0.22732
9	MS (3, 4)	0.00100	2.503	2.503	13	0.00250	0.00043	0.00040
10	Class DM	0.43000	2.080	2.080	13	0.89446	0.15510	0.14358

Table 15-7. Cumulative Bilinear Losses for Experts and for the Classical DM.

Study	5%	50%	95%	Total
ESTEC-1				
Exp. 1	0.31	2.37	1.49	4.17
Exp. 2	0.24	1.83	2.68	4.75
Exp. 3	0.24	1.71	1.38	3.33
Exp. 4	0.28	1.66	1.78	3.78
DM	0.25	1.48	1.59	3.32*
DSM-1				
Exp. 1	4.4 E05	4.3 E06	8.1 E06	1.3 E07
Exp. 2	1.5 E05	9.5 E06	1.5 E05	9.8 E06
Exp. 3	4.3 E05	4.3 E06	8.0 E06	1.3 E07
Exp. 4	4.4 E05	4.4 E06	8.2 E06	1.3 E07
Exp. 5	1.2 E05	3.2 E05	8.6 E04	5.2 E05
Exp. 6	7.4 E04	5.9 E05	8.8 E04	7.5 E05
Exp. 7	9.1 E04	6.1 E05	3.0 E04	7.3 E05
Exp. 8	9.1 E04	3.1 E05	1.5 E05	5.5 E05
Exp. 9	2.1 E05	1.6 E06	2.9 E06	4.7 E06
Exp. 10	2.4 E05	2.1 E06	3.6 E06	5.9 E06
DM	1.5 E05	7.4 E04	1.5 E05	3.7 E05*
DSM-2				
Exp. 1	2.7 E02	5.7 E02	1.5 E02	9.9 E02*
Exp. 2	4.3 E02	3.8 E03	6.7 E03	1.1 E04
Exp. 3	2.7 E03	3.1 E03	6.1 E02	6.4 E03
Exp. 4	1.3 E03	4.7 E03	6.1 E03	1.2 E04
Exp. 5	8.4 E01	4.7 E02	5.2 E02	1.1 E03
Exp. 6	5.2 E02	9.6 E02	7.1 E02	2.2 E03
Exp. 7	8.7 E02	1.6 E03	1.2 E03	3.7 E03
Exp. 8	4.2 E02	3.9 E03	6.4 E03	1.1 E04
DM	1.3 E02	1.0 E03	2.3 E02	1.4 E03
ESTEC-2				
Exp. 1	2.3 E03	2.0 E04	3.2 E04	5.4 E04
Exp. 2	6.9 E03	1.7 E04	1.3 E04	3.7 E04
Exp. 3	1.9 E04	1.8 E04	6.8 E03	3.9 E04
Exp. 4	6.1 E03	1.9 E04	2.4 E04	4.9 E04
Exp. 5	1.6 E04	1.9 E04	5.7 E03	3.5 E04
Exp. 6	4.8 E03	1.6 E04	3.7 E03	2.5 E04
Exp. 7	1.3 E04	2.0 E04	3.5 E03	3.6 E04
DM	2.2 E03	1.7 E04	3.9 E03	2.3 E04*
ESTEC-3				
Exp. 1	0.11	0.35	0.15	0.61
Exp. 2	0.02	0.07	0.11	0.20
Exp. 3	0.04	0.11	0.16	0.31
Exp. 4	0.007	0.14	0.13	0.28
Exp. 5	0.02	0.08	0.16	0.26
Exp. 6	0.04	0.06	0.04	0.14*
DM	0.02	0.11	0.04	0.17

An asterisk indicates the lowest loss in each study.

Table 15-8. Basic Model, Optimized DM, with Mendel-Sheridan
Updated Experts. Information is measured as in Table 2.

Study	Cal(DM)	Mean Seed Infom. (DM)	Unnormalized Weight (DM)	Normalized Weight (DM)
ESTEC1	0.20	0.67	0.13	0.05
DSM1	0.53	0.66	0.35	0.07
DSM2	0.70	0.59	0.41	0.50
ESTEC2	0.89	0.24	0.21	0.17
ESTEC3	0.47	0.74	0.35	0.50

FLANGE CONNECTIONS AT A PROCESS PLANT— PAIRED COMPARISONS

The following two case studies were undertaken at DSM, a large chemical process plant in the south of The Netherlands, to determine the probabilities of various failure causes of flange connections. A previous attempt by plant engineers to assess the probabilities of various failure modes via an internal questionnaire was unsuccessful, and it was decided to apply structured expert opinion to this problem. The present application utilized the psychological scaling model, using the method of paired comparisons, and a subsequent study used the classical model. It was hoped that the results of the classical model would generate reference probabilities for transforming the scale values emerging from the paired comparison study. It was decided to focus on one particular plant.

The Failure Causes

Different categories of failure causes can be distinguished for flange connections. Failures may arise from engineering failures or from operations failures. Engineering failures can be divided into design and construction flaws, and operations failures can be divided into environmental and procedural. This yields four subcategories of failure causes. After consultation with plant engineers, it was decided to concentrate on procedural and environmental causes. Hence two groups of failure causes were distinguished corresponding to environmental and procedural failures. Group one consisted of six environmental failures, and one procedural cause, "improper mounting" was added to the list to link to the procedural subgroup:

Group one: Environmental failures

1. Changes in temperature caused by process cycles
2. Temperature gradient on the contour of the flange
3. Corrosion of bolts/nuts
4. Corrosion of the flange
5. Aging of the gasket
6. Aging of bolts/nuts
7. Improper mounting

Group two subdivides the failure cause "improper mounting" into the following 10 specific procedural failure causes:

Group two; Procedural Failures:

7.1   Face of the flange not/insufficiently cleaned
7.2   Damaged face of the flange
7.3   Wrong gasket type
7.4   Bolts badly tightened due to difficult accessibility
7.5   Gasket stuck/greased while mounting
7.6   Bolts improperly tightened (flanges not parallel and pipe not in line; following a wrong tightening pattern)
7.7   Gasket not centered
7.8   Bolts/thread ends handled improperly (not made suitable, not greased)
7.9   Gasket damaged
7.10 Warm bolt connection not retightened during startup.

Preparation of the Elicitation

It was decided to apply the method of paired comparisons to obtain both a ranking and the relative scale values of the probabilities in groups one and two. At each comparison of two failure causes the following question was asked:

> "Which of these two causes has led more often to a failure of flanged connections at the plant?"

The paired comparison was split into two rounds. The first round involved comparisons of all pairs in group one, round two involved comparison of all pairs in group two. Both rounds were concluded by asking the experts for a direct ranking of the failure causes considered in each round.

In order to transform the paired comparison values into absolute values, reference values for one (in the case of Bradley-Terry) or two (in the case of Thurstone) scale values must be supplied. Since no reference values were available, it was decided to use the classical model to generate reference values. The classical model was applied to assess the probabilities for failure causes 1, 5, 7, 7.1, 7.6, and 7.10. More specifically, a subgroup of experts were asked how many times in the past 10 years each of these causes had occurred. The application of the classical model is described in the following section.

The Experts

The plant engineers selected 14 experts to participate in the study. They were operators, mechanics, maintenance engineers, mechanical engineers, and their chiefs. Ten of the experts had the equivalent of a bachelor of science degree, and these ten experts were selected for the classical model. All experts participated in the paired comparison study. The set of all experts is referred to as subgroup A. For each round a subgroup B of experts was selected, consisting of the experts having the fewest circular triads, that is, inconsistencies in preference (see Chap. 14). Each expert was interviewed individually. One plant representative and one member of the analysis team were present during each interview. Each interview consisted of the following parts:

- An introduction, intended to motivate the expert. The purpose of the interview and the type of questions in the method of paired comparisons were explained, and an overview of all failure causes was given.
- The actual paired comparisons data collection (rounds 1 and 2). The questionnaire was filled in by the expert himself. Direct rankings of the failure causes was elicited. For the 10 experts participating in the classical model study, the interview also included the following parts:
  —An introduction to the classical model, including a short training, intended to teach the experts how to quantify their uncertainty
  —The actual classical data collection

Results

The paired comparison data was analyzed with computer programs whose output results are included in supplement C to this chapter. Comparison of the results of the rankings based on paired comparisons with the direct rankings shows a high rank correlation, although the rankings typically differ with respect to the highest and lowest ranked items.

The experts' paired comparisons were analyzed for inconsistencies in preference, or circular triads. The average number of circular triads seems to be quite low, indicating stable preferences. The maximum number of circular triads for which the null hypothesis H_0 that the preferences are at random is still rejected at the significance level $\alpha = 0.05$ is 4 in the case of 7 objects (round 1) and 21 in the case of 10 objects (round 2). Only for expert e_{12} was H_0 not rejected in round one.

The coefficient of agreement u and the coefficient of concordance W were computed for subgroups A and B. Using the statistical tests described in Chapter 14, the null hypothesis H_0 that the preferences were at random was rejected at the 5% level for both groups in both rounds.

In order to transform the scale values to absolute values, assessments from the classical model were used as reference values. The following reference values were available:

Item Number in Classical Model	Failure Cause Number	Reference Value
9	1.0	1.13/year
10	5.0	0.64/year
11	7.0	2.24/year
12	7.1	0.55/year
13	7.6	1.62/year
14	7.10	0.45/year

Note that the value for failure number 7 should be greater than the sum of the values for failure numbers 7.1, 7.6, and 7.10. This need not point to a real inconsistency, as the reference values are medians of uncertainty distributions. On the other hand, it may be that detailed consideration of all subdivisions of failure cause 7 leads to higher assessments for "improper mounting" (number 7) than simply considering this cause in isolation.

Since the Thurstone model requires two reference values, there are three possible combinations of two reference values in each round. For all possible combinations, the scale values were transformed into absolute values for subgroup B. From the results it must be concluded that transformation is problematic in this case. The ranking of the reference values contradicts that of the scale values. The variation in the absolute values, given different reference values, is so large that determination of confidence intervals was deemed meaningless. Modeling uncertainty, that is, uncertainty with regard to the validity of the modeling assumptions, clearly swamps the sampling fluctuations represented by the confidence bounds.

Focusing on the untransformed values, the following results were communicated as representing a "group opinion" on the ranking of failure causes. The results of round one point to three important causes, cause 7 (improper mounting), cause 5 (aging of gasket), and cause 1 (changes in temperature caused by process cycles). Causes 2, 3, 4, and 6 are failure causes of lower order.

In the category of improper mounting, cause 7.4 (bolts badly tightened due to difficult accessibility) is deemed most important, followed by 7.7 (gasket not centered), 7.6 (bolts improperly tightened), and 7.10 (warm bolt connection not retightened during startup) in this order. Other causes are of lower order.

The following conclusions were drawn with respect to the preformance of the psychological scaling models and the method of paired comparisons:

- The paired comparison exercise went quite smoothly, and experts generally enjoyed having their expertise extracted in this manner. There was lively interest in the results.
- The rankings emerging from the paired comparison exercises appear meaningful and were accepted as such.
- The paired comparison scale values do not agree well with the results of the classical model, for those values assessed by both methods. This may not be surprising, as the paired comparison method treats all experts equally, whereas the classical model concentrates the weight on 2 of the 10 experts (see next case study).
- Three reference values for the Thurstone model are provided by the classical model, but these values are not affine related to the scale values emerging from the Thurstone model. Hence, the assumptions underlying the Thurstone model are not satisfied when the classical model's values play the role of true values in this case. The same conclusion applies to the Bradley-Terry (NEL) model, which requires only one reference value. Of course, it is not known whether these underlying assumptions would be satisfied with the true values themselves. The experience with this application teaches that it is advisable to have more than two reference values, in order to have a check on the modeling assumptions, before the transformation to absolute

values can be used with confidence. It is provisionally concluded that the psychological scaling models' primary strength lies in "consensus building" with regard to the ranking of objects. The transformation of scale values to absolute values must be approached with reservations.

FLANGE CONNECTIONS AT A PROCESS PLANT—THE CLASSICAL MODEL

As indicated above, the classical model was used to generate reference values for the paired comparison study of flange connection failure causes. Ten experts each assessed 14 items in total, of which the first 8 were seed variables. The elicitation was accomplished by directly assessing the 5%, 50%, and 95% quantiles of the uncertainty distributions for each variable. A brief training session preceded the elicitation.

The scores for the decision maker are shown in Table 15.9 for significance levels 0 and 0.14. Also shown are the decision maker's scores when the experts are assigned equal weights a priori. The decision maker's unnormalized weight is maximal at significance level 0.14. At this level, the decision maker is better than the weightiest expert, expert number 2, who receives weight 0.84. At significance level 0 all experts contribute to the decision maker, though unequally. The decision maker's calibration score is unaffected, but his information score is degraded. Substantial degradation results from assigning all experts equal weight. It is interesting that in this case the optimization is driven by the decision maker's information score, and not by his calibration score.

At significance level 0, experts 2 and 8 receive jointly about 80% of the weight. These experts are the best calibrated and among the least informative. At the

Table 15-9. Scores for Experts and DM at Significance Level $\alpha = 0.14$ (Optimal) and 0, and for the DM Gotten by Assigning All Experts Equal Weight

Expert	Calibration	Mean Seed Rel. Information	Unnormalized Weight
1	0.001	1.311	0.00131
2	0.54	1.549	0.83621
3	0.001	1.851	0.00185
4	0.02	1.152	0.02303
5	0.001	1.629	0.00163
6	0.0001	2.285	0.00023
7	0.001	2.125	0.00212
8	0.14	1.768	0.24758
9	0.0001	2.255	0.00023
10	0.02	2.275	0.04550
DM			
$\alpha = 0.14$	0.66	1.371	0.90463
$\alpha = 0.0$	0.66	1.091	0.72028
Equal weights	0.53	0.806	0.42740

optimal level of 0.14, their weights are, respectively, 0.74 and 0.26. "Range graphs" presented in Figure 15.7 give a visual appreciation of the results. These graphs show the 90% confidence bands and median assessments for all experts and for the optimal decision maker, for all variables. When present, the realization is also shown. From these graphs it is evident that the classical model produces very meaningful results for the optimal decision maker.

The following conclusions emerge from this application of the classical model:

- The elicitation for the classical model went quite smoothly. Experts were comfortable with the elicitation format.
- The numerical conclusions were accepted by the plant engineers as being quite meaningful. The numerical weights were found quite meaningful by the plant engineer who worked with the analysis team. In fact, he was one of the experts contributing to the optimal decision maker, and he was able to predict who the other contributer was.
- For reasons indicated in the previous section, the attempt to "bootstrap" the paired comparison model with the classical model, by using the latter to supply reference values, was not successful. Inherent differences in the two models caused the classical model's output to violate the assumptions underlying the psychological scaling models.

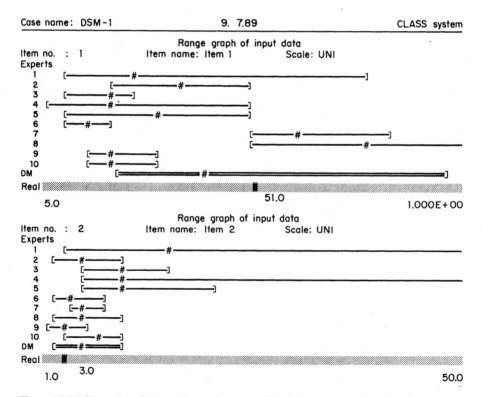

Figure 15.7 Bilinear loss for Bayesian estimates at 5%, 50%, and 95% quantiles, for experts 3, 4, Mendel-Sheridan combination of 3 and 4, and optimized decision maker.

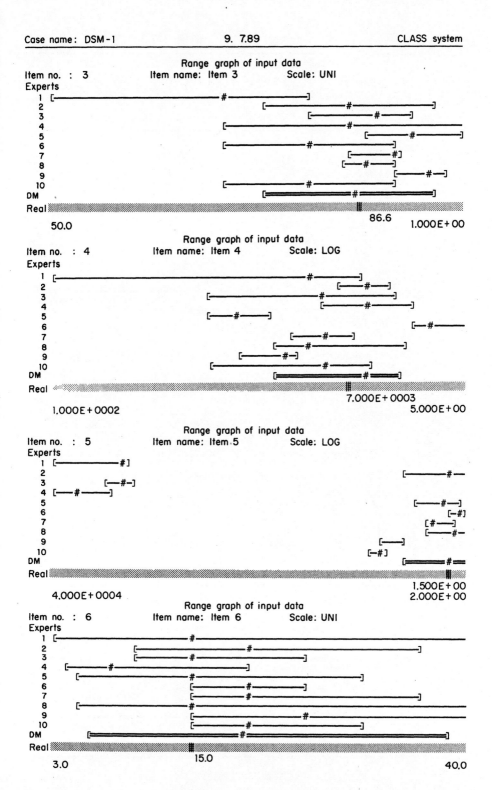

Range graph of input data
Item no. : 3   Item name: Item 3   Scale: UNI

Range graph of input data
Item no. : 4   Item name: Item 4   Scale: LOG

Range graph of input data
Item no. : 5   Item name: Item 5   Scale: LOG

Range graph of input data
Item no. : 6   Item name: Item 6   Scale: UNI

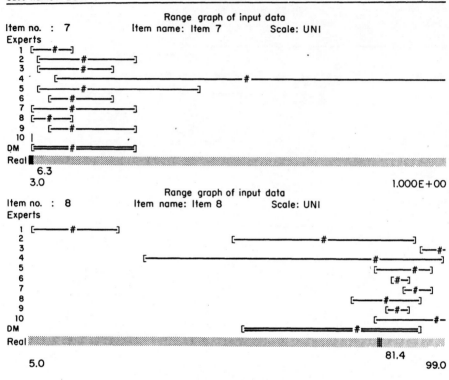

Range graph of input data
Item no. : 7   Item name: Item 7   Scale: UNI

6.3
3.0   1.000E+00

Range graph of input data
Item no. : 8   Item name: Item 8   Scale: UNI

81.4
5.0   99.0

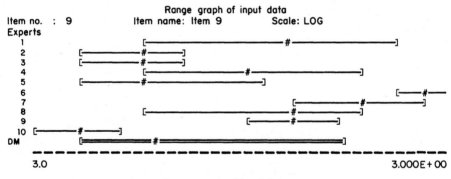

Range graph of input data
Item no. : 9   Item name: Item 9   Scale: LOG

3.0   3.000E+00

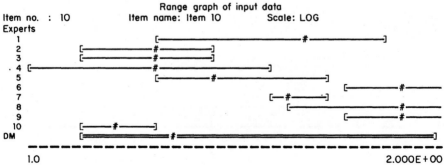

Range graph of input data
Item no. : 10   Item name: Item 10   Scale: LOG

1.0   2.000E+00

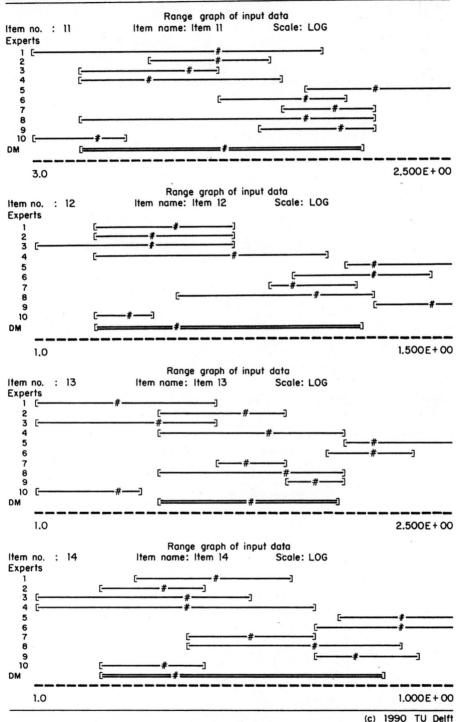

Range graph of input data
Item no. : 11   Item name: Item 11   Scale: LOG

3.0   2.500 E + 00

Range graph of input data
Item no. : 12   Item name: Item 12   Scale: LOG

1.0   1.500 E + 00

Range graph of input data
Item no. : 13   Item name: Item 13   Scale: LOG

1.0   2.500 E + 00

Range graph of input data
Item no. : 14   Item name: Item 14   Scale: LOG

1.0   1.000 E + 00

(c) 1990 TU Delft

SUPPLEMENT A

Results of Scoring Experts

Weights: global   DM optimization: yes
Significance Level: 0.040   Calibration power: 1.0

| Expert Name | Calibr. | Mean Rel. Infor. | | Number Realiz. | Unnorm. Weight | Normalized Weight | |
		Total	Realiz.			No DM	With DM
1	0.04000	2.845	2.141	13	0.08564	0.11000	0.05640
2	0.00100	5.736	4.515	13	0	0	0
3	0.14000	4.675	2.952	13	0.41328	0.53085	0.27220
4	0.14000	1.606	1.997	13	0.27961	0.35915	0.18416
DM	0.43000	2.279	1.720	13	0.73980		0.48725

Resulting Solution (combined DM distribution of values assessed by experts)

Weights: global   DM optimization: yes
Significance level: 0.040   Calibration power: 1.0

| Item Name | Quantiles of Solution | | | Scale |
	5%	50%	95%	
1	0.000461	0.017045	0.617775	U
2	0.001162	0.016693	0.652744	U
3	0.031861	0.174377	0.666953	U
4	0.005931	0.038161	0.186042	L
5	0.006499	0.034745	0.447966	U
6	0.009545	0.04356	0.507074	U
7	0.012795	0.074048	0.551059	U
8	3.597E-0010	3.357E-0009	0.222503	U
9	7.893E-0005	0.02238	0.912989	U
10	8.520E-0011	1.842E-0009	6.603E-0007	L
11	0.002101	0.028637	0.441022	U
12	9.022E-0008	0.029433	0.841162	U
13	5.604E-0010	5.761E-0008	5.922E-0005	L

List of parameters.   Quantile points (%):   5   50   95

No.	Name	Scale	Realization
1	Item 1	UNI	0.07
2	Item 2	UNI	0.15
3	Item 3	UNI	0.2
4	Item 4	LOG	1.00000E-0013
5	Item 5	UNI	0.07
6	Item 6	UNI	0.04
7	Item 7	UNI	0.04
8	Item 8	UNI	0.08
9	Item 9	UNI	0.005
10	Item 10	LOG	8.40000E-0007
11	Item 11	UNI	0.09
12	Item 12	UNI	0.041
13	Item 13	LOG	1.67000E-0007

Results of Scoring Experts

Weights: Global   DM optimization: yes
Significance level: 0.140   Calibration power: 1.0

Expert Name	Calibr.	Mean Rel. Infor.		Number Realiz.	Unnorm. Weight	Normalized Weight	
		Total	Realiz.			No DM	With DM
1	0.00100	1.714	1.311	8	0	0	0
2	0.54000	2.041	1.549	8	0.83621	0.77156	0.42054
3	0.00100	2.130	1.851	8	0	0	0
4	0.02000	1.578	1.152	8	0	0	0
5	0.00100	2.073	1.629	8	0	0	0
6	0.00010	2.547	2.285	8	0	0	0
7	0.00100	2.497	2.125	8	0	0	0
8	0.14000	2.001	1.768	8	0.24758	0.22844	0.12451
9	0.00010	2.577	2.255	8	0	0	0
10	0.02000	2.585	2.275	8	0	0	0
DM	0.66000	1.671	1.371	8	0.90463		0.45495

Resulting Solution (combined DM distribution of values assessed by experts)

Weights: global   DM optimization: yes
Significance level: 0.140   Calibration power: 1.0

| Item Name | Quantiles of Solution | | | Scale |
	5%	50%	95%	
1	20.293181	39.544299	93.189779	U
2	2.0	5.0	10.0	U
3	75.075281	85.993414	94.892696	U
4	2.511E+0003	9.257E+0003	1.528E+0004	L
5	1.000E+0006	1.522E+0006	2.000E+0006	L
6	6.364381	19.470554	37.405702	U
7	15.627639	97.106173	2.494E+0002	U
8	50.217407	75.499197	90.9	U

List of parameters.   Quantile points (%):   5   50   95

No.	Name	Scale	Realization
1	Item 1	UNI	51.0
2	Item 2	UNI	3.0
3	Item 3	UNI	86.6
4	Item 4	LOG	7.00000E+0003
5	Item 5	LOG	1.50000E+0006
6	Item 6	UNI	15.0
7	Item 7	UNI	6.3
8	Item 8	UNI	81.4

Results of Scoring Experts

Weights: global   DM optimization: yes
Significance level: 0.001   Calibration power: 1.0

| Expert Name | Calibr. | Mean Rel. Infor. | | Number Realiz. | Unnorm. Weight | Normalized Weight | |
		Total	Realiz.			No DM	With DM
1	0.00500	1.715	1.834	12	0.00917	0.37984	0.00782
2	0.00010	1.628	1.887	11	0	0	0
3	0.00010	1.331	1.313	12	0	0	0
4	0.00010	2.361	2.214	11	0	0	0
5	0.00100	2.436	2.679	12	0.00268	0.11099	0.00229
6	0.00010	3.547	3.278	12	0	0	0
7	0.00500	3.168	2.458	12	0.01229	0.50917	0.01048
8	0.00010	2.062	1.861	11	0	0	0
DM	0.84000	1.426	1.367	12	1.14824		0.97941

Resulting Solution (combined DM distribution of values assessed by experts)

Weights: global   DM optimization: yes
Significance level: 0.001   Calibration power: 1.0

	Quantiles of Solution			
Item Name	5%	50%	95%	Scale
1	69.48674	2.043E+0002	4.866E+0002	U
2	0.002498	0.005412	0.184828	U
3	51.203294	91.47458	1.993E+0002	U
4	25.671164	82.79404	1.963E+0002	U
5	0.002012	0.003908	0.005698	U
6	1.061E+0002	2.046E+0002	4.887E+0002	U
7	0.010026	0.016508	0.049704	U
8	60.132263	69.85794	88.7393	U
9	35.891794	45.890743	54.875431	U
10	0.005061	0.020157	0.04847	U
11	0.002014	0.003865	0.024214	U
12	0.010029	0.01561	0.049505	U

List of parameters.   Quantile points (%):   5   50   95

No.	Name	Scale	Realization
1	Item 1	UNI	67.3
2	Item 2	UNI	0.0047
3	Item 3	UNI	1.62600E+0002
4	Item 4	UNI	77.0
5	Item 5	UNI	0.0047
6	Item 6	UNI	3.26500E+0002
7	Item 7	UNI	0.0141
8	Item 8	UNI	67.5
9	Item 9	UNI	43.3
10	Item 10	UNI	0.0164
11	Item 11	UNI	0.0259
12	Item 12	UNI	0.0352

Results of Scoring Experts.

Weights: global   DM optimization: yes
Significance level: 0   Calibration power: 0.7

| Expert Name | Calibr. | Mean Rel. Infor. | | Number Realiz. | Unnorm. Weight | Normalized Weight | |
		Total	Realiz.			No DM	With DM
1	0.00010	1.655	1.848	26	0.00018	0.20105	0.00075
2	0.00010	2.269	1.337	26	0.00013	0.14550	0.00054
3	0.00010	1.747	1.084	26	0.00011	0.11800	0.00044
4	0.00010	2.901	2.285	26	0.00023	0.24870	0.00093
5	0.00010	1.465	0.831	26	0.00008	0.09045	0.00034
6	0.00050	1.793	0.299	26	0.00015	0.16280	0.00061
7	0.00010	1.452	0.308	26	0.00003	0.03350	0.00012
DM	0.78000	0.764	0.316	26	0.24610		0.99628

List of parameters.   Quantile points (%):   5   50   95

No.	Name	Scale	Realization
1	Item 1	UNI	2.92000E + 0002
2	Item 2	UNI	24.0
3	Item 3	UNI	1.50000E + 0002
4	Item 4	UNI	97.0
5	Item 5	UNI	8.23000E + 0002
6	Item 6	UNI	2.23000E + 0002
7	Item 7	UNI	27.0
8	Item 8	UNI	2.87000E + 0002
9	Item 9	UNI	3.56000E + 0002
10	Item 10	UNI	5.08000E + 0002
11	Item 11	UNI	1.87000E + 0002
12	Item 12	UNI	12.0
13	Item 13	UNI	5.56000E + 0002
14	Item 14	UNI	20.0
15	Item 15	UNI	5.85000E + 0002
16	Item 16	UNI	6.09000E + 0002
17	Item 17	UNI	5.52000E + 0002
18	Item 18	UNI	1.78000E + 0002
19	Item 19	UNI	87.0
20	Item 20	UNI	88.0
21	Item 21	UNI	5.78000E + 0002
22	Item 22	UNI	1.91000E + 0002
23	Item 23	UNI	84.0
24	Item 24	UNI	33.0
25	Item 25	UNI	5.46000E + 0002
26	Item 26	UNI	6.01000E + 0002

Resulting Solution (combined DM distribution of values assessed by experts)

Weights: global   DM optimization: yes
Significance level: 0   Calibration power: 0.7

| Item Name | Quantiles of Solution | | | Scale |
	5%	50%	95%	
1	1.453773	37.625713	5.018E+0002	U
2	3.274916	44.858171	5.059E+0002	U
3	3.437154	56.920514	5.258E+0002	U
4	3.547954	69.559114	5.520E+0002	U
5	3.606095	80.685451	5.977E+0002	U
6	3.6419	91.814675	6.038E+0002	U
7	3.6645	1.115E+0002	5.934E+0002	U
8	3.677556	1.321E+0002	5.759E+0002	U
9	3.684634	1.477E+0002	5.588E+0002	U
10	3.689713	1.614E+0002	5.505E+0002	U
11	15.08629	1.758E+0002	5.533E+0002	U
12	16.573554	1.881E+0002	5.673E+0002	U
13	17.161634	2.016E+0002	5.740E+0002	U
14	17.486284	2.047E+0002	5.784E+0002	U
15	17.694639	2.154E+0002	5.852E+0002	U
16	17.837223	2.414E+0002	5.919E+0002	U
17	17.939971	2.678E+0002	6.004E+0002	U
18	18.016218	2.940E+0002	6.116E+0002	U
19	18.955909	2.991E+0002	6.262E+0002	U
20	18.67528	3.106E+0002	6.484E+0002	U
21	18.781374	3.293E+0002	6.791E+0002	U
22	18.925831	3.394E+0002	7.120E+0002	U
23	19.906091	3.549E+0002	7.413E+0002	U
24	20.632502	3.786E+0002	7.649E+0002	U
25	21.172987	4.114E+0002	7.845E+0002	U
26	22.114121	4.483E+0002	8.092E+0002	U

Results of Scoring Experts

Weights: global   DM optimization: yes
Significance level: 0.005   Calibration power: 1.0

Expert Name	Calibr.	Mean Rel. Infor.		Number Realiz.	Unnorm. Weight	Normalized Weight	
		Total	Realiz.			No DM	With DM
1	0.00010	1.891	1.235	12	0	0	0
2	0.00010	2.192	2.154	12	0	0	0
3	0.00010	2.139	2.034	12	0	0	0
4	0.00500	2.308	2.549	12	0.01275	0.65518	0.03119
5	0.00010	2.685	3.253	12	0	0	0
6	0.00500	1.586	1.342	12	0.00671	0.34482	0.01641
DM	0.27000	1.295	1.442	12	0.38925		0.95240

Resulting Solution (combined DM distribution of values assessed by experts)

Weights: global   DM optimization: yes
Significance level: 0.005   Calibration power: 1.0

Item Name	Quantiles of Solution			Scale
	5%	50%	95%	
1	1.055E-0006	3.331E−0005	0.085273	U
2	1.169E−0009	8.172E−0007	4.604E−0005	L
3	0.907588	0.968812	0.99	U
4	5.545E−0007	0.000678	0.008778	L
5	1.450E−0005	0.000317	0.001	U
6	1.000E−0005	9.734E−0005	0.00463	L
7	1.027E−0005	0.000201	0.00463	L
8	1.048E−0005	0.000585	0.009957	U
9	0.000565	0.005	0.01	U
10	1.061E−0006	0.0002	0.001	L
11	1.052E−0006	7.206E−0005	0.004046	L
12	1.055E−0006	3.338E−0005	0.043346	U
13	5.230E−0008	3.370E−0006	8.760E−0005	L
14	1.526E−0007	6.262E−0006	0.022039	U

List of parameters.   Quantile points (%): 5   50   95

No.	Name	Scale	Realization
1	Item 1	UNI	
2	Item 2	LOG	$2.50000E-0006$
3	Item 3	UNI	0.95
4	Item 4	LOG	0.00077
5	Item 5	UNI	
6	Item 6	LOG	0.0005
7	Item 7	LOG	$9.80000E-0005$
8	Item 8	UNI	0.0036
9	Item 9	UNI	0.0059
10	Item 10	LOG	$1.00000E-0004$
11	Item 11	LOG	0.00035
12	Item 12	UNI	0.0023
13	Item 13	LOG	$2.30000E-0006$
14	Item 14	UNI	0.014

SUPPLEMENT B

ESTEC-1

Expert No.	Item No.	5%	50%	95%
1	1	$2.900E-0002$	$1.000E-0001$	$3.400E-0001$
1	2	$8.800E-0002$	$3.000E-0001$	$9.900E-0001$
1	3	$1.000E-0002$	$2.000E-0001$	$6.800E-0001$
1	4	$2.300E-0002$	$8.000E-0002$	$2.700E-0001$
1	5	$1.000E-0002$	$2.000E-0001$	$6.800E-0001$
1	6	$1.000E-0002$	$2.000E-0001$	$6.800E-0001$
1	7	$1.000E-0002$	$2.000E-0001$	$6.800E-0001$
1	8	$1.000E-0002$	$5.000E-0002$	$2.600E-0001$
1	9	$5.200E-0006$	$1.000E-0004$	$1.900E-0003$
1	10	$6.000E-0010$	$2.000E-0008$	$6.800E-0007$
1	11	$2.500E-0002$	$8.500E-0002$	$3.000E-0001$
1	12	$5.200E-0009$	$1.000E-0007$	$1.900E-0006$
1	13	$5.200E-0008$	$1.000E-0006$	$1.900E-0005$
2	1	$5.200E-0009$	$1.000E-0007$	$1.900E-0006$
2	2	$5.200E-0009$	$1.000E-0007$	$1.900E-0006$
2	3	$5.000E-0005$	$5.000E-0002$	$9.900E-0001$
2	4	$5.200E-0004$	$1.000E-0002$	$1.900E-0001$
2	5	$2.700E-0004$	$5.000E-0003$	$1.000E-0001$
2	6	$2.700E-0003$	$5.000E-0002$	$9.900E-0001$
2	7	$2.700E-0003$	$5.000E-0002$	$9.900E-0001$
2	8	$5.200E-0003$	$1.000E-0001$	$9.900E-0001$
2	9	$5.200E-0006$	$1.000E-0004$	$1.900E-0003$
2	10	$5.200E-0010$	$1.000E-0008$	$1.900E-0007$
2	11	$5.200E-0004$	$1.000E-0002$	$1.900E-0001$

Expert No.	Item No.	5%	50%	95%
2	12	5.200E − 0010	1.000E − 0008	1.900E − 0007
2	13	5.200E − 0008	1.000E − 0006	1.900E − 0005
3	1	3.400E − 0005	5.000E − 0003	6.700E − 0001
3	2	5.200E − 0004	1.000E − 0002	1.900E − 0001
3	3	5.900E − 0002	2.000E − 0001	6.800E − 0001
3	4	5.900E − 0003	3.000E − 0002	1.500E − 0001
3	5	5.900E − 0003	2.000E -- 0002	6.800E − 0002
3	6	1.200E − 0002	4.000E − 0002	1.300E − 0001
3	7	1.600E − 0002	8.000E − 0002	4.100E − 0001
3	8	2.900E − 0010	1.000−0009	3.400E − 0009
3	9	3.800E − 0004	5.000E − 0002	9.900E − 0001
3	10	2.000E − 0010	1.000E − 0009	5.100E − 0009
3	11	2.000E − 0003	1.000E − 0002	1.200E − 0001
3	12	2.300E − 0005	1.000E − 0003	3.400E − 0001
3	13	5.200E − 0010	1.000E − 0008	1.900E − 0007
4	1	2.900E − 0003	1.000E − 0002	3.400E − 0002
4	2	2.900E − 0003	1.000E − 0002	3.400E − 0002
4	3	2.900E − 0002	1.000E − 0001	3.400E − 0001
4	4	1.200E − 0002	4.000E − 0002	1.400E − 0001
4	5	1.200E − 0002	4.000E − 0002	1.400E − 0001
4	6	8.800E − 0003	3.000E − 0002	1.000E − 0001
4	7	1.200E − 0002	4.000E − 0002	1.400E − 0001
4	8	5.200E − 0005	1.000E − 0003	1.900E − 0002
4	9	2.900E − 0003	1.000E − 0002	3.400E − 0002
4	10	7.500E − 0011	1.000E − 0008	1.300E − 0006
4	11	2.000E − 0003	1.000E − 0002	5.100E − 0001
4	12	2.300E − 0003	5.000E − 0002	9.700E − 0001
4	13	7.500E − 0009	1.000E − 0006	1.300E − 0004

ESTEC-2

Expert No.	Item No.	5%	50%	95%
1	1	5.000E + 0000	3.750E + 0001	9.500E + 0001
1	2	5.000E + 0000	3.750E + 0001	9.600E + 0001
1	3	5.000E + 0000	3.750E + 0001	9.700E + 0001
1	4	5.000E + 0000	3.750E + 0001	9.750E + 0001
1	5	5.000E + 0000	3.750E + 0001	9.750E + 0001
1	6	5.000E + 0000	3.750E + 0001	9.750E + 0001
1	7	5.000E + 0000	3.750E + 0001	9.750E + 0001
1	8	5.000E + 0000	3.750E + 0001	9.750E + 0001
1	9	5.000E + 0000	3.750E + 0001	9.750E + 0001
1	10	5.000E + 0000	3.750E + 0001	9.750E + 0001
1	11	5.000E + 0000	3.750E + 0001	9.750E + 0001
1	12	5.000E + 0000	3.750E + 0001	9.750E + 0001
1	13	5.000E + 0000	3.750E + 0001	9.750E + 0001
1	14	5.000E + 0000	3.750E + 0001	9.750E + 0001
1	15	5.000E + 0000	3.750E + 0001	9.750E + 0001
1	16	5.000E + 0000	3.750E + 0001	9.750E + 0001

ESTEC-2. (Continued)

Expert No.	Item No.	5%	50%	95%
1	17	5.000E+0000	3.750E+0001	9.750E+0001
1	18	5.000E+0000	3.750E+0001	1.000E+0002
1	19	5.000E+0000	4.000E+0001	1.025E+0002
1	20	5.000E+0000	4.000E+0001	1.025E+0002
1	21	5.000E+0000	4.250E+0001	1.025E+0002
1	22	5.000E+0000	4.250E+0001	1.025E+0002
1	23	5.000E+0000	4.250E+0001	1.025E+0002
1	24	5.000E+0000	4.375E+0001	1.025E+0002
1	25	5.000E+0000	4.500E+0001	1.025E+0002
1	26	5.000E+0000	4.750E+0001	1.025E+0002
2	1	1.000E+0000	3.000E+0000	2.000E+0001
2	2	6.000E+0000	1.000E+0001	4.400E+0001
2	3	1.000E+0001	2.300E+0001	6.800E+0001
2	4	1.750E+0001	4.400E+0001	1.000E+0002
2	5	2.800E+0001	6.800E+0001	1.350E+0002
2	6	4.300E+0001	9.600E+0001	1.720E+0002
2	7	6.400E+0001	1.270E+0002	2.140E+0002
2	8	8.750E+0001	1.610E+0002	2.570E+0002
2	9	1.080E+0002	1.940E+0002	3.020E+0002
2	10	1.280E+0002	2.270E+0002	3.440E+0002
2	11	1.500E+0002	2.700E+0002	3.860E+0002
2	12	1.750E+0002	2.880E+0002	4.250E+0002
2	13	2.010E+0002	3.100E+0002	4.580E+0002
2	14	2.275E+0002	3.300E+0002	4.825E+0002
2	15	2.510E+0002	3.475E+0002	4.970E+0002
2	16	2.700E+0002	3.620E+0002	5.000E+0002
2	17	2.880E+0002	3.750E+0002	4.990E+0002
2	18	2.950E+0002	3.890E+0002	4.870E+0002
2	19	2.140E+0002	3.840E+0002	4.610E+0002
2	20	8.600E+0001	3.600E+0002	4.300E+0002
2	21	3.600E+0001	3.375E+0002	3.920E+0002
2	22	3.900E+0001	3.275E+0001	3.740E+0002
2	23	1.325E+0002	3.375E+0002	3.825E+0002
2	24	2.950E+0002	3.630E+0002	4.100E+0002
2	25	3.770E+0002	4.070E+0002	4.460E+0002
2	26	3.750E+0002	4.375E+0002	4.725E+0002
3	1	1.025E+0002	2.420E+0002	3.350E+0002
3	2	1.175E+0002	2.440E+0002	3.440E+0002
3	3	1.325E+0002	2.450E+0002	3.460E+0002
3	4	1.475E+0002	2.500E+0002	3.475E+0002
3	5	1.625E+0002	2.525E+0002	3.475E+0002
3	6	1.800E+0002	2.575E+0002	3.475E+0002
3	7	1.975E+0002	2.800E+0002	3.475E+0002
3	8	2.225E+0002	3.000E+0002	3.475E+0002
3	9	2.380E+0002	3.180E+0002	3.525E+0002
3	10	2.600E+0002	3.370E+0002	3.800E+0002
3	11	2.790E+0002	3.575E+0002	4.080E+0002
3	12	2.980E+0002	3.750E+0002	4.375E+0002
3	13	3.180E+0002	3.960E+0002	4.600E+0002
3	14	3.375E+0002	4.150E+0002	4.850E+0002
3	15	3.575E+0002	4.350E+0002	5.090E+0002

Expert No.	Item No.	5%	50%	95%
3	16	3.750E + 0002	4.580E + 0002	5.340E + 0002
3	17	3.925E + 0002	4.775E + 0002	5.600E + 0002
3	18	4.105E + 0002	5.000E + 0002	5.860E + 0002
3	19	4.300E + 0002	5.200E + 0002	6.175E + 0002
3	20	4.475E + 0002	5.400E + 0002	6.490E + 0002
3	21	4.625E + 0002	5.580E + 0002	6.800E + 0002
3	22	4.775E + 0002	5.780E + 0002	7.100E + 0002
3	23	5.010E + 0002	5.975E + 0002	7.410E + 0002
3	24	5.220E + 0002	6.175E + 0002	7.700E + 0002
3	25	5.400E + 0002	6.380E + 0002	8.000E + 0002
3	26	5.560E + 0002	6.800E + 0002	8.320E + 0002
4	1	1.000E − 0011	1.400E + 0001	3.750E + 0001
4	2	1.000E − 0011	1.400E + 0001	3.750E + 0001
4	3	1.000E − 0011	1.400E + 0001	3.750E + 0001
4	4	1.000E − 0011	1.400E + 0001	3.750E + 0001
4	5	1.000E − 0011	1.400E + 0001	3.750E + 0001
4	6	1.000E − 0011	1.400E + 0001	3.750E + 0001
4	7	1.000E − 0011	1.400E + 0001	3.750E + 0001
4	8	1.000E − 0011	1.400E + 0001	3.750E + 0001
4	9	1.000E − 0011	1.400E + 0001	3.750E + 0001
4	10	1.000E − 0011	1.400E + 0001	3.750E + 0001
4	11	2.000E + 0001	4.260E + 0001	7.375E + 0001
4	12	4.000E + 0001	7.120E + 0001	1.100E + 0002
4	13	6.000E + 0001	9.980E + 0001	1.463E + 0002
4	14	8.000E + 0001	1.284E + 0002	1.825E + 0002
4	15	1.000E + 0002	1.570E + 0002	2.188E + 0002
4	16	1.200E + 0002	1.856E + 0002	2.550E + 0002
4	17	1.400E + 0002	2.142E + 0002	2.913E + 0002
4	18	1.600E + 0002	2.428E + 0002	3.275E + 0002
4	19	1.800E + 0002	2.714E + 0002	3.638E + 0002
4	20	2.000E + 0002	3.000E + 0002	4.000E + 0002
4	21	2.205E + 0002	3.300E + 0002	4.365E + 0002
4	22	2.410E + 0002	3.600E + 0002	4.730E + 0002
4	23	2.615E + 0002	3.900E + 0002	5.095E + 0002
4	24	2.820E + 0002	4.200E + 0002	5.460E + 0002
4	25	3.025E + 0002	4.500E + 0002	5.825E + 0002
4	26	3.230E + 0002	4.800E + 0002	6.190E + 0002
5	1	1.000E + 0001	9.000E + 0001	2.200E + 0002
5	2	4.000E + 0001	1.150E + 0002	2.490E + 0002
5	3	6.500E + 0001	1.400E + 0002	2.780E + 0002
5	4	9.200E + 0001	1.620E + 0002	3.100E + 0002
5	5	1.140E + 0002	1.810E + 0002	3.375E + 0002
5	6	1.375E + 0002	2.000E + 0002	3.620E + 0002
5	7	1.600E + 0002	2.150E + 0002	3.870E + 0002
5	8	1.775E + 0002	2.325E + 0002	4.130E + 0002
5	9	1.970E + 0002	2.525E + 0002	4.400E + 0002
5	10	2.150E + 0002	2.780E + 0002	4.625E + 0002
5	11	2.330E + 0002	3.040E + 0002	4.870E + 0002
5	12	2.520E + 0002	3.270E + 0002	5.110E + 0002
5	13	2.700E + 0002	3.530E + 0002	5.360E + 0002
5	14	2.890E + 0002	3.810E + 0002	5.600E + 0002

ESTEC-2. (Continued)

Expert No.	Item No.	5%	50%	95%
5	15	3.100E+0002	4.110E+0002	5.860E+0002
5	16	3.320E+0002	4.370E+0002	6.100E+0002
5	17	3.520E+0002	4.660E+0002	6.300E+0002
5	18	3.740E+0002	4.950E+0002	6.525E+0002
5	19	3.970E+0002	5.280E+0002	6.750E+0002
5	20	4.225E+0002	5.580E+0002	7.000E+0002
5	21	4.475E+0002	5.900E+0002	7.260E+0002
5	22	4.710E+0002	6.150E+0002	7.520E+0002
5	23	4.950E+0002	6.480E+0002	7.770E+0002
5	24	5.190E+0002	6.810E+0002	8.025E+0002
5	25	5.410E+0002	7.150E+0002	8.300E+0002
5	26	5.630E+0002	7.510E+0002	8.560E+0002
6	1	9.200E+0002	2.350E+0002	6.490E+0002
6	2	1.000E+0002	2.390E+0002	6.500E+0002
6	3	1.070E+0002	2.460E+0002	6.510E+0002
6	4	1.120E+0002	2.490E+0002	6.510E+0002
6	5	1.140E+0002	2.510E+0002	6.525E+0002
6	6	1.160E+0002	2.520E+0002	6.525E+0002
6	7	1.175E+0002	2.520E+0002	6.510E+0002
6	8	1.175E+0002	2.520E+0002	6.500E+0002
6	9	1.175E+0002	2.520E+0002	6.475E+0002
6	10	1.175E+0002	2.520E+0002	6.475E+0002
6	11	1.175E+0002	2.520E+0002	6.470E+0002
6	12	1.175E+0002	2.520E+0002	6.440E+0002
6	13	1.175E+0002	2.520E+0002	6.425E+0002
6	14	1.175E+0002	2.510E+0002	6.400E+0002
6	15	1.175E+0002	2.490E+0002	6.390E+0002
6	16	1.175E+0002	2.490E+0002	6.380E+0002
6	17	1.180E+0002	2.490E+0002	6.375E+0002
6	18	1.180E+0002	2.490E+0002	6.375E+0002
6	19	1.200E+0002	2.490E+0002	6.375E+0002
6	20	1.200E+0002	2.490E+0002	6.370E+0002
6	21	1.190E+0002	2.490E+0002	6.325E+0002
6	22	1.175E+0002	2.490E+0002	6.325E+0002
6	23	1.150E+0002	2.490E+0002	6.320E+0002
6	24	1.130E+0002	2.490E+0002	6.310E+0002
6	25	1.120E+0002	2.490E+0002	6.330E+0002
6	26	1.080E+0002	4.350E+0002	6.350E+0002
7	1	1.200E+0002	1.960E+0002	3.600E+0002
7	2	1.900E+0002	3.150E+0002	4.500E+0002
7	3	2.430E+0002	4.200E+0002	5.620E+0002
7	4	2.930E+0002	4.890E+0002	6.610E+0002
7	5	3.125E+0002	5.730E+0002	7.880E+0002
7	6	3.270E+0002	6.000E+0002	8.150E+0002
7	7	3.075E+0002	5.470E+0002	8.100E+0002
7	8	2.925E+0002	4.920E+0002	7.500E+0002
7	9	2.730E+0002	4.670E+0002	6.830E+0002
7	10	2.520E+0002	3.800E+0002	6.720E+0002
7	11	2.500E+0002	4.000E+0002	6.650E+0002
7	12	2.425E+0002	4.500E+0002	6.950E+0002
7	13	2.400E+0002	4.360E+0002	7.170E+0002

ESTEC-2. (Continued)

Expert No.	Item No.	5%	50%	95%
7	14	2.425E+0002	4.530E+0002	7.180E+0002
7	15	2.480E+0002	5.030E+0002	7.080E+0002
7	16	2.420E+0002	5.070E+0002	6.780E+0002
7	17	2.125E+0002	4.600E+0002	6.870E+0002
7	18	2.020E+0002	3.870E+0002	7.000E+0002
7	19	2.025E+0002	3.820E+0002	7.125E+0002
7	20	2.170E+0002	4.270E+0002	7.450E+0002
7	21	2.420E+0002	4.810E+0002	7.990E+0002
7	22	2.720E+0002	5.340E+0002	8.800E+0002
7	23	2.770E+0002	5.970E+0002	9.150E+0002
7	24	2.630E+0002	6.000E+0002	9.000E+0002
7	25	2.510E+0002	5.500E+0002	8500E+0002
7	26	2.370E+0002	4.690E+0002	7.670E+0002

ESTEC-3

Expert No.	Item No.	5%	50%	95%
1	1	1.000E−0008	1.000E−0007	1.000E−0006
1	2	1.000E−0008	1.000E−0006	1.000E−0005
1	3	9.000E−0001	9.500E−0001	9.900E−0001
1	4	5.000E−0007	1.000E−0006	1.000E−0004
1	5	1.000E−0005	1.000E−0004	1.000E−0003
1	6	5.000E−0004	1.000E−0003	1.000E−0002
1	7	5.000E−0004	1.000E−0003	1.000E−0002
1	8	1.000E−0002	5.000E−0002	1.000E−0001
1	9	5.000E−0004	5.000E−0003	5.000E−0002
1	10	1.000E−0004	1.000E−0003	5.000E−0003
1	11	1.000E−0002	5.000E−0002	1.000E−0001
1	12	1.000E−0003	5.000E−0003	1.000E−0002
1	13	1.000E−0007	1.000E−0006	1.000E−0005
1	14	1.000E−0004	1.000E−0003	5.000E−0003
2	1	1.000E−0008	1.000E−0007	1.000E−0006
2	2	1.000E−0009	1.000E−0008	1.000E−0007
2	3	9.000E−0001	9.500E−0001	9.900E−0001
2	4	1.000E−0005	1.000E−0004	1.000E−0003
2	5	1.000E−0005	1.000E−0004	1.000E−0003
2	6	1.000E−0008	1.000E−0007	1.000E−0006
2	7	1.000E−0006	1.000E−0004	1.000E−0002
2	8	1.000E−0004	1.000E−0003	1.000E−0002
2	9	1.000E−0004	1.000E−0003	1.000E−0002
2	10	1.000E−0007	1.000E−0006	1.000E−0005
2	11	1.000E−0004	1.000E−0003	1.000E−0002
2	12	1.000E−0007	1.000E−0006	1.000E−0005
2	13	1.000E−0009	1.000E−0008	1.000E−0007
2	14	1.000E−0007	1.000E−0005	1.000E−0003
3	1	1.000E−0007	5.000E−0006	1.000E−0004
3	2	1.000E−0006	1.000E−0005	1.000E−0004
3	3	9.000E−0001	9.500E−0001	9.900E−0001
3	4	5.000E−0007	5.000E−0006	1.000E−0005

ESTEC-3. (Continued)

Expert No.	Item No.	5%	50%	95%
3	5	1.000E − 0005	5.000E − 0005	2.500E − 0004
3	6	1.000E − 0008	1.000E − 0007	1.000E − 0006
3	7	7.500E − 0007	5.000E − 0006	5.000E − 0005
3	8	1.000E − 0003	7.500E − 0003	2.500E − 0002
3	9	1.000E − 0005	5.000E − 0005	1.000E − 0004
3	10	5.000E − 0007	5.000E − 0006	2.500E − 0005
3	11	2.500E − 0003	7.500E − 0003	5.000E − 0002
3	12	2.500E − 0003	7.500E − 0003	2.500E − 0002
3	13	2.500E − 0007	1.000E − 0006	5.000E − 0006
3	14	5.000E − 0006	1.000E − 0005	5.000E − 0005
4	1	5.000E − 0007	1.000E − 0005	5.000E − 0005
4	2	1.000E − 0007	1.000E − 0005	5.000E − 0005
4	3	9.500E − 0001	9.750E − 0001	9.900E − 0001
4	4	5.000E − 0004	1.000E − 0003	1.000E − 0002
4	5	1.000E − 0005	1.000E − 0004	1.000E − 0003
4	6	1.000E − 0005	1.000E − 0003	5.000E − 0003
4	7	5.000E − 0005	1.000E − 0003	5.000E − 0003
4	8	5.000E − 0006	1.000E − 0004	1.000E − 0003
4	9	5.000E − 0004	5.000E − 0003	1.000E − 0002
4	10	1.000E − 0006	1.000E − 0004	1.000E − 0003
4	11	1.000E − 0006	5.000E − 0005	1.000E − 0004
4	12	5.000E − 0007	1.000E − 0005	5.000E − 0005
4	13	5.000E − 0008	1.000E − 0006	1.000E − 0005
4	14	1.000E − 0007	1.000E − 0006	1.000E − 0005
5	1	1.000E − 0008	1.000E − 0006	5.000E − 0006
5	2	1.000E − 0008	1.000E − 0006	1.000E − 0005
5	3	9.000E − 0001	9.500E − 0001	9.900E − 0001
5	4	1.000E − 0006	5.000E − 0006	1.000E − 0005
5	5	1.000E − 0004	5.000E − 0004	1.000E − 0003
5	6	1.000E − 0004	5.000E − 0004	1.000E − 0001
5	7	1.000E − 0004	5.000E − 0004	1.000E − 0003
5	8	1.000E − 0006	5.000E − 0006	1.000E − 0001
5	9	1.000E − 0005	5.000E − 0005	1.000E − 0004
5	10	1.000E − 0005	5.000E − 0005	1.000E − 0004
5	11	1.000E − 0005	5.000E − 0005	1.000E − 0004
5	12	1.000E − 0005	5.000E − 0005	1.000E − 0004
5	13	1.000E − 0007	1.000E − 0006	1.000E − 0004
5	14	1.000E − 0008	1.000E − 0006	5.000E − 0005
6	1	1.000E − 0003	1.000E − 0002	1.000E − 0001
6	2	1.000E − 0009	1.000E − 0008	1.000E − 0007
6	3	9.000E − 0001	9.500E − 0001	9.900E − 0001
6	4	5.000E − 0007	2.500E − 0006	1.000E − 0005
6	5	1.000E − 0004	5.000E − 0004	1.000E − 0003
6	6	1.000E − 0005	5.000E − 0005	1.000E − 0004
6	7	1.000E − 0005	5.000E − 0005	1.000E − 0004
6	8	7.500E − 0004	7.500E − 0003	1.000E − 0002
6	9	7.500E − 0004	5.000E − 0003	1.000E − 0002
6	10	1.000E − 0004	5.000E − 0004	1.000E − 0003
6	11	1.000E − 0004	7.500E − 0004	5.000E − 0003
6	12	5.000E − 0003	1.000E − 0002	5.000E − 0002
6	13	5.000E − 0006	5.000E − 0005	1.000E − 0004
6	14	2.500E − 0003	7.500E − 0003	2.500E − 0002

Expert No.	Item No.	5%	50%	95%
1	1	1.000E − 0001	2.500E + 0001	7.500E + 0001
1	2	3.000E − 0000	1.500E + 0001	5.000E + 0001
1	3	5.000E + 0001	7.000E + 0001	8.000E + 0001
1	4	1.000E + 0002	4.500E + 0003	9.000E + 0003
1	5	4.000E + 0004	7.500E + 0004	8.000E + 0004
1	6	3.000E + 0000	1.500E + 0001	4.000E + 0001
1	7	1.000E + 0001	6.000E + 0001	1.000E + 0002
1	· 8	5.000E + 0000	1.400E + 0001	2.500E + 0001
2	1	2.000E + 0001	3.500E + 0001	5.000E + 0001
2	2	2.000E + 0000	5.000E + 0000	1.000E + 0001
2	3	7.500E + 0001	8.500E + 0001	9.500E + 0001
2	4	6.500E + 0003	1.000E + 0004	1.350E + 0004
2	5	1.000E + 0006	1.500E + 0006	2.000E + 0006
2	6	1.000E + 0001	2.000E + 0001	3.500E + 0001
2	7	2.500E + 0001	1.250E + 0002	2.500E + 0002
2	8	5.000E + 0001	7.000E + 0001	9.000E + 0001
3	1	1.000E + 0001	2.000E + 0001	2.500E + 0001
3	2	5.000E + 0000	1.000E + 0001	1.500E + 0001
3	3	8.000E + 0001	8.800E + 0001	9.200E + 0001
3	4	1.000E + 0003	5.000E + 0003	1.500E + 0004
3	5	6.500E + 0004	7.500E + 0004	8.580E + 0004
3	6	1.000E + 0001	1.500E + 0001	2.500E + 0001
3	7	2.500E + 0001	1.250E + 0002	2.000E + 0002
3	8	9.000E + 0001	9.500E + 0001	9.900E + 0001
4	1	5.000E + 0000	2.000E + 0001	5.000E + 0001
4	2	5.000E + 0000	1.000E + 0001	5.000E + 0001
4	3	7.000E + 0001	8.500E + 0001	1.000E + 0002
4	4	5.000E + 0003	1.000E + 0004	2.000E + 0004
4	5	4.000E + 0004	5.000E + 0004	7.000E + 0004
4	6	4.000E + 0000	8.000E + 0000	2.000E + 0001
4	7	6.300E + 0001	5.000E + 0002	1.000E + 0003
4	8	3.000E + 0001	8.000E + 0001	9.500E + 0001
5	1	1.000E + 0001	3.000E + 0001	5.000E + 0001
5	2	5.000E + 0000	1.000E + 0001	2.000E + 0001
5	3	8.700E + 0001	9.200E + 0001	9.800E + 0001
5	4	1.000E + 0003	1.500E + 0003	2.500E + 0003
5	5	1.100E + 0006	1.400E + 0006	1.700E + 0006
5	6	5.000E + 0000	1.500E + 0001	3.000E + 0001
5	7	2.500E + 0001	1.250E + 0002	4.000E + 0002
5	8	8.000E + 0001	8.800E + 0001	9.200E + 0001
6	1	1.000E + 0001	1.500E + 0001	2.000E + 0001
6	2	2.000E + 0000	4.000E + 0000	8.000E + 0000
6	3	7.000E + 0001	8.000E + 0001	9.000E + 0001
6	4	2.000E + 0004	2.500E + 0004	5.000E + 0004
6	5	1.500E + 0006	1.680E + 0006	1.750E + 0006
6	6	1.500E + 0001	].£££E + 0001	2.500E + 0001
6	7	5.000E + 0001	1.000E + 0002	2.000E + 0002
6	8	8.300E + 0001	8.500E + 0001	8.700E + 0001
7	1	5.000E + 0001	6.000E + 0001	8.000E + 0001
7	2	4.000E + 0000	5.000E + 0000	8.000E + 0000

DSM-1. (Continued)

Expert No.	Item No.	5%	50%	95%
7	3	8.500E + 0001	9.000E + 0001	9.100E + 0001
7	4	3.000E + 0003	5.000E + 0003	8.000E + 0003
7	5	1.200E + 0006	1.300E + 0006	1.600E + 0006
7	6	1.500E + 0001	2.000E + 0001	3.500E + 0001
7	7	1.000E + 0001	1.000E + 0002	2.500E + 0002
7	8	8.600E + 0001	8.800E + 0001	9.200E + 0001
8	1	5.000E + 0001	7.500E + 0001	1.000E + 0002
8	2	2.000E + 0000	5.000E + 0000	1.000E + 0001
8	3	8.400E + 0001	8.700E + 0001	9.000E + 0001
8	4	2.500E + 0003	4.000E + 0003	1.600E + 0004
8	5	1.200E + 0006	1.600E + 0006	2.000E + 0006
8	6	5.000E + 0000	1.500E + 0001	4.000E + 0001
8	7	1.000E + 0001	5.000E + 0001	1.000E + 0002
8	8	7.500E + 0001	8.200E + 0001	9.000E + 0001
9	1	1.500E + 0001	2.000E + 0001	3.000E + 0001
9	2	1.000E + 0000	3.000E + 0000	6.000E + 0000
9	3	9.000E + 0001	9.400E + 0001	9.600E + 0001
9	4	1.500E + 0003	2.800E + 0003	3.500E + 0003
9	5	8.000E + 0005	9.800E + 0005	1.000E + 0006
9	6	1.500E + 0001	2.500E + 0001	4.000E + 0001
9	7	5.000E + 0001	1.000E + 0002	2.500E + 0002
9	8	8.200E + 0001	8.400E + 0001	8.700E + 0001
10	1	1.500E + 0001	2.000E + 0001	3.000E + 0001
10	2	3.000E + 0000	7.000E + 0000	1.000E + 0001
10	3	7.000E + 0001	8.000E + 0001	9.000E + 0001
10	4	1.000E + 0003	5.000E + 0003	1.000E + 0004
10	5	7.200E + 0005	8.160E + 0005	8.640E + 0005
10	6	1.500E + 0001	2.000E + 0001	3.000E + 0001
10	7	3.000E + 0000	4.000E + 0000	1.000E + 0001
10	8	8.000E + 0001	9.300E + 0001	9.700E + 0001

DSM-2

Expert No.	Item No.	5%	50%	95%
1	1	1.000E + 0002	1.600E + 0002	2.000E + 0002
1	2	2.000E − 0002	1.000 − 0001	2.000E − 0001
1	3	1.000E + 0002	1.600E + 0002	2.000E + 0002
1	4	2.000E + 0001	8.000E + 0001	2.000E + 0002
1	5	2.000E − 0003	3.300E − 0003	5.000E − 0003
1	6	3.000E + 0002	4.000E + 0002	5.000E + 0002
1	7	1.000E − 0002	2.000E − 0002	5.000E − 0002
1	8	7.500E + 0001	8.000E + 0001	9.000E + 0001
1	9	4.000E + 0001	5.000E + 0001	5.500E − 0001
1	10	5.000E − 0003	5.600E − 0003	1.000E − 0002
1	11	5.000E − 0003	5.600E − 0003	1.000E − 0002
1	12	1.000E − 0002	1.670E − 0002	2.000E − 0002

Expert No.	Item No.	5%	50%	95%
2	1	1.000E + 0001	2.000E + 0001	5.000E + 0001
2	2	1.000E − 0002	2.000E − 0002	5.000E − 0002
2	3	1.000E + 0001	2.000E + 0001	5.000E + 0001
2	4	1.000E + 0001	2.000E + 0001	5.000E + 0001
2	5	5.000E − 0002	1.000E − 0001	2.000E − 0001
2	6	2.000E + 0002	3.120E + 0002	5.000E + 0002
2	7	1.250E − 0001	2.000E − 0001	5.000E − 0001
2	8	6.000E + 0001	6.500E + 0001	7.000E + 0001
2	9	2.500E + 0001	3.000E + 0001	3.500E + 0001
2	10	2.000E − 0004	1.000E − 0003	5.000E − 0003
2	11	−9.996E + 0002	−9.996E + 0002	−9.996E + 0002
2	12	1.000E − 0004	5.000E − 0004	2.000E − 0003
3	1	1.500E + 0002	2.000E + 0002	5.000E + 0002
3	2	5.000E − 0003	1.000E − 0001	2.000E − 0001
3	3	5.000E + 0002	8.500E + 0002	1.000E + 0003
3	4	1.000E + 0002	2.500E + 0002	6.000E + 0002
3	5	1.000E − 0005	1.000E − 0004	2.000E − 0004
3	6	2.000E + 0002	3.200E + 0002	5.000E + 0002
3	7	5.000E − 0004	2.000E − 0003	1.000E − 0002
3	8	7.500E + 0001	8.000E + 0001	9.000E + 0001
3	9	6.000E + 0001	7.000E + 0001	8.000E + 0001
3	10	5.000E − 0003	1.000E − 0002	2.000E − 0002
3	11	2.000E − 0003	2.000E − 0002	1.000E − 0001
3	12	5.000E − 0002	1.000E − 0001	5.000E − 0001
4	1	2.000E + 0002	3.000E + 0002	5.000E + 0002
4	2	5.000E − 0002	1.000E − 0001	2.000E − 0001
4	3	5.000E + 0001	1.000E + 0002	2.000E + 0002
4	4	1.000E + 0002	2.400E + 0002	5.000E + 0002
4	5	2.000E − 0006	5.000E − 0006	1.000E − 0005
4	6	1.000E + 0002	4.000E + 0002	5.000E + 0002
4	7	5.000E − 0004	1.000E − 0003	5.000E − 0003
4	8	2.500E + 0001	3.500E + 0001	5.000E + 0001
4	9	3.500E + 0001	5.000E + 0001	7.000E + 0001
4	10	1.000E − 0004	5.000E − 0004	5.000E − 0003
4	11	−9.996E + 0002	−9.996E + 0002	−9.996E + 0002
4	12	2.000E − 0003	5.000E − 0003	1.000E − 0002
5	1	6.500E + 0001	7.500E + 0001	8.500E + 0001
5	2	2.000E − 0003	2.900E − 0003	5.000E − 0003
5	3	7.500E + 0001	8.500E + 0001	1.000E + 0002
5	4	8.000E + 0001	9.000E + 0001	1.000E + 0002
5	5	4.400E − 0003	5.000E − 0003	5.700E − 0003
5	6	2.500E + 0002	2.750E + 0002	3.000E + 0002
5	7	1.670E − 0002	2.000E − 0002	2.500E − 0002
5	8	7.000E + 0001	7.250E + 0001	7.500E + 0001
5	9	3.500E + 0001	4.000E + 0001	5.000E + 0001
5	10	5.000E − 0003	6.700E − 0003	1.000E − 0002
5	11	1.670E − 0002	2.000E − 0002	2.500E − 0002
5	12	2.000E − 0002	4.000E − 0002	5.000E − 0002
6	1	5.000E + 0001	6.000E + 0001	7.000E + 0001

DSM-2. (Continued

Expert No.	Item No.	5%	50%	95%
6	2	1.000E−0003	2.000E−0003	5.000E−0003
6	3	4.500E+0001	5.000E+0001	5.500E+0001
6	4	8.000E+0001	9.000E+0001	1.000E+0002
6	5	2.000E−0006	6.700E−0006	2.000E−0005
6	6	4.000E+0002	5.000E+0002	6.000E+0002
6	7	2.000E−0003	5.000E−0003	1.000E−0002
6	8	6.000E+0001	7.000E+0001	8.000E+0001
6	9	5.000E+0001	5.500E+0001	6.000E+0001
6	10	5.000E−0004	1.000E−0003	2.000E−0003
6	11	1.000E−0003	2.000E−0003	5.000E−0003
6	12	1.000E−0003	2.000E−0003	5.000E−0003
7	1	2.000E+0002	3.500E+0002	5.000E+0002
7	2	4.400E−0003	5.000E−0003	5.700E−0003
7	3	5.000E+0001	7.500E+0001	1.000E+0002
7	4	5.000E+0001	8.000E+0001	1.000E+0002
7	5	2.000E−0003	4.000E−0003	5.000E−0003
7	6	1.000E+0002	1.750E+0002	2.000E+0002
7	7	1.000E−0002	1.330E−0002	2.000E−0002
7	8	6.000E+0001	6.500E+0001	7.000E+0001
7	9	4.000E+0001	4.500E+0001	5.000E+0001
7	10	2.000E−0002	3.330E−0002	5.000E−0002
7	11	2.000E−0003	2.200E−0003	2.500E−0003
7	12	1.000E−0002	1.330E−0002	2.000E−0002
8	1	1.000E+0001	3.000E+0001	5.000E+0001
8	2	5.000E−0003	1.000E−0002	2.000E−0002
8	3	2.000E+0001	5.000E+0001	1.000E+0002
8	4	1.000E+0001	2.000E+0001	5.000E+0001
8	5	1.000E−0002	2.000E−0002	5.000E−0002
8	6	2.000E+0002	4.000E+0002	5.000E+0002
8	7	2.000E−0003	1.000E−0002	2.000E−0002
8	8	6.000E+0001	7.000E+0001	7.500E+0001
8	9	4.000E+0001	6.000E+0001	6.500E+0001
8	10	1.000E−0003	2.000E−0003	5.000E−0003
8	11	2.000E−0003	5.000E−0003	1.000E−0002
8	12	−9.996E+0002	−9.996E+0002	−9.996E+0002

Elicitation Format

Round 1

Question: Which of the following two failure causes has led more often to the failure of a flanged connection in the plant?

⋮

10.   Ageing of gasket   0
         or
         Wrong mounting   0
11.   Ageing of bolts/nuts   0
         or
         Changes in temperature caused by process cycles   0

⋮

Input Paired Comparisons (Round 1)

Round 1, subgroup A (all experts):
(a)   Number of failure causes: 7
(b)   Number of experts: 14
(c)   Confidence intervals via simulation (Y/N):   N
(d)   Confidence level (in %):
(e)   Number of simulations:
(f)   Transformation (Y/N):   N
(g)   Reference cause 1 = cause no:
(h)   Reference cause 2 = cause no:
(i)   Reference value 1:
(j)   Reference value 2:
(k)   Order in which the pairs were presented to the experts:

Pair	Cause A	Cause B
1.	1	2
2.	3	7
3.	4	6
4.	5	1
5.	2	3
6.	7	4
7.	6	5
8.	1	3
9.	4	2
10.	5	7
11.	6	1
12.	3	4
13.	2	5
14.	7	6
15.	1	4
16.	5	3
17.	6	2
18.	7	1
19.	4	5
20.	3	6
21.	2	7

(1) answers experts:
     Notation: 1 if Obj. A > Obj. B
                2   Obj. A < Obj. B
                0   Obj. A = Obj. B
                9   no knowledge of
Obj. A or Obj. B

Pair no.	12345678901234567 8901
Expert 1	111122211112222112211
Expert 2	122111212221211121212
Expert 3	122111212221211111212
Expert 4	122221212221211111212
Expert 5	221111212121111121222
Expert 6	121111211221211111222
Expert 7	122121212221211121222
Expert 8	121211212222111121222
Expert 9	222111212121111122212
Expert 10	121211211222111121122
Expert 11	112222122111221112222
Expert 12	112212211121221211212
Expert 13	222111212121211121211
Expert 14	222111212221211121222

Output Paired Comparisons (Round 1)

Round 1, subgroup A (all experts):
Circular Triads Experts:
Notation circular triads of type A > B, B > C and C > A: C1
                                  A = B, A > C and B < C: C2
                                  A = B, A = C and B < C: C3

Experts	Objects	Circular Triad
Expert 1	1, 3, 6	C1
Total		1 C1 + 0 C2 + 0 C3 = 1
Expert 3	2, 3, 6	C1
Total		1 C1 + 0 C2 + 0 C3 = 1
Expert 5	2, 5, 7	C1
	3, 4, 6	C1
Total		2 C1 + 0 C2 + 0 C3 = 2
Expert 6	2, 3, 4	C1
	3, 4, 6	C1
Total		2 C1 + 0 C2 + 0 C3 = 2
Expert 7	2, 3, 6	C1
Total		1 C1 + 0 C2 + 0 C3 = 1
Expert 9	1, 2, 7	C1
	2, 5, 7	C1
Total		2 C1 + 0 C2 + 0 C3 = 2
Expert 11	1, 3, 5	C1
	2, 4, 7	C1

Output Paired Comparisons (Round 1). (Continued)

Experts	Objects	Circular Triad
Total		2 C1 + 0 C2 + 0 C3 = 2
Expert 12	1, 3, 7	C1
	1, 4, 7	C1
	1, 5, 7	C1
	1, 6, 7	C1
	2, 3, 4	C1
	2, 3, 5	C1
	2, 3, 6	C1
	2, 3, 7	C1
Total		8 C1 + 0 C2 + 0 C3 = 8

Round 1, subgroup B (expert 12 excluded)

Coefficients of agreement:	$u = 0.324$
Chi square approx.	$u' = 123.438$
Coefficient of concordance:	$W = 0.454$
Chi square approx.	$W' = 35.374$
Σ (mean rank deviation)^2	$S = 2146.000$

Data-Proportion Matrix

Obj	2	3	4	5	6	7
1	9.0/13	12.0/13	12.0/13	4.0/13	11.0/13	3.0/13
2		9.0/13	10.0/13	4.0/13	8.0/13	2.0/13
3			10.0/13	0.0/13	6.0/13	2.0/13
4				1.0/13	5.0/13	2.0/13
5					12.0/13	5.0/13
6						2.0/13

Object Values and Goodness of Fit

Object	NEL (Bradley-Terry)	Thurstone
1	0.1734	0.4480
2	0.0784	−0.0704
3	0.0350	−0.5874
4	0.0208	−0.8055
5	0.3041	0.7666
6	0.0395	−0.4815
7	0.3487	0.7302

Goodness of fit: NEL (Bradley-Terry) model, $F = 4.0141$
Thurstone model, $D = 11.5310$

Transformed Object Values (I) reference values;
obj., 7: 1.13, obj., 5: 0.64

Obj.	NEL (Bradley-Terry)	Thurstone (lin)
1	1.13E+00	1.13E+00
2	5.11E−01	1.93E+00
3	2.28E−01	2.72E+00
4	1.35E−01	3.06E+00
5	1.98E+00	6.40E−01
6	2.58E−01	2.56E+00
7	2.27E+00	6.96E−01

Transformed Object Values (II), reference values;
obj., 5: 0.64; obj. 7: 2.24

Obj.	NEL (Bradley-Terry)	Thurstone (lin)
1	3.65E−01	1.46E+01
2	1.65E−01	3.74E+01
3	7.37E−02	6.02E+01
4	4.37E−02	6.97E+01
5	6.40E−01	6.40E−01
6	8.32E−02	5.55E+01
7	7.34E−01	2.24E+00

Thurstone model, reference values and/or expert opinions are contradictory.

Transformed Object Values (III): reference values;
obj., 7: 2.24; obj., 1; 1.13.

Obj.	NEL (Bradley-Terry)	Thurstone (lin)
1	1.11E+00	1.13E+00
2	5.04E−01	0.00E+00
3	2.25E−01	0.00E+00
4	1.34E−01	0.00E+00
5	1.95E+00	2.38E+00
6	2.54E−01	0.00E+00
7	2.24E+00	2.24E+00

Thurstone model, reference values and/or expert opinions are contradictory.

16
Conclusions

This final brief chapter attempts to draw together conclusions from the preceding parts of this study, and indicates the important open problems.

PART I

Part I addressed the data on expert opinion and the methods for using expert opinion currently found in practice. The overall conclusion to be drawn from this material is twofold:

- Expert opinions can contain useful data for rational decision support.
- Considered as a source of data, expert opinion has certain characteristics that require new techniques of data collection and processing.

The most experience to date with quantified expert opinion is found in the field of risk analysis. Here we confront all the problems and potential payoffs of expert opinion. Expert opinions typically show a very wide spread, they may be poorly calibrated, and experts tend to cluster into optimists and pessimists. On the other hand, there are dramatic examples of successful application of expert opinion in this field.

The field of artificial intelligence has invited massive applications of expert opinion, in the form of input into expert systems. The use of expert opinion in this field is highly unstructured, and the representations of uncertainty tend to be ad hoc, lacking a firm axiomatic basis. In particular, these representations contradict the axioms of probability. When uncertainty is represented in an ad hoc manner, it is impossible to evaluate and combine expert opinions in an intelligent way. On the other hand, the most cursory acquaintance with expert data demonstrates the need for evaluation and combination.

Psychometric studies have thrown some light on factors that may adversely affect the quality of subjective probability assessments. However, the data relating specifically to experts are sparse, and the analytical tools used to analyze these data could be improved.

The most important problem emerging from the material reviewed in Part I is the need for an overarching methodology for using expert opinion in science. The methodological rules for collecting and processing "objective data" have been developed over a great many years. "Subjective data" in the form of expert opinion may be a useful new for form of information, but rules for its collection and processing must be developed. Such rules must serve the fundamental aim of science, namely, building rational consensus, and they must also take account of the peculiar features of subjective data. The conclusions of Part I outlined the hesitant first steps in this direction, which have guided the development in the remainder of this study.

PART II

Part II studied the representation of uncertainty as subjective probability. Savage's classical theory of rational decision is taken as an axiomatic basis for the representation of uncertainty. Though the normative basis for the "logic of partial belief" emerging from this theory is not as strong as that of logic per se, it is quite ample, in this author's opinion, to support the type of applications described in this book. Moreover, there is no serious competitor, that is, no alternative representation of uncertainty with an axiomatic basis rivaling that which Savage has provided. While aspects of his theory have been under debate for some 30 years, this debate has not produced anything resembling stable coalitions, let alone consensus, on any alternatives.

Of course, this is not to say that some future theory will not rival Savage's theory in normative power, while at the same time providing a better model of how people actually behave. Perhaps such theories will also support the structured use of expert opinion. However, given the wealth of techniques from probability theory that Savage's representation of uncertainty places at our disposal, it seems unreasonable to expect a serious competitor to emerge overnight.

Subsequent chapters in Part II attempt to capitalize on the representation of uncertainty as subjective probability and derive tools for evaluating and combining expert opinion. The idea of these developments might be put simply as follows. If we are simply given point estimates of unknown quantities, either from experts or from some other source, then the statistical tools at our disposal for analyzing these estimates on the basis of observed values are extremely limited. The quantities may have different physical dimensions, and the uncertainty distributions over the possible ranges are different, and perhaps unknown. When an expert quantifies his uncertainty over the (continuously many) possible values by giving his subjective probability distributions, he effectively transformations each quantity into quantities measured in units of probability and having, for him, a uniform distribution over the unit interval. Indeed if X is an uncertain quantity with continuous cumulative distribution function F, then the transformed variable $F(X)$ is uniformly distributed over the unit interval. With these transformed quantities a much richer body of statistical methods, both classical and Bayesian, becomes available. In the final analysis, this is the reason for quantifying expert opinion as subjective probability.

We may hope that many more techniques from statistics and probability theory will be "downloaded" onto the expert problem in the future. The techniques for testing the calibration hypotheses can surely be improved and extended. Perhaps more pressing is the need for more and refined techniques for eliciting subjective probability distributions. These techniques must be user friendly and should allow us encode more information easily. Quantile assessment is somewhat crude in this respect. Parametric elicitation techniques constitute a promising direction for future work. Closely related to elicitation is the problem of communicating probabilistic information to decision makers. This requires, as it were, inverting the elicitation procedure. Little has been done in this area to date, but no one will deny its importance. Finally, many interesting mathematical problems remain regarding the asymptotic behavior of weights and scoring rules. The results in Chapter 9 represent a "first pass."

For practical applications the most pressing need is to develop techniques for identifying meaningful seed variables easily. This requires experience in ferreting out such variables, but might also profit from the use of "laboratory seed variables" if these could be demonstrated to predict performance on variables of interest.

PART III

In Part III, three models for combining expert opinions are developed. These models are all operational, have been applied, and have proved to be of value. It is appropriate to conclude by offering a summary assessment of these models.

The classical model is easy to understand, easy to apply, and can be applied whenever a modest number of seed variables are available. The experience to date indicates that its results are relatively robust—it has never yielded strange or intuitively unacceptable results. The two most arbitrary features are the choice of seed variables and the measure of informativeness. With regard to the latter, experience to date indicates that the results are not highly sensitive to the choice of information measures. With regard to the former, little can be said at present. However, one can anticipate that variations of the classical model, with more or less equal normative appeal, will yield somewhat differing results, and a choice between these will be very difficult on normative grounds. In short, the classical model is probably substantially better than doing nothing, but is probably not the "best" solution to any given problem.

The Bayesian model described in Chapter 13 is probably most useful when applied to a single expert. The results reported in Chapter 15 are encouraging in this respect. Even in this case, however, the results are highly sensitive to decisions of the analyst. For multiple experts, the solution presented in Chapter 13 is rather speculative, and better solutions, particularly for a small number of seed variables, may be anticipated. The results of the application in Chapter 15 confirm these conclusions. Of the three models developed in Part III, the Bayesian models have the strongest mathematical foundation. Translating this theoretical advantage into a practical advantage, while at the same time reducing dependence on ad hoc prior information, is a challenge for the future.

The psychological scaling models are quite different from the other two. The primary virtues of these models lie in their user-friendliness and in their role as consensus builders. The transformations to absolute values remains problematic. The modeling assumptions underlying the transformations are quite strong, and little is known about their validity in practical applications. Here again, the variety of models, and the difficulty in making a reasoned choice among them, must figure as a disadvantage.

Appendix A

Mathematical Framework and Interpretation

This appendix is devoted to the mathematical and conceptual preliminaries that underly the mathematical modeling of expert opinion. The discussion of interpretation is extremely brief, and serves only to remind readers that such discussions exist. The mathematics introduced here is quite elementary, but enables the reader to follow the main line through Parts II and III. Specific subjects requiring more mathematical detail are placed in supplements of the relevant chapters. We assume acquaintence with, or access to, the elementary facts and notions of probability and statistics, as contained in any undergraduate introductory course.

MATHEMATICAL FRAMEWORK

We denote the real numbers as **R** and the integers as **N**. A *set* is indicated by curly brackets, with the generic element separated from the defining property by a vertical stroke:

$A = \{x \mid x$ satisfies the defining property of A$\}$ for example;

Bachelor $= \{x \mid x$ is an unmarried man$\}$

A′ denotes the complement of A, the set of elements not belonging to A. If A and B are sets, $A \cup B$ denotes the union of A and B, and $A \cap B$ denotes the intersection of A and B. $A \supset B$ says that B is a subset of A, which may also be written $B \subset A$. $x \in A$ says that x is an element of A. $\emptyset$ denotes the empty set, the set without elements.

The mathematical framework and notation used in this study is the following.

The Set of Possible Worlds as a Probability Space

$S =$ the set of *states*, or equivalently the set of possible worlds.

A *state* is regarded as a complete description of the world. Hence, a single state may be identified with a single possible world. **S** is the set of all such. Of course, in all practical situations, the descriptions of the possible worlds will be "truncated" and the possible worlds will be described only up to a preassigned level. Such truncation is merely a convenience of the analyst and it is implicitly assumed that each state description could be indefinitely refined.

A set of subsets of S is called *field* if it contains the empty set $\varnothing$ (which is a subset of every set), and is closed under unions and complementation. More exactly,

A set $\mathscr{F}$ of subsets of S is a *field* of subsets of S if

(i) $\varnothing \in \mathscr{F}$

(ii) If $A \in \mathscr{F}$ and $B \in \mathscr{F}$,
    then $A \cup B \in \mathscr{F}$

(iii) If $A \in \mathscr{F}$, then $A' \in \mathscr{F}$ (A.1)

If $\mathscr{F}$ is a field of subsets of S, one also says that $\mathscr{F}$ is a *field* over S. There are many different fields over S. The smallest field over S is the trivial field containing only $\varnothing$ and S itself. The largest field over S is the set of all subsets of S. The set of all subsets of S is denoted[1] 2^S. Any set of subsets of S generates a unique field, namely the intersection of all fields that contain that set of subsets (which intersection is itself a field).

The pair $(S, \mathscr{F})$ is called a *measurable space*; it is the sort of mathematical object over which probabilities can be defined.

A (finitely additive) *probability measure p* over the measurable space $(S, \mathscr{F})$ is a function $p : \mathscr{F} \to [0, 1]$ from $\mathscr{F}$ to the unit interval $[0, 1]$ satisfying

(i) For all $A \in \mathscr{F}$, $0 \leqslant p(A) \leqslant 1$

(ii) $p(S) = 1$

(iii) If $A \in \mathscr{F}$, and $B \in \mathscr{F}$, and if $A \cap B = \varnothing$, then $p(A \cup B) = p(A) + p(B)$

The object $(S, \mathscr{F}, p)$ is called a *probability space*. In most mathematical contexts it is customary to assume as well that p is *countably additive*; that is, if $A_i \in \mathscr{F}$, $i = 1, 2, \ldots$, with $A_i \cap A_j = \varnothing$ when $i \neq j$, then $p(\cup A_i) = \Sigma p(A_i)$. It is then necessary to assume that $\mathscr{F}$ is closed under countable unions, in which case $\mathscr{F}$ is called a *σ-field*. We do not insist on this assumption, as subjectivists have strong philosophical reservations against countable additivity (see below, and see De Finetti, 1974, pp 116 f.f.).

Unless otherwise stated, the field $\mathscr{F}$ will always be the field of all subsets of S. For countably additive probability measures this would involve mathematical indiscretions, but for finitely additive measures it is quite harmless.

The Field of Events

Abstractly, a state, or a possible world, may be represented as a complete description of all that is the case in that possible world. A state will be denoted by a lowercase letter, typically "s." An *event* is a set of states. If A is an event, then we may also think of A as a *proposition*, namely, the proposition asserting that the event in question occurs. For example, the event "It rains on February 9, 2010" is

---

[1] 2^S denotes the set of maps from S to $\{0, 1\}$, and each such map corresponds to a unique subset of S.

the set of possible worlds in which it rains on February 9, 2010, that is, the set of possible worlds in which the quoted proposition holds. The following statements are therefore equivalent:

"Event A occurs"
"Proposition A holds"
"The real world belongs to A"

The boolean operations on events, union, intersection and complement, correspond to logical operations on propositions. Thus,

$A \cap B$ corresponds to "A and B"
$A \cup B$ corresponds to "A or B"
A' corresponds to "not A"

The event/proposition S corresponds to the trivial proposition (the proposition true in all possible worlds) and the empty set $\varnothing$ corresponds to the impossible proposition (the proposition that is true in no possible world).

To be utterly precise we must distinguish between propositions and statements. The statements "$2 + 2 = 5$" and "$2 + 2 = 6$" both express the same proposition, namely, the proposition corresponding to $\varnothing$. All true mathematical statements express the same proposition, namely, the proposition corresponding to the event S. Still being utterly precise, we should distinguish between a state s and the event $\{s\}$ whose single element is the state s. A degree of partial belief is not assigned to s, but to $\{s\}$. These distinctions will be made only if necessary to avoid ambiguity.

In any concrete application we obviously don't want to consider the set of *all* possible worlds. It is customary to introduce a reduced set of states and a reduced field of events. We can do this by isolating first a set of propositions in which we are interested. Suppose our interest is confined to the following propositions:

A: It rains tomorrow.
B: My car won't start tomorrow.
C: I have a cold the day after tomorrow.

A reduced state, or equivalently a reduced possible world, is a complete description in terms of the above propositions. In other words, it specifies whether each of the above events occurs; that is, specifies the truth value of each of the above propositions. In the present example there are $2^3 = 8$ elements in the set of reduced states.

If we start with n propositions, then we generate in this way 2^n reduced states, assuming that the propositions are logically disjoint (i.e., that no proposition or its negation is entailed by other propositions or their negations). An event is a set of states. If follows that there are

$$2^{2^n}$$

events to which probabilities must be assigned. Hence, the number of events grows very quickly. If $n = 5$, then we shall have to assign probabilities to more than 4 trillion events. Of course, these assignments are not independent. The probability of an event is the sum of the probabilities of the states comprising the event. For $n = 5$, we have to assign probabilities to each of the $2^5 = 32$ states, and the probabilities of all other events can be calculated.

It is sometimes convenient to speak of *atoms* of a field of events. A is an atom of $\mathscr{F}$ if $A \in \mathscr{F}$, $A \neq \emptyset$, and if $B \in \mathscr{F}$, $B \subset A$, then $B = A$ or $B = \emptyset$. States are atoms of the field of all subsets of S. The trivial field $\{\emptyset, S\}$ has S as its only atom. The atoms of a reduced field of events are the reduced states.

INTERPRETATION

The above formalism is more or less accepted as the mathematical description of probability. Differences arise regarding the question how this formalism should be interpreted. We outline very briefly the main interpretations.

The Classical Interpretation

The classical interpretation is generally attributed to Laplace (see Laplace, in Madden (1960)). It holds that probability is to be interpreted via the principle of indifference: If n alternatives are physically indistinguishable, then they all have the same probability of being realized. S is chosen as a finite set whose elements are physically indistinguishable, for example, the faces of a fair die. If S contains n elements, then the probability of each element is $1/n$. If A is a subset of S containing m elements, then $p(A) = m/n$. The classical interpretation never gave a satisfactory analysis of the meaning of "physically indistinguishable." If the faces of a fair die were really indistinguishable, then we could not tell them apart. How indistinguishable must they then be?

The Logical Interpretation

The logical interpretation (Keynes, 1921, Carnap, 1950) holds that conditional probability is a generalization of logical entailment. If $p(A) \neq 0$, then the *conditional probability of B given A* is by definition

$$P(B|A) = \frac{p(A \cap B)}{p(A)}$$

If the occurrence of B is logically entailed by the occurrence of A, then $p(B|A) = 1$. This interpretation holds that the occurrence of A can *partially entail* the occurrence of B, and this is expressed by a value for $p(B|A)$ strictly less than 1. Just as logical entailment is an "objective" relation between events, partial entailment is also held to be an objective relation, the same for all observers. If we substitute for A the trivial event S (i.e., the event that always occurs), then $p(B) = p(B|S)$ should be the same for all observers. Hence, the logical interpretation holds that the probability of each event is uniquely determined. A satisfactory definition of partial entailment has never been given.

The Frequency Interpretation

The frequency interpretation (von Mises, 1919, 1981; Reichenbach, 1932) holds that $p(A)$ is the limit relative frequency of the event A in an infinite sequence of "independent trials for A." Let $1_A(i)$ be the indicator function for "A on the ith trial," that is $1_A(i) = 1$ if A occurs on the ith trial, and $=0$ otherwise. By definition

$$p(A) = \lim_{n \to \infty} \frac{1}{n} \sum_{i=1}^{n} 1_A(i)$$

The frequency interpretation has always stumbled over the definition of "independent trials for A." The definition of independence cannot appeal to the notion of probability, as independence is used here in defining probability. Von Mises attempted to characterize the sort of sequences of trials whose limiting relative frequencies could be interpreted as probabilities, calling such sequences "collectives." His attempts are now generally regarded as unsuccessful, although there have been notable attempts to rehabilitate this notion (see Martin-Lof, 1970; Schnorr, 1970; van Lambalgen, 1987).

Even if frequencies cannot be used to *define* probability, they can be used to *measure* probabilities.[3] The main mechanism for doing this is the

Weak Law of Large Numbers. Let events $\{A_i\}_{i=1,\ldots,\infty}$ be independent with $p(A_i) = p$, $i = 1, \ldots$; then for all $d > 0$, (A.3)

$$p\left(\left|\frac{1}{n} \sum_{i=1}^{n} 1_A(i) - p\right| > d\right) \to 0 \qquad \text{as } n \to \infty$$

A sequence $\{A_i\}_{i=1,\ldots,\infty}$ of independent events with $p(A_i) = p$ is called a *Bernoulli sequence*, or a *Bernoulli process*, with parameter p. The weak law says that in a long finite sequence of independent trials for A, the relative frequency of occurence of A will very likely be very close to the probability of A. Of course, in order to apply this law we must know that the trials for A are independent, and that the probability of observing A is the same on each trial.

The Subjective Interpretation

The subjective interpretation is the principal focus of Part II. The discussion here will therefore be brief. Although its orgins are somewhat obscure, the right reverend Thomas Bayes (1702–1761) was an early protagonist, and his followers are sometimes called Bayesians. The mathematicians Emil Borel (1924) and Bruno De Finetti (1937) contributed to the formal development, and the latter was an especially forceful advocate. The philospher Frank Ramsey (1931) made the essential connection with the theory of utility, which was rediscovered by Von Neumann and Morgenstern in their classic *Theory of Games and Economic Behavior* (1944). L. J. Savage's *The Foundations of Statistics* (1954) presented a

[3]Philosophers of the positivist persuasion will wonder what the difference is between defining and measuring. There is this difference: Measurements can apply under special circumstances, where definitions are general. Thus we can measure the momentum of a charged particle in a magnetic field by observing its deflection, but we cannot define momentum in general in this way.

comprehensive theory in which the subjective interpretation of probability is embedded in a general theory of rational behavior.

The subjective interpretation holds that $p(A)$ represents the degree of belief that a subject invests in the occurrence of event A. The theory is *normative* in the same sense as logic. Under this interpretation, the laws of probability are sometimes said to comprise the *logic of partial belief.* Logic does not prescribe the events in whose occurrence we should believe. However, if we believe in A and if we believe that "If A then B," then logic does prescribe that we should also believe in B. It is easy to find many examples in which people do not conform to the laws of logic. This does not mean that logic is incorrect, it simply means that logic does not always *describe* the way people actually think.

In the same vein, the subjective interpretation does not prescribe the degree of belief a subject should invest in the occurrence of an event A. However, it does prescribe, for example, that his belief in $A \cup B$ should be at least as great as his belief in A. People often violate this and other prescriptions of the theory. From Chapter 3 we know that probabilistic reasoning is much more subtle and treacherous than elementary logical reasoning. Immoral people cannot invalidate the laws of morality (if such there be), illogical people cannot invalidate the laws of logic, and in the same way, poor probabilistic reasoning cannot invalidate the interpretation of probability as the logic of partial belief.

Of course, we must give some account of this normative mandate for the logic of partial belief. Why should partial belief conform to the laws of probability? In Chapter 6 one answer to this question is given: Savage's normative decision model derives the representation of uncertainty via probability from a deeper theory of rational preference.

De Finetti has given another argument, not reproduced here, with more cash value. He argues that uncertainty should be expressed in personal betting rates and shows that these rates conform to the laws of probability if and only if the person in question avoids a "Dutch book." A Dutch book is a combination of bets in which the person can only lose.

It is well to emphasize that the logic of partial belief in the context of subjective probability has nothing to do with the logical interpretation of probability mentioned above. The logic of partial belief does not prescribe *ab initio* degrees of partial belief in events, whereas the logical interpretation would.

Finally, it is well to emphasize that subjective probabilities can sometimes be interpreted as relative frequencies. The notion of a well-calibrated assessor is studied in detail in Chapters 8 and 9. Under appropriate circumstances, the subjective probabilities of a well-calibrated assessor correspond to relative frequencies. Further, De Finetti's theory of exchangeability gives general conditions under which subjective probabilities approach relative frequencies.

CONDITIONAL PROBABILITY, INDEPENDENCE, BAYES' THEOREM

If $P(A) \neq 0$, then we define the *conditional probability* of B given A as

$$p(B|A) = \frac{p(B \cap A)}{p(A)}$$

A set $\{A_i\}_{i=1,\ldots,n}$ of events is a *partition* of S if $A_i \cap A_j = \emptyset$ whenever $i \neq j$, and if $\cup A_i = S$. A useful fact is

$$p(B) = \sum_{i=1}^{n} p(B|A_i)p(A_i) \tag{A.4}$$

whenever $\{A_i\}_{i=1,\ldots,n}$ is a partition with $p(A_i) \neq 0$, $i = 1,\ldots,n$.

Events A and B are said to be *independent* (with respect to probability measure p) if

$$p(A \cap B) = p(A)p(B)$$

If A and B are independent, then clearly $p(A|B) = p(A)$ $[p(B) \neq 0]$.

A simple though important result is known as *Bayes' theorem:*

$$p(A|B) = \frac{p(B|A)p(A)}{p(B)} \tag{A.5}$$

If A is a "hypothesis," and B is observed data, then $p(A)$ is called the *prior probability* of the hypothesis, $p(A|B)$ is called the *posterior probability* of the hypothesis given the observation B, and $p(B|A)$ is called the *likelihood of the hypothesis* given the observation B. $p(A)$ and $p(B)$ are also called the *base rates.*

EXPECTATION, MOMENTS, MOMENT CONVERGENCE

If f is a real valued function defined on S, with S finite, then the *expectation* of f, $E(f)$, is

$$E(f) = \sum_{s \in S} f(s)p(s)^1$$

Expectation is linear, that is, for all real numbers a and b, $E(af + bg) = a\,E(f) + b\,E(g)$.

If S is infinite the definition of the expectation of a function f on S is more delicate, especially when the probability p is only finitely additive. When we write the expectation of f as

$$E(f) = \int_{s \in S} f(s)\,dp(s)$$

we shall assume that the measure p is countably additive, and that this expression is defined in the usual way.

The *conditional expectation* of f given $B[p(B) > 0)]$, for S finite is defined as

$$E(f|B) = \sum_{s \in S} f(s)p(s|B)$$

and a similar definition applies for S infinite. The *indicator function* for an event A, 1_A, is defined as

$$1_A(s) = 1 \quad \text{if } s \in A \quad \text{and} \quad = 0 \quad \text{otherwise}$$

[1] This expression is meaningful if $\{s\} \in \mathscr{F}$, for all $s = S$. This of course holds if $\mathscr{F} = 2^S$. It need not hold in general, and then we should say that it was not measurable with respect to $\mathscr{F}$.

A very useful fact is

$$E(1_A) = p(A)$$

The *variance* of f, $V(f)$, is defined to be $E([f - E(f)]^2) = E(f^2) - E^2(f)$. The *standard deviation* of f is $[V(f)]^{1/2}$. If f and g are functions on S, the *covariance* of f and g, $\text{cov}(f, g)$, is $E(fg) - E(f)E(g)$. The *product moment correlation* of f and g, $\rho(f, g)$ is defined to be $\text{cov}(f, g)/[V(f)V(g)]^{1/2}$.

If p is a probability over the real numbers R (that is, $S = R$, and $\mathscr{F}$ is a σ-field over R containing all intervals), then the rth *moment* of p is

$$E(x^r) = \int_R x^r \, dp(x)$$

the *expectation* of p is $E(x)$ and the *variance* $V(p)$ of p is $E([x - E(x)]^2) = E(x^2) - [E(x)]^2$. In Chapter 7 we invoke the important

Moment Convergence Theorem. Let $S = [0, 1]$, let $\mathscr{F}$ be a σ-field over S, let $\{p_n\}_{n=1,\ldots,\infty}$ be a sequence of countably additive probabilities on $(S, \mathscr{F})$ and let E_n denote expectation with respect to p_n. Then p_n converges[4] to a probability p on $(S, \mathscr{F})$ as $n \to \infty$ if and only if for $r = 1, 2, \ldots$, $E_n(x^r)$ converges to some limit m_r as $n \to \infty$. In this case m_r is the rth moment of p, and p is uniquely determined by its moments (Feller, 1971, p. 251).

The number of different ways of choosing m objects from a set of n objects, $n \geqslant m \geqslant 0$, is given by the *binomial coefficient*:

$$\binom{n}{m} = \frac{n!}{m!(n-m)!} \tag{A.7}$$

The above expression is defined to be 0 if $m > n$ or $m < 0$.

The *binomial theorem* states that, for real numbers a and b,

$$(a + b)^n = \sum_{m=0}^{n} \binom{n}{m} a^m b^{n-m} \tag{A.8}$$

DISTRIBUTIONS

A *random variable* X on the probability space $(S, \mathscr{F}, p)$ is a real-valued function S such that for all $r \in R$, $\{s \in S \mid X(s) \leqslant r\} \in \mathscr{F}$. The latter requirement ensures that X is measurable with respect to $\mathscr{F}$. Convention dictates that random variables are written with capital letters, and values of the random variables with lowercase letters, thus "$X = x$" means that random variable X takes value x. A function $F: R \to R$ is the cumulative distribution function of X if

$$F(r) = p\{s \mid X(s) \leqslant r\}$$

Distribution functions are continuous from the right, and hence differentiable except on a set that is at most countable. If F is everywhere differentiable, then

---

[4]Specifically, $p_n(A) \to p(A)$ for every set A whose boundary has probability 0 under p. An equivalent definition is that $E_n(f) \to E(f)$ for every real continuous function f on $[0, 1]$.

dF/dr is the *density* corresponding to F.

For example, the *normal distribution* with mean μ and standard deviation σ is

$$\Phi(r) = \int_{-\infty}^{r} (2\pi\sigma^2)^{-1/2} e^{-1/2[(x-\mu)/\sigma]^2} \, dx$$

The integrand is the normal density with mean μ and standard deviation σ. If X is a normally distributed random variable with mean μ and standard deviation σ, then $Y = (X - \mu)/\sigma$ is a *standard normal* variable having mean 0 and unit variance. Values for the standard normal cumulative distribution function are given in table B.1. The inverse distribution is given in table B.2. Any *positive affine transformation* $f(Y)$:

$$f(Y) = aY + b \qquad a, b \in \mathbf{R}, \, a > 0$$

is normal with mean b and standard deviation a. The constants a and b are sometimes called the *scale* and *location* parameters of the transformation f. Affine transformations (positive or otherwise) preserve the ratios of intervals, that is, for all $y_1, y_2, y_3, y_4, y_3 \neq y_4$,

$$\frac{y_1 - y_2}{y_3 - y_4} = \frac{f(y_1) - f(y_2)}{f(y_3) - f(y_4)}$$

We say that a function is (positive) affine unique if it is determined up to an arbitrary (positive) affine transformation. The product moment correlation is invariant under positive affine transformations, that is, $\rho(af + b, g) = \rho(f, g)$.

For X normal with mean μ and standard deviation σ, the variable $Z = e^X$ is *lognormal*, and $\ln Z$ is normal with mean and standard deviation μ and σ. As the lognormal distribution is of special interest in estimating small probabilities, we develop some of its properties. Let Y be a standard normal variable; then

$$\frac{\ln Z - \mu}{\sigma} = {}_d Y$$

where "$=_d$" means "has the same distribution as." Let z_a and y_a be the ath quantiles of the distributions of Z and Y; that is, the unique numbers z_a, y_a such that

$$p(Z \leqslant z_a) = a = p(Y \leqslant y_a)$$

Filling z_a and y_a into the previous equation, we see

$$z_a = e^{(\sigma y_a + \mu)}$$

Since $0 = y_{0.5}$ we deduce that the median of z is e^μ. By direct integration we can calculate the mean and variance of Z:

$$E(Z) = e^{\mu + \sigma^2/2}$$

$$E(Z - E(Z))^2 = (e^{\sigma^2} - 1)e^{2\mu + \sigma^2}$$

Further, writing $k_a = z_a/e^\mu$, we derive

$$\frac{\ln k_a}{y_a} = \sigma$$

Since the standard normal variable is symmetric about the origin, $y_{1-a} = -y_a$, and hence $k_{1-a} = 1/k_a$. This means that the quantiles z_a and z_{1-a} can be derived by respectively multiplying and dividing the median by k_a. It is therefore convenient to give the median 5% and 95% quantiles of a lognormal distribution by giving the median and the factor $k_{.95}$. $k_{.95}$ is called the *error factor* or *range factor* of the distribution of z. Moreover, given k_a for one value of a, we can derive k_b for $b \neq a$ from the last equation, and easily compute other quantiles. Finally, given two symmetric quantiles, z_a and z_{1-a}, the median is just their geometric mean:

$$z_{0.5} = (z_a z_{1-a})^{1/2}$$

Another important distribution is the chi square distribution, denoted χ^2. The chi square distributions are parametrized by their "degrees of freedom." χ_B^2 denotes the cumulative chi square distribution function with B degrees of freedom. Its mean is B and its variance is $2B$. If X is a standard normal variable, then X^2 is a chi square variable with one degree of freedom. The sum of independent chi square variables is chi square, whose number of degrees of freedom is the sum of the numbers of degrees of freedom of the summands. Hence χ_B^2 is the distribution of the sum of B independent squared standard normal variables. $\chi_B^2(r)$ denotes the value of the chi square cumulative distribution function with B degrees of freedom at point r. Table B.3 gives quantiles of chi square distributions with the number of degrees of freedom ranging from 1 to 200.

ENTROPY, INFORMATION, RELATIVE INFORMATION

Let $p = (p_1, \ldots, p_n)$, $p_i > 0, i = 1, \ldots, n$, be a probability distribution over the outcomes $x_1, \ldots, x_n$. Where ln denotes the natural logarithm, the *entropy* of p, $H(p)$, is

$$H(p) = -\sum_{i=1}^{n} p_i \ln p_i \tag{A.9}$$

The *information* of p, $I(p)$, is the negative of the entropy; $I(p) = -H(p)$. $H(p)$ is always nonnegative. Its minimal value, 0, is attained if and only if $p_i = 1$ for some i. Its maximal value, $\ln n$, is attained if and only if $p_i = 1/n$, $i = 1, \ldots, n$. $H(P)$ is a measure for the lack of information in the distribution P.

For $n = 2$ we may graph $H(p)$ as a function of p_1, as shown in Figure A.1.

Let $q = q_1, \ldots, q_m$, $q_i > 0$, $i = 1, \ldots, m$, be a distribution over the outcomes $y_1, \ldots, y_m$. If p and q are independent, then the *joint distribution* (p, q) is given by

$$(p, q)(x_i, y_j) = p_i q_j \qquad i = 1, \ldots, n; j = 1, \ldots, m$$

In this case, the entropy of (p, q) satisfies

$$H(p, q) = H(p) + H(q) \tag{A.10}$$

Whereas entropy is a natural, dimensionless, measure for lack of information in finite distributions, there is no wholly satisfactory generalization for probability

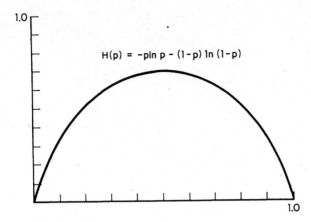

Figure A.1 $H(p)$ shown as a function of p_1, where $p = (p_1, 1 - p_1)$.

densities. For a probability density $f(x)$, we can consider the "analog":

$$H(f) = -\int [\ln f(x)] f(x)\, dx \qquad (A.11)$$

By direct integration, we can calculate the entropies for the normal and lognormal distributions with parameters μ, σ:

$$\text{Entropy of normal} = \ln(\sqrt{2\pi\sigma^2}) + \tfrac{1}{2}$$

and

$$\text{Entropy of lognormal} = \mu + \ln(\sqrt{2\pi\sigma^2}) + \tfrac{1}{2}$$

We see that both these expressions can be negative and go to $-\infty$ as $\sigma \to 0$. In general, probability densities, as opposed to mass functions, can have negative entropy, as $f(x)$ can be greater than 1.

These problems are caused by the fact that (A.11) is not really a continuous analog of (A.10). In going from finite probability mass functions to continuous density functions, we should replace summation by integration and replace "p_i" by $f(x)\, dx$. However, the expression $\ln [f(x)\, dx]$ would not make sense. The "missing dx" in (A.11) causes $H(f)$ to behave quite differently than $H(p)$. For example, suppose that a digital weight scale distinguishes 200 different kilogram readings, and suppose we consider a distribution over possible scale readings for a given object. Since there are 200 possible outcomes, the entropy of this distribution is a dimensionless number reflecting the lack of information in the distribution. If we convert the scale from kilograms to grams, the value of the entropy remains the same. Consider now a normal random variable X with standard deviation σ_x reflecting uncertainty in weight, measured in grams. If we now express weight in kilograms instead of grams, we transform the variable X into the variable $Y = X/1000$. This is a positive affine transformation with scale parameter $\frac{1}{1000}$; hence, Y is normally distributed with the same mean as X, but with standard

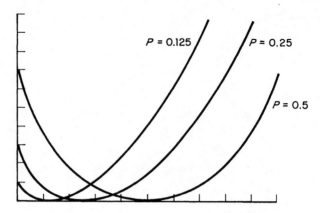

Figure A.2 $I(s, p)$ for $n = 2$, $s = (s_1, 1 - s_1)$, $p = (p_1, 1 - p_1)$, shown as a function of s_1, for $p_1 = 0.125$, $p_1 = 0.25$, and $p_1 = 0.5$.

deviation $\sigma_y = \sigma_x/1000$. Consulting the above formula, we see that the entropy drops when we change the scale from grams to kilograms.

There is no universally accepted measure for information for continuous distributions. We discussed practical solutions for this problem in Chapter 8.

Let p be as above, and let $s = (s_1, \ldots, s_n)$ be another distribution over $x_1, \ldots, x_n$. The *relative information* of s with respect to p is

$$I(s, p) = \sum_{i=1}^{n} s_i \ln \frac{s_i}{p_i} \qquad (A.12)$$

$I(s, p)$ is always nonnegative (Kullback, 1959, p. 15), and its minimal value, 0, is attained if and only if $s = p$. Note that $s_i = 0$ is allowed but $p_i > 0$ must be assumed. If u denotes the uniform distribution over $\{1, \ldots, n\}$, such that $u_i = 1/n, i = 1, \ldots, n$; then an easy calculation shows that

$$I(p, u) = \ln n - H(p) \qquad (A.12)$$

from which it follows that the maximal value of $I(p, u)$, $\ln n$ is attained when $H(p) = 0$.

Let s be the sample distribution generated by N independent samples from the distribution p. Then as N goes to infinity, the quantity $2NI(s, p)$ becomes χ^2 distributed with $n - 1$ degrees of freedom.

If again we set $n = 2$, and we fix p_1, then we can graph $I(s, p)$ as a function of s_1, as shown in Figure A.2.

We note that the relative information, in contrast with the entropy, does have a natural generalization for probability densities. Let f_1 and f_2 be two continuous densities which are nowhere equal to zero. Then

$$I(f_1, f_2) = \int f_1(x) \ln \frac{f_1(x)}{f_2(x)} dx \qquad (A.13)$$

is always nonnegative, and equals zero if and only if $f_1 = f_2$ (Kullback, 1959, p. 15). The "missing dx" of (A.11) drops out in the argument of the natural logarithm in (A.13).

Appendix B
Tables

Table B.1 The c.d.f. of the Standard Normal Distribution $\Phi(z) = \displaystyle\int_t^z \frac{1}{\sqrt{2\pi}} e^{-(1/2)t^2}\, dt$

z	0	1	2	3	4	5	6	7	8	9
−3.9	0.0000	0000	0000	0000	0000	0000	0000	0000	0000	0000
−3.8	0.0001	0001	0001	0001	0001	0001	0001	0001	0001	0001
−3.7	0.0001	0001	0001	0001	0001	0001	0001	0001	0001	0001
−3.6	0.0002	0002	0001	0001	0001	0001	0001	0001	0001	0001
−3.5	0.0002	0002	0002	0002	0002	0002	0002	0002	0002	0002
−3.4	0.0003	0003	0003	0003	0003	0003	0003	0003	0003	0002
−3.3	0.0005	0005	0005	0004	0004	0004	0004	0004	0004	0003
−3.2	0.0007	0007	0006	0006	0006	0006	0006	0005	0005	0005
−3.1	0.0010	0009	0009	0009	0008	0008	0008	0008	0007	0007
−3.0	0.0013	0013	0013	0012	0012	0011	0011	0011	0010	0010
−2.9	0.0019	0018	0018	0017	0016	0016	0015	0015	0014	0014
−2.8	0.0026	0025	0024	0023	0023	0022	0021	0021	0020	0019
−2.7	0.0035	0034	0033	0032	0031	0030	0029	0028	0027	0027
−2.6	0.0047	0045	0044	0043	0041	0040	0039	0038	0037	0036
−2.5	0.0062	0060	0059	0057	0055	0054	0052	0051	0049	0048
−2.4	0.0082	0080	0078	0075	0073	0071	0069	0069	0066	0064
−2.3	0.0107	0104	0102	0096	0096	0094	0091	0089	0087	0084
−2.2	0.0139	0136	0132	0129	0125	0122	0119	0116	0113	0110
−2.1	0.0179	0174	0170	0166	0162	0158	0154	0150	0146	0143
−2.0	0.0228	0222	0217	0212	0207	0202	0197	0192	0188	0183
−1.9	0.0287	0281	0274	0268	0262	0256	0250	0244	0239	0233
−1.8	0.0359	0351	0344	0336	0329	0322	0314	0307	0301	0294
−1.7	0.0446	0436	0427	0418	0409	0401	0392	0384	0375	0367
−1.6	0.0548	0537	0526	0516	0505	0495	0485	0475	0465	0455
−1.5	0.0668	0655	0643	0630	0618	0606	0594	0582	0571	0559
−1.4	0.0808	0793	0778	0764	0749	0735	0721	0708	0694	0681
−1.3	0.0968	0951	0934	0918	0901	0885	0869	0853	0838	0823
−1.2	0.1151	1131	1112	1093	1075	1056	1038	1020	1003	0985

Table B.1 Continued).

z	0	1	2	3	4	5	6	7	8	9
−1.1	0.1357	1335	1314	1292	1271	1251	1230	1210	1190	1170
−1.0	0.1587	1562	1539	1515	1492	1469	1446	1423	1401	1379
−0.9	0.1841	1814	1788	1762	1736	1711	1685	1660	1635	1611
−0.8	0.2119	2090	2061	2033	2005	1977	1949	1922	1894	1867
−0.7	0.2420	2389	2358	2327	2296	2266	2236	2206	2177	2148
−0.6	0.2743	2709	2676	2643	2611	2578	2546	2514	2483	2451
−0.5	0.3085	3050	3015	2981	2946	2912	2877	2843	2810	2776
−0.4	0.3446	3409	3372	3336	3300	3264	3228	3192	3156	3121
−0.3	0.3821	3783	3745	3707	3669	3632	3594	3557	3520	3483
−0.2	0.4207	4168	4129	4090	4052	4013	3974	3936	3897	3859
−0.1	0.4602	4562	4522	4483	4443	4404	4364	4325	4286	4247
−0.0	0.5000	4960	4920	4880	4840	4801	4761	4721	4681	4641
0.0	0.5000	5040	5080	5120	5160	5199	5239	5279	5319	5359
0.1	0.5398	5438	5478	5517	5557	5596	5636	5675	5714	5753
0.2	0.5793	5832	5871	5910	5948	5987	6026	6064	6103	6141
0.3	0.6179	6217	6255	6293	6331	6368	6406	6443	6480	6517
0.4	0.6554	6591	6628	6664	6700	6736	6772	6808	6844	6879
0.5	0.6915	6950	6985	7019	7054	7088	7123	7157	7190	7224
0.6	0.7257	7291	7324	7357	7389	7422	7454	7486	7517	7549
0.7	0.7580	7611	7642	7673	7704	7734	7764	7794	7823	7852
0.8	0.7881	7910	7939	7967	7995	8023	8051	8078	8106	8133
0.9	0.8159	8186	8212	8238	8264	8289	8315	8340	8365	8389
1.0	0.8413	8438	8461	8485	8508	8531	8554	8577	8599	8621
1.1	0.8643	8665	8686	8708	8729	8749	8770	8790	8810	8830
1.2	0.8849	8869	8888	8907	8925	8944	8962	8980	8997	9015
1.3	0.9032	9049	9066	9082	9099	9115	9131	9147	9162	9177
1.4	0.9192	9207	9222	9236	9251	9265	9279	9292	9306	9319
1.5	0.9332	9345	9357	9370	9382	9394	9406	9418	9429	9441
1.6	0.9452	9463	9474	9484	9495	9505	9515	9525	9535	9545
1.7	0.9554	9564	9573	9582	9591	9599	9608	9616	9625	9633
1.8	0.9641	9649	9656	9664	9671	9678	9686	9693	9699	9706
1.9	0.9713	9719	9726	9732	9738	9744	9750	9756	9761	9767
2.0	0.9772	9778	9783	9788	9793	9798	9803	9808	9812	9817
2.1	0.9821	9826	9830	9834	9838	9842	9846	9850	9854	9857
2.2	0.9861	9864	9868	9871	9875	9878	9881	9884	9887	9890
2.3	0.9893	9896	9898	9901	9904	9906	9909	9911	9913	9916
2.4	0.9918	9920	9922	9925	9927	9929	9931	9932	9934	9936
2.5	0.9938	9940	9941	9943	9945	9946	9948	9949	9951	9952
2.6	0.9953	9955	9956	9957	9959	9960	9961	9962	9963	9964
2.7	0.9965	9966	9967	9968	9969	9970	9971	9972	9973	9974
2.8	0.9974	9975	9976	9977	9977	9978	9979	9979	9980	9981
2.9	0.9981	9982	9982	9983	9984	9984	9985	9985	9986	9986
3.0	0.998650	.998694	.998736	.998777	.998817	.998856	.998893	.998930	.998965	.998999
3.1	0.999032	.999065	.999096	.999126	.999155	.999184	.999211	.999238	.999264	.999289
3.2	0.999313	.999336	.999359	.999381	.999402	.999423	.999443	.999462	.999481	.999499
3.3	0.999517	.999534	.999550	.999566	.999581	.999596	.999610	.999624	.999638	.999651
3.4	0.999663	.999675	.999687	.999698	.999709	.999720	.999730	.999740	.999749	.999758

Table B.1. (Continued)

z	0	1	2	3	4	5	6	7	8	9
3.5	0.999767	.999776	.999784	.999792	.999800	.999807	.999815	.999822	.999828	.999835
3.6	0.999841	.999847	.999853	.999858	.999864	.999869	.999874	.999879	.999883	.999888
3.7	0.999892	.999896	.999900	.999904	.999908	.999912	.999915	.999918	.999922	.999925
3.8	0.999928	.999931	.999933	.999936	.999938	.999941	.999943	.999946	.999948	.999950
3.9	0.999952	.999954	.999956	.999958	.999959	.999961	.999963	.999964	.999966	.999967
4.0	0.999968	.999970	.999971	.999972	.999973	.999974	.999975	.999976	.999977	.999978
4.1	0.999979	.999980	.999981	.999982	.999983	.999983	.999984	.999985	.999985	.999986
4.2	0.999987	.999987	.999988	.999988	.999989	.999989	.999990	.999990	.999991	.999991
4.3	0.999991	.999992	.999992	.999993	.999993	.999993	.999993	.999994	.999994	.999994
4.4	0.999995	.999995	.999995	.999995	.999996	.999996	.999996	.999996	.999996	.999996
4.5	0.999997	.999997	.999997	.999997	.999997	.999997	.999997	.999998	.999998	.999998
4.6	0.999998	.999998	.999998	.999998	.999998	.999998	.999998	.999998	.999999	.999999
4.7	0.999999	.999999	.999999	.999999	.999999	.999999	.999999	.999999	.999999	.999999
4.8	0.999999	.999999	.999999	.999999	.999999	.999999	.999999	.999999	.999999	.999999
4.9	1.000000	1.00000	1.00000	1.00000	1.00000	1.00000	1.00000	1.00000	1.00000	1.00000

z	0	1	2	3	4	5	6	7	8	9

Table B.2. The Inverse Normal Function $z = \Phi^{-1}(q) = \Phi^{-1}(1 - p)$

p	.000	.001	.002	.003	.004	.005	.006	.007	.008	.009	.010	
.00	∞	3.0902	2.8782	2.7478	2.6521	2.5758	2.5121	2.4573	2.4089	2.3656	2.3263	.99
.01	2.3263	2.2904	2.2571	2.2262	2.1973	2.1701	2.1444	2.1201	2.0969	2.0749	2.0537	.98
.02	2.0537	2.0335	2.0141	1.9954	1.9774	1.9600	1.9431	1.9268	1.9110	1.8957	1.8808	.97
.03	1.8808	1.8663	1.8522	1.8384	1.8250	1.8119	1.7991	1.7866	1.7744	1.7624	1.7507	.96
.04	1.7507	1.7392	1.7279	1.7169	1.7060	1.6954	1.6849	1.6747	1.6646	1.6546	1.6449	.95
.05	1.6449	1.6352	1.6258	1.6164	1.6072	1.5982	1.5893	1.5805	1.5718	1.5632	1.5548	.94
.06	1.5548	1.5464	1.5382	1.5301	1.5220	1.5141	1.5063	1.4985	1.4909	1.4833	1.4758	.93
.07	1.4758	1.4684	1.4611	1.4538	1.4466	1.4395	1.4325	1.4255	1.4187	1.4118	1.4051	.92
.08	1.4051	1.3984	1.3917	1.3852	1.3787	1.3722	1.3658	1.3595	1.3532	1.3469	1.3408	.91
.09	1.3408	1.3346	1.3285	1.3225	1.3165	1.3106	1.3047	1.2988	1.2930	1.2873	1.2816	.90
.10	1.2816	1.2759	1.702	1.2646	1.2591	1.2536	1.2481	1.2426	1.2372	1.2319	1.2265	.89
.11	1.2265	1.2212	1.2160	1.2107	1.2055	1.2004	1.1952	1.1901	1.1850	1.1800	1.1750	.88
.12	1.1750	1.1700	1.1650	1.1601	1.1552	1.1503	1.1455	1.1407	1.1359	1.1311	1.1264	.87
.13	1.1264	1.1217	1.1170	1.1123	1.1077	1.1031	1.0985	1.0939	1.0893	1.0848	1.0803	.86
.14	1.0803	1.0758	1.0714	1.0669	1.0625	1.0581	1.0537	1.0494	1.0450	1.0407	1.0364	.85
.15	1.0364	1.0322	1.0279	1.0237	1.0194	1.0152	1.0110	1.0069	1.0027	0.9986	0.9945	.84
.16	0.9945	0.9904	0.9863	0.9822	0.9782	0.9741	0.9701	0.9661	0.9621	0.9581	0.9542	.83
.17	0.9542	0.9502	0.9463	0.9424	0.9385	0.9346	0.9307	0.9269	0.9230	0.9192	0.9154	.82
.18	0.9154	0.9116	0.9078	0.9040	0.9002	0.8965	0.8927	0.8890	0.8853	0.8816	0.8779	.81
.19	0.8779	0.8742	0.8705	0.8669	0.8633	0.8596	0.8560	0.8524	0.8488	0.8452	0.8416	.80
.20	0.8416	0.8381	0.8345	0.8310	0.8274	0.8239	0.8204	0.8169	0.8134	0.8099	0.8064	.79
.21	0.8064	0.8030	0.7995	0.7961	0.7926	0.7892	0.7858	0.7824	0.7790	0.7756	0.7722	.78
.22	0.7722	0.7688	0.7655	0.7621	0.7588	0.7554	0.7521	0.7488	0.7454	0.7421	0.7388	.77
.23	0.7388	0.7356	0.7323	0.7290	0.7257	0.7225	0.7192	0.7160	0.7128	0.7095	0.7063	.76
.24	0.7063	0.7031	0.6999	0.6967	0.6935	0.6903	0.6871	0.6840	0.6808	0.6776	0.6745	.75
.25	0.6745	0.6713	0.6682	0.6651	0.6620	0.6588	0.6557	0.6526	0.6495	0.6464	0.7433	.74

Table B.2. The Inverse Normal Function $z = \Phi^{-1}(q) = \Phi^{-1}(1 - p)$

p	.000	.001	.002	.003	.004	.005	.006	.007	.008	.009	.010	
.26	0.6433	0.6403	0.6372	0.7341	0.6311	0.6280	0.6250	0.6219	0.6189	0.6158	0.6128	.73
.27	0.6128	0.6098	0.6068	0.6038	0.6008	0.5978	0.5948	0.5918	0.5888	0.5858	0.5828	.72
.28	0.5828	0.5799	0.5769	0.5740	0.5710	0.5681	0.5651	0.5622	0.5592	0.5563	0.5534	.71
.29	0.5534	0.5505	0.5476	0.5446	0.5417	0.5388	0.5359	0.5330	0.5302	0.5273	0.5244	.70
.30	0.5244	0.5215	0.5187	0.5158	0.5129	0.5101	0.5072	0.5044	0.5115	0.4987	0.4959	.69
.31	0.4959	0.4930	0.4902	0.4874	0.4845	0.4817	0.4789	0.4761	0.4733	0.4705	0.4677	.68
.32	0.4677	0.4649	0.4621	0.4593	0.4565	0.4538	0.4510	0.4482	0.4454	0.4427	0.4399	.67
.33	0.4399	0.4372	0.4344	0.4316	0.4289	0.4261	0.4234	0.4207	0.4179	0.4152	0.4125	.66
.34	0.4125	0.4097	0.4070	0.4043	0.4016	0.3989	0.3961	0.3934	0.3907	0.3880	0.3853	.65
.35	0.3853	0.3826	0.3799	0.3772	0.3745	0.3719	0.3692	0.3665	0.3638	0.3611	0.3585	.64
.36	0.3585	0.3558	0.3531	0.3505	0.3478	0.3451	0.3425	0.3398	0.3372	0.3345	0.3319	.63
.37	0.3319	0.3292	0.3266	0.3239	0.3213	0.3186	0.3160	0.3134	0.3107	0.3081	0.3055	.62
.38	0.3055	0.3029	0.3002	0.2976	0.2950	0.2924	0.2898	0.2871	0.2845	0.2819	0.2793	.61
.39	0.2793	0.2767	0.2741	0.2715	0.2689	0.2663	0.2637	0.2611	0.2585	0.2559	0.2533	.60
.40	0.2533	0.2508	0.2482	0.2456	0.2430	0.2404	0.2378	0.2353	0.2327	0.2301	0.2275	.59
.41	0.2275	0.2250	0.2224	0.2198	0.2173	0.2147	0.2121	0.20096	0.2070	0.2045	0.2019	.58
.42	0.2019	0.1993	0.1968	0.1942	0.1917	0.1891	0.1866	0.1840	0.1815	0.1789	0.1764	.57
.43	0.1764	0.1738	0.1713	0.1687	0.1662	0.1637	0.1611	0.1586	0.1560	0.1535	0.1510	.56
.44	0.1510	0.1484	0.1459	0.1434	0.1408	0.1383	0.1358	0.1332	0.1307	0.1282	0.1257	.55
.45	0.1257	0.1231	0.1206	0.1181	0.1156	0.1130	0.1105	0.1080	0.1055	0.1030	0.1004	.54
.46	0.1004	0.0979	0.0954	0.0929	0.0904	0.0878	0.0853	0.0828	0.0803	0.0778	0.0753	.53
.47	0.0753	0.728	0.0702	0.0677	0.0652	0.0627	0.0602	0.0577	0.0552	0.0527	0.0502	.52
.48	0.0502	0.0476	0.0451	0.0426	0.0401	0.0376	0.0351	0.0326	0.0301	0.0276	0.0251	.51
.49	0.0251	0.0226	0.0201	0.0175	0.0150	0.0125	0.0100	0.0075	0.0050	0.0025	0.0000	.50
	.010	.009	.008	.007	.006	.005	.004	.003	.002	.001	.000	q

Table B.3. The Chi-Squared Distribution

ν \ q	0.500	0.600	0.700	0.800	0.850	0.900	0.925	0.950	0.975	0.990	0.995	0.999	0.9995
1	0.455	0.708	1.074	1.642	2.072	2.706	3.170	3.841	5.024	6.635	7.879	10.83	12.12
2	1.386	0.833	2.408	3.219	3.794	4.605	5.181	5.991	7.378	9.210	10.60	13.82	15.20
3	2.366	2.946	3.665	4.642	5.317	6.251	6.905	7.815	9.348	11.34	12.84	16.27	17.73
4	3.357	4.045	4.878	5.989	6.745	7.779	8.496	9.488	11.14	13.28	14.86	18.47	20.00
5	4.351	5.132	6.064	7.289	8.115	9.236	10.01	11.07	12.83	15.09	16.75	20.52	22.11
6	5.348	6.211	7.231	8.558	9.446	10.64	11.47	12.59	14.45	16.81	18.55	22.46	24.10
7	6.346	7.283	8.383	9.803	10.75	12.02	12.88	14.07	16.01	18.48	20.28	24.32	26.02
8	7.344	8.351	9.524	11.03	12.03	13.36	14.27	15.51	17.53	20.09	21.95	26.12	27.87
9	8.343	9.414	10.66	12.24	13.29	14.68	15.63	16.92	19.02	21.67	23.59	27.88	29.67
10	9.342	10.47	11.78	13.44	14.53	15.99	16.97	18.31	20.48	23.21	25.19	29.59	31.42
11	10.34	11.53	12.90	14.63	15.77	17.28	18.29	19.68	21.92	24.72	26.76	31.26	33.14
12	11.34	12.58	14.01	15.81	16.99	18.55	19.60	21.03	23.34	26.22	28.30	32.91	34.82
13	12.34	13.64	15.12	16.98	18.20	19.81	20.90	22.36	24.74	27.69	29.82	34.53	36.48
14	13.34	14.69	16.22	18.15	19.41	21.06	22.18	23.68	26.12	29.14	31.32	36.12	38.11
15	14.34	15.73	17.32	19.31	20.60	22.31	23.45	25.00	27.49	30.58	32.80	37.70	39.72
16	15.34	16.78	18.42	20.47	21.79	23.54	24.72	26.30	28.85	32.00	34.27	39.25	41.31
17	16.34	17.82	19.51	21.61	22.98	24.77	25.97	27.59	30.19	33.41	35.72	40.79	42.88
18	17.34	18.87	20.60	22.76	24.16	25.99	27.22	28.87	31.53	34.81	37.16	42.31	44.43
19	18.34	19.91	21.69	23.90	25.33	27.20	28.46	30.14	32.85	36.19	38.58	43.82	45.97
20	19.34	20.95	22.77	25.04	26.50	28.41	29.69	31.41	34.17	37.57	40.00	45.31	47.50
21	20.34	21.99	23.86	26.17	27.66	29.62	30.92	32.67	35.43	38.93	41.40	46.80	49.01
22	21.34	23.03	24.94	27.30	28.82	30.81	32.14	33.92	36.78	40.29	42.80	48.27	50.51
23	22.34	24.07	26.02	28.43	29.98	32.01	33.36	35.17	38.08	41.64	44.18	49.73	52.00
24	23.34	25.11	27.10	29.55	31.13	33.20	34.57	36.42	39.36	42.98	45.56	51.18	53.48
25	24.34	26.14	28.17	30.68	32.28	34.38	35.78	37.65	40.65	44.31	46.93	52.62	54.95
26	25.34	27.18	29.25	31.79	33.43	35.56	36.98	38.89	41.92	45.64	48.29	54.05	56.41
27	26.34	28.21	30.32	32.91	34.57	36.74	38.18	40.11	43.19	46.96	49.64	55.48	57.86
28	27.34	29.25	31.39	34.03	35.71	37.92	39.38	41.34	44.46	48.28	50.99	56.89	59.30
29	28.34	30.28	32.46	35.14	36.85	39.09	40.57	42.56	45.72	49.59	52.34	58.30	60.73
30	29.34	31.32	33.53	36.25	37.99	40.26	41.76	43.77	46.98	50.89	53.67	59.70	62.16

Table B.3. The Chi-Squared Distribution

ν	|	q												
	0.500	0.600	0.700	0.800	0.850	0.900	0.925	0.950	0.975	0.990	0.995	0.999	0.9995
31	30.34	32.35	34.60	37.36	39.12	41.42	42.95	44.99	48.23	52.19	55.00	61.10	63.58
32	31.34	33.38	35.66	38.47	40.26	42.58	44.13	46.19	49.48	53.49	56.33	62.49	65.00
33	32.34	34.41	36.73	39.57	41.39	43.75	45.31	47.40	50.73	54.78	57.65	63.87	66.40
34	33.34	35.44	37.80	40.68	42.51	44.90	46.49	48.60	51.97	56.06	58.96	65.25	67.80
35	34.34	36.47	38.86	41.78	43.64	46.06	47.66	49.80	53.20	57.34	60.27	66.62	69.20
36	35.34	37.50	39.92	42.88	44.76	47.21	48.84	51.00	54.44	58.62	61.58	67.99	70.59
37	36.34	38.53	40.98	43.98	45.89	48.36	50.01	52.19	55.67	59.89	62.88	69.35	71.97
38	37.34	39.56	42.05	45.08	47.01	49.51	51.17	53.38	56.90	61.16	64.18	70.70	73.35
39	38.34	40.59	43.11	46.17	48.13	50.66	52.34	54.57	58.12	62.43	65.48	72.05	74.73
40	39.34	41.62	44.16	47.27	49.24	51.81	53.50	55.76	59.34	63.69	66.77	73.40	76.09
45	44.34	46.76	49.45	52.73	54.81	57.51	59.29	61.66	65.41	69.96	73.17	80.08	82.88
50	49.33	51.89	54.72	58.16	60.35	63.17	65.03	67.50	71.42	76.15	79.49	86.66	89.56
60	59.33	62.13	65.23	68.97	71.34	74.40	76.41	79.08	83.30	88.38	91.95	99.61	102.7
70	69.33	72.36	75.69	79.71	82.26	85.53	87.68	90.53	95.02	100.4	104.2	112.3	115.6
80	79.33	82.57	86.12	90.41	93.11	96.58	98.86	101.9	106.6	112.3	116.3	124.8	128.3
90	89.33	92.76	96.52	101.1	103.9	107.6	110.0	113.1	118.1	124.1	128.3	137.2	140.8
100	99.33	102.9	106.9	111.7	114.7	118.5	121.0	124.3	129.6	135.8	140.2	149.4	153.2
120	119.3	123.3	127.6	132.8	136.1	140.2	143.0	146.6	152.2	159.0	163.6	173.6	177.6
150	149.3	153.8	158.6	164.3	168.0	172.6	175.6	179.6	185.8	193.2	198.4	209.3	213.6
200	199.3	204.4	210.0	216.6	220.7	226.0	229.5	234.0	241.1	249.4	255.3	267.5	272.4

ν = degrees of freedom; q = value of cumulative distribution function.

Table 5.3. (Continued)

ν	q 0.500	0.600	0.700	0.800	0.850	0.900	0.925	0.950	0.975	0.990	0.995	0.999	.9995
1	0.455	0.708	1.074	0.642	2.072	2.706	3.170	3.841	5.024	6.635	7.879	10.83	12.12
2	1.386	1.833	2.408	3.219	3.794	4.605	5.181	5.991	7.378	9.210	10.60	13.82	15.20
3	2.366	2.946	3.665	4.642	5.317	6.251	6.905	7.815	9.348	11.34	12.84	16.27	17.73
4	3.357	4.045	4.878	5.989	6.745	7.779	8.496	9.488	11.14	13.28	14.86	18.47	20.00
5	4.351	5.132	6.064	7.289	8.115	9.236	10.01	11.07	12.83	15.09	16.75	20.52	22.11
6	5.348	6.211	7.231	8.558	9.446	10.64	11.47	12.59	14.45	16.81	18.55	22.46	24.10
7	6.346	7.283	8.383	9.803	10.75	12.02	12.88	14.07	16.01	18.48	20.28	24.32	26.02
8	7.344	8.351	9.524	11.03	12.03	13.36	14.27	15.51	17.53	20.09	21.95	26.12	27.87
9	8.343	9.414	10.66	12.24	13.29	14.68	15.63	16.92	19.02	21.67	23.59	27.88	29.67
10	9.342	10.47	11.78	13.44	14.53	15.99	16.97	18.31	20.48	23.21	25.19	29.59	31.42
11	10.34	11.53	12.90	14.63	15.77	17.28	18.29	19.68	21.92	24.72	26.76	31.26	33.14
12	11.34	12.58	14.01	15.81	16.99	18.55	19.60	21.03	23.34	26.22	28.30	32.91	34.82
13	12.34	13.64	15.12	16.98	18.20	19.81	20.90	22.36	24.74	27.69	29.82	34.53	36.48
14	13.34	14.69	16.22	18.15	19.41	21.06	22.18	23.68	26.12	29.14	31.32	36.12	38.11
15	14.34	15.73	17.32	19.31	20.60	22.31	23.45	25.00	27.49	30.58	32.80	37.70	39.72
16	15.34	16.78	18.42	20.47	21.79	23.54	24.72	26.30	28.85	32.00	34.27	39.25	41.31
17	16.34	17.82	19.51	21.61	22.98	24.77	25.97	27.59	30.19	33.41	35.72	40.79	42.88
18	17.34	18.87	20.60	22.76	24.16	25.99	27.22	28.87	31.53	34.81	37.16	42.31	44.43
19	18.34	19.91	21.69	23.90	25.33	27.20	28.46	30.14	32.85	36.19	38.58	43.82	45.97
20	19.34	20.95	22.77	25.04	26.50	28.41	29.69	31.41	34.17	37.57	40.00	45.31	47.50
21	20.34	21.99	23.86	26.17	27.66	29.62	30.92	32.67	35.48	38.93	41.40	46.80	49.01
22	21.34	23.03	24.94	27.30	28.82	30.81	32.14	33.92	36.78	40.29	42.80	48.27	50.51
23	22.34	24.07	26.02	28.43	29.98	32.01	33.36	35.17	38.08	41.64	44.18	49.73	52.00
24	23.34	25.11	27.10	29.55	31.13	33.20	34.57	36.42	39.36	42.98	45.56	51.18	53.48
25	24.34	26.14	28.17	30.68	32.28	34.38	35.78	37.65	40.65	44.31	46.93	52.62	54.95

Table 5.3. (Continued)

ν	q 0.500	0.600	0.700	0.800	0.850	0.900	0.925	0.950	0.975	0.990	0.995	0.999	.9995
26	25.34	27.18	29.25	31.79	33.43	35.56	36.98	38.89	41.92	45.64	48.29	54.05	56.41
27	26.34	28.21	30.32	32.91	34.57	36.74	38.18	40.11	43.19	46.96	49.64	55.48	57.86
28	27.34	29.25	31.39	34.03	35.71	37.92	39.38	41.34	44.46	48.28	50.99	56.89	59.30
29	28.34	30.28	32.46	35.14	36.85	39.09	40.57	42.56	45.72	49.59	52.34	58.30	60.73
30	29.34	31.32	33.53	36.25	37.99	40.26	41.76	43.77	46.98	50.89	53.67	59.70	62.16
31	30.34	32.35	34.60	37.36	39.12	41.42	42.95	44.99	48.23	52.19	55.00	61.10	63.58
32	31.34	33.38	35.66	38.47	40.26	42.58	44.13	46.19	49.48	53.49	56.33	62.49	65.00
33	32.34	34.41	36.73	39.57	41.39	43.75	45.31	47.40	50.73	54.78	57.65	63.87	66.40
34	33.34	35.44	37.80	40.68	42.51	44.90	46.49	48.60	51.97	56.06	58.96	65.25	67.80
35	34.34	36.47	38.86	41.78	43.64	46.06	47.66	49.80	53.20	57.34	60.27	66.62	69.20
36	35.34	37.50	39.92	42.88	44.76	47.21	48.84	51.00	54.44	58.62	61.58	67.99	70.59
37	36.34	38.53	40.98	43.98	45.89	48.36	50.01	52.19	55.67	59.89	62.88	69.35	71.97
38	37.34	39.56	42.05	45.08	47.01	49.51	51.17	53.38	56.90	61.16	64.18	70.70	73.35
39	38.34	40.59	43.11	46.17	48.13	50.66	52.34	54.57	58.12	62.43	65.48	72.05	74.73
40	39.34	41.62	44.16	47.27	49.24	51.81	53.50	55.76	59.34	36.69	66.77	73.40	76.09
45	44.34	46.76	49.45	52.73	54.81	57.51	59.29	61.66	65.41	69.96	73.17	80.08	82.88
50	49.33	51.89	54.72	58.16	60.35	63.17	65.03	67.50	71.42	76.15	79.49	86.66	89.56
60	59.33	62.13	65.23	68.97	71.34	74.40	76.41	79.08	83.30	88.38	91.95	99.61	102.7
70	69.33	72.36	75.69	79.71	82.26	85.53	87.68	90.53	95.02	100.4	104.2	112.3	115.6
80	79.33	82.57	86.12	90.41	93.11	96.58	98.86	101.9	106.6	112.3	116.3	124.8	128.3
90	89.33	92.76	96.52	101.1	103.9	107.6	110.0	113.1	118.1	124.1	128.3	137.2	140.8
100	99.33	102.9	106.9	111.7	114.7	118.5	121.0	124.3	129.6	135.8	140.2	149.4	153.2
120	119.3	123.3	127.6	132.8	136.1	140.2	143.0	146.6	152.2	159.0	163.6	173.6	177.6
150	149.3	153.8	158.6	164.3	168.0	172.6	175.6	179.6	185.8	193.2	198.4	209.3	213.6
200	199.3	204.4	210.0	216.6	220.7	226.0	229.5	234.0	241.1	249.4	255.3	267.5	272.4

Table B.4. Frequencies (f) for Values of the Number d of Circular Triads in Paired Comparisons, and the Probability (P) That These Values Will Be Attained or Exceeded

Value of d	n=2 f	n=2 P	n=3 f	n=3 P	n=4 f	n=4 P	n=5 f	n=5 P	n=6 f	n=6 P	n=7 f	n=7 P
0	2	1.000	6	1.000	24	1.000	120	1.000	720	1.000	5,040	1.000
1			2	0.250	16	0.623	120	0.883	960	0.978	8,400	0.908
2					24	0.375	240	0.766	2,240	0.949	21,840	0.994
3							240	0.531	2,880	0.880	33,600	0.983
4							280	0.297	6,240	0.792	75,600	0.967
5							24	0.023	3,648	0.602	90,384	0.931
6									8,640	0.491	179,760	0.888
7									4,800	0.227	188,160	0.802
8									2,640	0.081	277,200	0.713
9											280,560	0.580
10											384,048	0.447
11											244,160	0.263
12											233,520	0.147
13											72,240	0.036
14											2,640	0.001
Total	2		8		64		1,024		37,768		2,097,152	

Value of d	n=8 f	n=8 P	n=9 f	n=9 P	n=10 f	n=10 P	n=10 P′
0	40,320	1.000	362,880	1.000	3,628,800	1.0^5	1.000
1	80,640	$0.9^3 85$	846,720	0.9^3	9,676,800	1.0^5	1.000
2	228,480	$0.9^3 55$	2,580,480	$0.9^4 8$	31,449,600	1.0^5	1.000
3	403,200	$0.9^3 87$	5,093,760	$0.9^4 4$	68,275,200	1.0^5	1.000
4	954,240	$0.9^3 72$	12,579,840	$0.9^3 87$	175,392,000	1.0^3	1.000
5	1,304,576	$0.9^3 36$	19,958,400	$0.9^5 69$	311,592,960	0.9^5	1.000
6	3,042,816	0.989	44,698,752	$0.9^3 40$	711,728,640	0.9^4	1.000
7	3,870,720	0.977	70,785,792	$0.9^3 87$	1,193,794,560	0.9^4	1.000
8	6,926,080	0.963	130,032,000	$0.9^3 77$	2,393,475,840	0.9^4	1.000
9	8,332,800	0.937	190,834,560	$0.9^3 58$	3,784,596,480	0.9^3	1.000
10	15,821,568	0.906	361,525,248	$0.9^3 21$	7,444,104,192	0.9^3	1.000
11	14,755,328	0.847	443,931,264	0.988	10,526,745,600	0.9^3	1.000
12	24,487,680	0.792	779,950,080	0.981	19,533,696,000	0.9^3	1.000
13	24,514,560	0.701	1,043,763,840	0.970	27,610,168,320	$0.9^3 87$	1.000
14	34,762,240	0.610	1,529,101,440	0.955	47,107,169,280	$0.9^3 79$	0.999
15	29,288,448	0.480	1,916,619,264	0.933	64,016,040,960	$0.9^3 66$	0.997
16	37,188,480	0.371	2,912,257,152	0.905	107,446,832,640	$0.9^3 47$	0.996
17	24,487,680	0.232	3,078,407,808	0.862	134,470,425,600	$0.9^3 17$	0.995
18	24,312,960	0.141	4,506,485,760	0.817	218,941,470,720	0.988	0.993
19	10,402,560	0.051	4,946,417,280	0.752	272,302,894,080	0.982	0.988
20	3,230,080	0.012	6,068,256,768	0.680	417,512,148,480	0.974	0.980
21			6,160,876,416	0.592	404,080,834,560	0.962	0.968

Table B.4. (Continued)

For the case when the number (n) of objects is ten and approximate probability (P') is given based on the χ^2 approximation.

Value of d	$n=8$ f	$n=8$ P	$n=9$ f	$n=9$ P	$n=10$ f	$n=10$ P	P'
22			7,730,384,256	0.502	743,278,970,880	0.948	0.952
23			6,292,581,120	0.389	829,743,344,640	0.927	0.930
24			6,900,969,600	0.298	1,202,317,401,600	0.903	0.905
25			5,479,802,496	0.197	1,334,577,484,800	0.869	0.874
26			4,327,787,520	0.118	1,773,862,272,000	0.831	0.834
27			2,399,241,600	0.055	1,878,824,586,240	0.781	0.786
28			1,197,020,160	0.020	2,496,636,103,680	0.727	0.727
29			163,094,400	$0.0^3 24$	2,406,981,104,640	0.656	0.659
30			3,230,080	$0.0^3 5$	3,032,021,672,960	0.588	0.583
31					2,841,072,675,840	0.502	0.500
32					3,166,378,709,760	0.421	0.413
33					2,743,311,191,040	0.331	0.325
34					2,877,794,035,200	0.253	0.243
35					2,109,852,702,720	0.171	0.170
36					1,840,136,336,640	0.111	0.109
37					1,109,253,196,800	0.059	0.063
38					689,719,564,800	0.028	0.032
39					230,683,084,800	$0.0^3 70$	0.014
40					48,251,508,430	$0.0^3 14$	0.005

Source: Kendal, M., Rank Correlation Methods, Ch. Griffin & Co. London, 1962.
Notation: d = number of circular triads
n = number of objects

Table B.5. Agreement in Paired Comparisons. The Probability P That a Value of Σ (for u) Will Be Attained or Exceeded, for $m=8$, $n=2$ to 8

$n=2$ Σ	P	$n=3$ Σ	P	$n=4$ Σ	P	$n=5$ Σ	P	$n=6$ Σ	P	$n=7$ Σ	P	$n=8$ Σ	P
1	1.000	3	1.000	6	1.000	10	1.000	15	1.000	21	1.000	28	1.000
3	0.250	5	0.578	8	0.822	12	0.944	17	0.987	23	0.998	30	1.000
		7	0.156	10	0.466	14	0.756	19	0.920	25	0.981	82	0.997
		9	0.016	12	0.169	16	0.474	21	0.764	27	0.925	84	0.983
				14	0.038	18	0.224	23	0.539	29	0.808	36	0.945
				16	0.0046	20	0.078	25	0.314	31	0.633	38	0.805
				18	$0.0^3 24$	22	0.020	27	0.148	33	0.433	40	0.736
						24	0.0035	29	0.057	35	0.256	42	0.572
						26	$0.0^3 42$	31	0.017	37	0.130	44	0.400
						28	$0.0^4 30$	33	0.0042	39	0.056	46	0.250
						30	$0.0^4 95$	35	$0.0^3 79$	41	0.021	48	0.138
								37	$0.0^3 12$	43	0.0064	50	0.068
								39	$0.0^4 12$	45	0.0017	52	0.029

Table B.5.

n=2		n=3		n=4		n=5		n=6		n=7		n=8	
Σ	P	Σ	P	Σ	P	Σ	P	Σ	P	Σ	P	Σ	P
								41	0.0^692	47	0.0^337	54	0.011
								43	0.0^743	49	0.0^468	56	0.0038
								45	0.0^903	51	0.0^410	58	0.0011
										53	0.0^512	60	0.0^329
										55	0.0^612	62	0.0^466
										57	0.0^886	64	0.0^413
										59	0.0^944	66	0.0^522
										61	$0.0^{10}15$	68	0.0^432
										63	$0.0^{13}23$	70	0.0^740
												72	0.0^842
												74	0.0^936
												76	$0.0^{10}24$
												78	$0.0^{11}13$
												80	$0.0^{13}48$
												82	$0.0^{14}12$
												84	$0.0^{10}14$

Agreement in Paired Comparisons. The Probability P that a Value of Σ (for u) will be attained or exceeded, for m = 4 and n = 2 to 6 (for n = 6 only Values beyond the 1 per cent Point are given)

n=2		n=3		n=4		n=5		n=5		n=6		n=6	
Σ	P	Σ	P	Σ	P	Σ	P	Σ	P	Σ	P	Σ	P
2	1.000	6	1.000	12	1.000	20	1.000	42	0.0048	57	0.014	79	0.0^542
3	0.625	7	0.947	13	0.997	21	1.000	43	0.0030	58	0.0092	80	0.0^828
6	0.125	8	0.786	14	0.975	22	0.999	44	0.0017	59	0.0058	81	0.0^998
		9	0.455	15	0.901	23	0.995	45	0.0^373	60	0.0037	82	0.0^915
		10	0.330	16	0.769	24	0.979	46	0.0^341	61	0.0022	83	0.0^912
		11	0.277	17	0.632	25	0.942	47	0.0^324	62	0.0013	84	$0.0^{10}51$
		12	0.137	18	0.524	26	0.882	48	0.0^490	63	0.0^376	86	$0.0^{11}30$
		14	0.048	19	0.410	27	0.803	49	0.0^437	64	0.0^344	87	$0.0^{11}17$
		15	0.025	20	0.278	28	0.719	50	0.0^425	65	0.0^323	90	$0.0^{13}28$
		18	0.0020	21	0.185	29	0.621	51	0.0^593	66	0.0^313		
				22	0.137	30	0.514	52	0.0^521	67	0.0^472		
				23	0.088	31	0.413	53	0.0^517	68	0.0^436		
				24	0.044	32	0.327	54	0.0^674	69	0.0^418		
				25	0.027	33	0.249	56	0.0^766	70	0.0^597		
				26	0.019	34	0.179	57	0.0^738	71	0.0^547		
				27	0.0079	35	0.127	60	0.0^993	72	0.0^520		
				28	0.0030	36	0.090			73	0.0^510		
				29	0.0025	37	0.060			74	0.0^631		
				30	0.0011	38	0.038			75	0.0^618		
				32	0.0^316	39	0.024			76	0.0^778		
				33	0.0^495	40	0.016			77	0.0^744		
				36	0.0^538	41	0.0088			78	0.0^715		

The Probability P that a Value of Σ (for u) will be attained or exceeded, for m = 5 and n = 2 to 5

Table B.5. (Continued)

n=2 Σ	P	n=3 Σ	P	n=4 Σ	P	n=5 Σ	P	n=5 Σ	P
4	1.000	12	1.000	24	1.000	40	1.000	76	$0.0^4 50$
6	0.375	14	0.756	26	0.940	42	0.991	78	$0.0^4 16$
10	0.063	16	0.390	28	0.762	44	0.945	80	$0.0^5 50$
		18	0.207	30	0.538	46	0.843	82	$0.0^5 15$
		20	0.103	32	0.353	48	0.698	84	$0.0^6 39$
		22	0.030	34	0.208	50	0.537	86	$0.0^6 10$
		24	0.011	36	0.107	52	0.384	88	$0.0^7 23$
		26	0.0039	38	0.053	54	0.254	90	$0.0^8 53$
		30	$0.0^3 24$	40	0.024	56	0.158	92	$0.0^6 12$
				42	0.0093	58	0.092	94	$0.0^9 14$
				44	0.0036	60	0.050	96	$0.0^{10} 46$
				46	0.0012	62	0.026	100	$0.0^{13} 91$
				48	$0.0^3 36$	64	0.012		
				50	$0.0^3 12$	66	0.0057		
				52	$0.0^4 28$	68	0.0025		
				54	$0.0^5 54$	70	0.00:0		
				56	$0.0^5 18$	72	$0.0^3 39$		
				60	$0.0^7 60$	74	$0.0^3 14$		

The probability P that a Value of Σ (for u) will be attained or exceeded, for $m=6$ and $n=2$ to 4

n=2 Σ	P	n=3 Σ	P	n=4 Σ	P	n=4 min	P	n=4 Σ	P
6	1.000	18	1.000	36	1.000	55	0.043	74	$0.0^4 12$
7	0.688	19	0.969	37	0.999	56	0.029	75	$0.0^5 89$
10	0.219	20	0.332	38	0.991	57	0.020	76	$0.0^5 49$
15	0.031	21	0.626	39	0.959	58	0.016	77	$0.0^5 32$
		22	0.523	40	0.896	59	0.011	80	$0.0^6 58$
		23	0.468	41	0.822	60	0.0072	81	$0.0^4 17$
		24	0.303	42	0.755	61	0.0049	82	$0.0^6 12$
		26	0.180	43	0.669	62	0.0034	85	$0.0^7 34$
		27	0.147	44	0.556	63	0.0025	90	$0.0^8 93$
		28	0.088	45	0.466	64	0.0016		
		29	0.061	46	0.409	65	$0.0^3 83$		
		30	0.040	47	0.337	66	$0.0^3 66$		
		31	0.034	48	0.257	67	$0.0^3 48$		
		32	0.023	49	0.209	68	$0.0^3 26$		
		35	0.0062	50	0.175	69	$0.0^3 16$		
		36	0.0029	51	0.133	70	$0.0^4 86$		
		37	0.0020	52	0.097	71	$0.0^4 68$		
		40	$0.0^3 58$	53	0.073	72	$0.0^4 48$		
		45	$0.0^4 31$	54	0.057	73	$0.0^4 16$		

Source: Kendall (1962).
Notation: n = number of objects
m = number of experts
Σ = sum of the number of agreements between pairs of experts

Table B-6. Between-Expert Agreement Table (Coefficient of Concordance)*

		N				Additional Values for $N=3$	
k	3†	4	5	6	7 .	k	s
Values at the .05 level of significance							
3			64.4	103.9	157.3	9	54.0
4		49.5	88.4	143.3	217.0	12	71.9
5		62.6	112.3	182.4	276.2	14	83.8
6		75.7	136.1	221.4	335.2	16	95.8
8	48.1	101.7	183.7	299.0	453.1	18	107.6
10	60.0	127.8	231.2	376.7	571.0		
15	89.8	192.9	349.8	570.5	864.9		
20	119.7	258.0	468.5	764.4	1,158.7		
Values at the .01 level of significance							
3			75.6	122.8	185.6	9	75.9 .
4		61.4	109.3	176.2	265.0	12	103.5
5		80.5	142.8	229.4	343.8	14	121.9
6		99.5	176.1	282.4	422.6	16	140.2
8	66.8	137.4	242.7	388.3	579.9	18	158.6
10	85.1	175.3	309.1	494.0	737.0		
15	131.0	269.8	475.2	758.2	1,129.5		
20	177.0	364.2	641.2	1,022.2	1,521.9		

*Adapted from M. Friedman, "A Comparison of Alternative Tests of Significance for the Problem of m rankings," *Ann. Math. Statist.*, vol. 11, pp. 86–92, 1940, with the kind persmission of the author and the publisher.
†Notice that additional critical values of s for $N = 3$ are given in the right-hand column of this table.
Source: Siegel, S., *Nonparametric Statistics for the Behavioral Sciences*, McGraw-Hill, New York, 1956.

Notation: N = number of objects
k = number of experts

References

CHAPTER 1

Brockhoff, K., "The Performance of Forecasting Groups in Computer Dialogue and Face to Face Discussion," in H. A. Linstone and M. Turoff (eds.), *The Delphi Method, Techniques and Applications*, Addison Wesley, Reading, Mass., 1975, pp. 291–321.

Dalkey, N., Brown, B., and Cochran, S., "Use of Self-Ratings to Improve Group Estimates," *Technological Forecasting*, vol. 1, no. 3, pp. 283–291, 1970.

Delbecq, A., Van de Ven, A. and Gusstafson, D., *Group Techniques for Program Planning*, Scott, Foresman, Glenview, Ill., 1975.

Federation of American Scientists, Public Interest Report, vol. 33, no. 8, October 1980.

Fischer, G., "An Experimental Study of Four Procedures for Aggregating Subjective Probability Assessments," Technical Report 75-7, Decisions and Designs, Inc., McLean, Va., 1975.

Gofman, J., and Tamplin, A., *Population Control Through Nuclear Pollution*, Nelson-Hall Co. Chicago, 1970.

Gough, R., "The Effect of Group Format on Aggregate Subjective Probability Distributions," in D. Wendt and C. Viek (eds.), Utility, Probability and Human Decision Making, Dordrecht, Reidel, 1975.

Gustafson, D., Shulka, R., Delbecq, A., and Walster, A., "A Comparative Study of Differences in Subjective Likelihood Estimates Made by Individuals, Interacting Groups, Delphi Groups, and Nominal Groups," *Organizational Behaviour and Human Performance*, vol. 9, pp. 280–291, 1973.

Helmer, Olaf, "Analysis of the Future: The Delphi Method" and "The Delphi Method–An Illustration," in J. Bright (ed.), *Technological Forecasting for Industry and Government*, Prentice-Hall, Englewood Cliffs, N.J., 1968, pp. 116–134.

Helmer, Olaf, *Social Technology*, Basic Books, New York, 1966.

Kahn, Herman, *On Thermonuclear War*, Free Press, New York, 1960.

Kahn, Herman, and Wiener, Anthony J., *The Year 2000, A Framework for Speculation*, Macmillan, New York, 1967.

Kevles, Daniel, *The Physicists*, Alfred Knopf, New York, 1978.

Linstone, H. A., and Turoff, M., *The Delphi Method, Techniques and Applications*, Addison Wesley, Reading, Mass., 1975.

Mazur, Allen, "Opinion Pool Measurements of American Confidence in Science," *Science, Technology and Human Values*, vol. 6, no. 36, pp. 16–19, 1981.

Newman, J. R., "Thermonuclear War," *Scientific American*, March 1961. See: *Readings from Sci. Amer. Science, Conflict and Society*, W. M. Freeman, San Francisco, 1969, pp. 282–286.

Parenté, F. J., and Anderson-Parenté, J. K., "Delphi Inquiry Systems," in G. Wright and P. Ayton (eds.), *Judgmental Forecasting*, Wiley, Chichester, 1987.

Reichenbach, H., *The Rise of Scientific Philosophy*, University of California Press, 1968; first edition, 1951.

Sackman, H., *Delphi Critique, Expert Opinion, Forecasting and Group Processes*, Lexington Books, Lexington, Mass., 1975.

Seaver, D., "How Groups Can Assess Uncertainty" *Proc. Int. Conf. on Cybernetics and Society*, Wash. D.C., Sept. 19–21, 1977, pp. 185–190.

Science p. 171, March 5, 1971.

Toekomstonderzoek, suppl. 10, pp. 6.6.2-01–6.6.4–07, November 1974.

CHAPTER 2

Amendola, A., "Systems Reliability Benchmark Exercise Parts I and II," EUR-10696, EN/I, 1986.

American Physical Society, "Study Group on Light Water Reactor Safety; Report to the American Physical Society," *Review of Modern Physics*, vol. 47, suppl. no. 1, 1975.

Beaver, W. M. *Financial Reporting: An Accounting Revolution*. Prentice Hall, Englewood Cliffs, N. J., 1981.

Bell, T. E., and Esch, K., "The Space Shuttle: A Case of Subjective Engineering," *IEEE Spectrum*, pp. 42–46, June 1989.

Bernreuter, D. L., Savy, J. B., Mensing, R. W., and Chung, D. H., Seismic Hazard Characterization of the Eastern United States: Methodology and Interim Results for Ten Sites," NUREG/CR-3756, 1984.

Brune, R., Weinstein, M. and Fitzwater, M., Peer Review Study of the Draft Handbook for Human Reliability Analysis with Emphasis on Nuclear Power Plant Applications," NUREG/CR-1278, Human Performance Technologies, Inc., Thousand Oaks, Calif., 1983.

Christensen-Szalanski, J. J. J., and Beach, L. R., "The Citation Bias: Fad and Fashion in the Judgment and Decision Literature," *American Psychologist*, vol. 39, pp. 75–78, 1984.

Clemen, R., and Winkler, R., "Combining Economic Forecasts," *J. of Business and Economic Statistics*, vol. 4, pp. 39–46, January 1986.

Colglazier, E. W., and Weatherwax, R. K., "Failure Estimates for the Space Shuttle," Abstracts for Society for Risk Analysis, Annual Meeting 1986, Boston, Mass., p. 80, Nov. 9–12, 1986.

Cooke, R. "Problems with Empirical Bayes," *Risk Analysis*, vol. 6, no. 3, pp. 269–272, 1986b.

Cooke, R., and Waij, R., "Monte Carlo Sampling for Generalized Knowledge Dependence with Application to Human Reliability," *Risk Analysis*, vol. 6, no. 3, pp. 335–343, 1986.

Cottrell W., and Minarick, C., "Precursors to Potential Severe Core Damage Accidents: 1980–1982, a Status Report," NUREG/CR-3591, 1984.

Covello, V. T., and Mumpower, J. "Risk Analysis and Risk Management: An Historical Perspective," *Risk Analysis*, vol. 5, no. 2, pp. 103–120, 1985.

Dalrymple, G. J., and Willows, M., "DoE Disposal Assessments, vol. 4: Expert Elicitation," TR-DR3-4, Yard, London, July 1990.

Electric Power Research Institute, "*Seismic Hazard Methodology for the Central* and Eastern United States," vol. 1, *Methodology*, NP-4/26, 1986.

Environmental Protection Agency, *Reactor Safety Study Oversight Hearings Before the Subcommittee on Energy and the Environment of the Committee on Interior and Insular Affairs*. House of Representatives, 94th Congress, second session, serial no. 94—61, Washington, D.C., June 11, 1976.

Feynman, R. P., "Mr. Feynman Goes to Washington," *Engineering and Science*, pp. 6–22, Fall 1987.

Flavin, C., "Electricity's Future: The Shift to Efficiency and Small Scale Power," *Worldwatch paper 61*, 1984; reprinted in *Bull. Sci. Tech. Soc.*, vol. 5, 55–103, 1985.

Granger, C. W. J., *Forecasting in Business and Economics*, Academic Press, New York, 1980.

Granger Morgan, M., and Henrion, M., *Uncertainty; a Guide to Dealing with Uncertainty in Quantitative Risk and Policy Analysis*, Department of Engineering and Public Policy, Carnegie Mellon University, 1988.

Granger Morgan, M., Amaral, D., Henrion, M., and Morris, S., "Technological Uncertainty in Quantitative Policy Analysis—A Wulfur Pollution Example," *Risk Analysis*, vol. 3, pp. 201–220, 1984.

Hofer, E., Javeri, V., and Loffler, H., "A Survey of Expert Opinion and Its Probabilistic Evaluation for Specific Aspects of the SNR-300 Risk Study," *Nuclear Technology*, vol. 68, pp. 180–225, 1985.

Honano, E. J., Hora, S. C., Keeney, R. L., and Von Winterfeldt, D., *Elicitation and Use of Expert Judgment in Performance Assessment for High-level Radioactive Waste Repositories*, NUREG/CR-5411, Washington, D.C., May 1990.

Humphreys, P., *Human Reliability Assessors Guide*, Safety and Reliability Directorate, United Kingdom Atomic Energy Authority, 1988.

IEEE, *IEEE Guide to the Collection and Presentation of Electrical, Electronic and Sensing Component Reliability Data of Nuclear Power Generation Stations*, IEEE st-500, 1977.

Jungermann, H., and Thuring, M., "The Use of Mental Models for Generating Scenarios," in G. Wright and P. Ayton (eds.), *Judgmental Forecasting*, Wiley, New York, 1987.

Kaplan, S., and Garrick, B., "On the Quantitative Definition of Risk," *Risk Analysis*, vol. 1, pp. 11–27, 1981.

Kemeny J., *Report of the President's Commission on the Accident at Three Mile Island*, Washington, D.C., 1979.

Kok, M., "Multiple Objective Energy Modeling: Experimental Results with Interactive Methods in The Netherlands," Ph.D. thesis, Department of Mathematics, Delft University of Technology, report 85–49, 1985.

Lee, Y. T., Orkent, D. and Apostolakis, G., "A Comparison of Background Seismic Risks and the Incremental Seismic Risk Due to Nuclear Power Plants," *Nuclear Engineering & Design*, vol. 53, pp. 141–154, 1979.

Levine, S., and Rasmussen, N., "Nuclear Plant PRA: How Far Has It Come," *Risk Analysis*, vol. 4, 247–255, 1984.

Lewis, H. W., Budnitz, R. J., Kouts, H. J. C., Lowenstein, W. B., Rowe, W. D., Von Hippel, F., and Zachariasen, F., *Risk Assessment Review Group Report to the U.S. Nuclear Regulatory Commission*, NUREG/CR-0400, 1979.

Merkhofer, M., and Keeney, R., "A Multiattribute Utility Analysis of Alternative Sites for the Disposal of Nuclear Waste," *Risk Analysis*, vol. 7, no. 2, pp. 173–194, 1987.

Minarick, J., and Kukielka, C., "Precursors to Potential Severe Core Damage Accidents: 1969–1979," NUREG/CR-2497, 1982.

Morris, J. M., and D'Amore, R. J., "Aggregating and Communicating Uncertainty," Pattern Analysis and Recognition Corp., 228 Liberty Plaza, Rome, New York, 1980.

Mosleh, A., Bier, V., and Apostolakis, G., Critique of Current Practice for the Use of Expert Opinions in Probabilistic Risk Assessment," *Reliability Engineering and System Safety*, vol. 20, pp. 63–85, 1988.

Mosleh, A., Bier, V. M., and Apostolakis, G., "Methods for the Elicitation and Use of Expert Opinion in Risk Assessment," NUREG/CR-4962, 1987.

Office of Nuclear Regulatory Research, "Reactor Risk Reference Document," NUREG-1150, 1987.

Orkent, D., "A Survey of Expert Opinion on Row Probability Earthquakes," in *Annals of Nuclear Energy*, Pergamon Press, pp. 601–614, 1975.

Poucet, A., "The European Benchmark Exercise on Human Reliability Analysis," Proceedings PSA '89, International Topical Meeting on Probability, Reliability and Safety Assessment, Pittsburgh, pp. 103–110, April 2–7, 1989.

Preyssl, C., and Cooke, R., "Expert Judgment; Subjective and Objective Data for Risk Analysis of Spaceflight Systems," *Proceedings PSA '89, International Topical Meeting on Probability, Reliability and Safety Assessment*, Pittsburgh, pp. 603–612, April 2–7, 1989.

Rogovin, M., and Frampton, G. T., *Three Mile Island, a Report to the Commissioners and to the Public*, Government Printing Office, 1980.

Samet, M. G., "Quantitative Interpretation of Two Qualitative Scales Used to Rate Military Intelligence," *Human Factors*, vol. 17, no. 2, pp. 192–202, 1975.

Shooman, M., and Sinkar, S., "Generation of Reliability and Safety Data by Analysis of Expert Opinion," Proc. 1977 Annual Reliability and Maintainability Symposium, pp. 186–193, 1977.

Snaith, E. R., "The Correlation Between the Predicted and Observed Reliabilities of Components, Equipment, and Systems," National Center of Systems Reliability, U.K. Atomic Energy Authority, NCSR R18, 1981.

Sui, N., and Apostolakis, G., "Combining Data and Judgment in Fire Risk Analysis," 8th International Conference on Structural Mechanics in Reactor Technology, Brussels, Belgium, August 26–27, 1985.

Swain, A., and Guttman, H., *Handbook of Human Reliability, Analysis with Emphasis on Nuclear Power Plant Applications*, NUREG/CR-1278, 1983.

Union of Concerned Scientists, "The Risks of Nuclear Power Reactors: A Review of the NRC Reactor Safety Study," WASH-1400, 1977.

U.S. AEC, "Theoretical Possibilities and Consequences of Major Accident in Large Nuclear Power Plants," U.S. Atomic Energy Commission, WASH-740, 1957.

U.S. NRC, *PRA Procedures Guide*, U.S. Nuclear Regulatory Commission, NUREG/CR-2300, 1983.

U.S. NRC, "Nuclear Regulatory Commission Issues Policy Statement on Reactor Safety Study and Review by the Lewis Panel," *NRC Press Release*, no. 79–19, January 19, 1979.

U.S. NRC, "Reactor Safety Study," U.S. Nuclear Regulatory Commission, WASH-1400, NUREG-751014, 1975.

Vlek, C., "Rise, Decline and Aftermath of the Dutch 'Societal Discussion on (Nuclear) Energy Policy (1981–1983)'," in H. A. Becker and A. Porter (eds.), *Impact Assessment Today*, Van Arkel, Utrecht, 1986.

Vlek, C., and Otten, W., "Judgmental Handling of Energy Scenarios: A Psychological Analysis and Experiment," in G. Wright and P. Ayton (eds.), *Judgmental Forecasting*, Wiley, New York, 1987.

Wheeler, T. A., Hora, S. C., Cramond, W. R., and Unwin, S. D., "Analysis of Core Damage Frequency from Internal Events: Expert Judgment Elicitation," NUREG/CR-4550, vol. 2, Sandia National Laboratories, 1989.

Wiggins, J., "ESA Safety Optimization Study," Hernandez Engineering, HEI-685/1026, Houston, Texas, 1985.

Woo, G., "The Use of Expert Judgment in Risk Assessment; Draft Report for Her Majesty's Inspectorate of Pollution," Yard, London, October 1990.

CHAPTER 3

Adams, J. B., "A Probability Model of Medical Reasoning and the MYCIN Model," *Mathematical Biosciences*, vol. 32, pp. 177–186, 1976.

Carnap, R., *Logical Foundations of Probability*, Routledge and Kegan Paul, Chicago, 1950.

Cendrowska, J., and Bramer, M., "Inside an Expert Systems: A Rational Reconstruction of the MYCIN Consultation System," in O'Shea T. and Eisenstadt, M. (eds.), *Artificial Intelligence, Tools Techniques and Applications*, Harper and Row, New York, 1984, pp. 453–497.

Cooke, R. M., "Probabilistic Reasoning in Expert Systems Reconstructed in Probability Semantics," *Philosophy of Science Association*, vol. 1, pp. 409—421, 1986.

Dubois, D., and Prade, H., *Fuzzy Sets and Systems: Theory and Applications*, Academic Press, New York, 1980.

French, S., "Fuzzy Sets: The Unanswered Questions," Manchester-Sheffield School of Probability and Statistics Research Report, November 1987.

French, S., "Fuzzy Decision Analysis, Some Problems," in Zimmermann, H., Zadeh, L., and Gains (eds.), *Fuzzy Sets and Decision Analysis*, Elsevier North-Holland, Amsterdam, 1984.

Gordon, J., and Shortliffe, E., "The Dempster-Shafer Theory of Evidence," in E. Shortliffe and B. Buchanan (eds.), *Rule Based Expert Systems*, Reading, Mass., 1984.

Johnson, R. W., "Independence and Bayesian Updating Methods," in L. Kanal and J. Lemmer (eds.), *Uncertainty in Artificial Intelligence*, Elsevier North Holland, Amsterdam, 1986.

Los, J., "Semantic Representation of the Probability of Formulas in Formalized Theories," *Studia Logika* vol. 14, pp. 183–196, 1963.

Shortliffe, E., and Buchanan, B., *Rule Based Expert Systems*, Reading, Mass., 1984.

Shortliffe, E., and Buchanan, B., "A Model of Inexact Reasoning in Medicine," *Mathematical Biosciences*, vol. 23, pp. 351–379, 1975.

Stefik, M., "Strategic Computing at DARPA: Overview and Assessment," *Comm. of the ACM*, July, vol. 28, no. 7, pp. 690–704, July 1985.

Szolovits, P., and Pauker, S., "Categorical and Probabilistic Reasoning in Medical Diagnoses," *Artificial Intelligence*, vol. 11, pp. 115–144, 1978.

Yu, V. L., Fagan, L., Wraith, S., Clancey, W., Scott, A., Hannigan, J., Blum, R., Buchanan, B., Cohen, S., "Antimicrobial Selection by a Computer," *JAMA*, vol. 242, no. 12, pp. 1279–1282, Sept. 21, 1979.

Zadeh, L., "Is Probability Theory Sufficient for Dealing with Uncertainty in AI: A Negative View," in Kanal, L. and Lemmer, J. (eds.), *Uncertainty in Artificial Intelligence*, Elsevier North Holland, Amsterdam, 1986, pp. 103–116.

Zadeh, L., "Probability Measures of Fuzzy Events," *J. Math. Anal. Appl.*, vol. 23, pp. 421–427, 1968.

Zimmermann, H., "Fuzzy set theory and inference mechanism," in G. Mitra (eds), *Mathematical Models for Decision Support*, Springer-Verlag, Berlin pp. 727–743, 1987.

CHAPTER 4

Apostolakis, G., "The Broadening of Failure Rate Distributions in Risk Analysis: How Good Are the Experts?" *Risk Analysis*, vol. 5, no. 2, pp. 89–95, 1985.

Christensen-Szalanski, J., and Bushyhead, J., "Physicians; Use of Probabilistic Information in a Real Clinical Setting," *Journal of Experimental Psychology: Human Perception and Performance*, vol. 7, pp. 928–935, 1981.

Cooke, R., "Problems with Empirical Bayes," *Risk Analysis*, vol. 6, no. 3, pp. 269–272, 1986.

Eddy, D., "Probabilistic Reasoning in Clinical Medicine," in D. Kahneman, P. Slovic, and A. Tversky (eds.), *Judgment under Uncertainty*, Cambridge University Press, Cambridge, 1982.

Hynes, M. E, and Vanmarcke, E. H., "Reliability of Enbankment Performance Predictions," in *Mechanics in Engineering*, Proc. 1st ASCE-EMD Specialty Conf. University of Waterloo, May 26–28, 1976, pp. 367–384.

Kahneman, D., Slovic, P., and Tversky, A., (eds.) *Judgment under Uncertainty, Heuristics and Biases*, Cambridge University Press, Cambridge, 1982.

Langer, E., "The Illusion of Control," *The Journal of Personality and Social Psychology*, vol. 32, pp. 311–328, 1975.

Lichtenstein, S., Fischhoff, B., and Phillips, L., "Calibration of Probabilities: The State of the Art to 1980", in D. Kahneman, P. Slovic, and A. Tversky (eds.), *Judgment under Uncertainty*, pp. 306–334.

Martz, H., "Response to 'Problems with Empirical Bayes'," *Risk Analysis*, vol. 6, no. 3, pp. 273–274, 1986.

Martz, H., "On Broadening Failure Rate Distributions in PRA Uncertainty Analysis," *Risk Analysis*, vol. 4, no. 1, pp. 15–23, 1984.

Martz, H., and Bryson, M., "On Combining Data for Estimating the Frequency of Low-Probability Events with Application to Sodium Value Failure Rates," *Nuclear Science and Engineering*, vol. 83, pp. 267–280, 1983.

Murphy, A., and Daan, H., "Impacts of Feedback and Experience on the Quality of Subjective Probability Forecasts: Comparison of Results from the First and Second Years of the Zierikzee Experiment," *Monthly Weather Review*, vol. 112, pp. 413–423, 1984.

Murphy, A., and Winkler, R., "Can Weather Forecasters Formulate Reliable Probability Forecasts of Precipitation and Temperature?" *National Weather Digest*, vol. 2, pp. 2–9, 1977.

Oskamp, S., "Overconfidence in case-study judgments," in D. Kahneman, P. Slovic, and A. Tversky (eds.), *Judgment under Uncertainty*, Cambridge University Press, Cambridge, 1982, pp. 287–293.

Slovic, P., Fischhoff, B., and Lichtenstein, S., "Facts versus Fears: Understanding Perceived Risk," in D. Kahneman, P. Slovic, and A. Tversky (eds.), *Judgment under Uncertainty*, Cambridge University Press, Cambridge, 1982.

Tversky, A., and Kahneman, D., "Availability: A Heuristic for Judging Frequency and Probability," in D. Kahneman, P. Slovic, and A. Tversky (eds.), *Judgment under Uncertainty*, Cambridge University Press, Cambridge, 1982a.

Tversky, A., and Kahneman, D., "Causal Schemas in Judgments Under Uncertainty," in D. Kahneman, P. Slovic, and A. Tversky (eds.), *Judgment under Uncertainty*, Cambridge University Press, Cambridge, 1982b.

Tversky, A., and Kahneman, D., "Judgment under Uncertainty: Heuristics and Biases," in D. Kahneman, P. Slovic, and A. Tversky (eds.), *Judgment under Uncertainty*, Cambridge University Press, Cambridge, 1982c.

Thys, W., "Fault Management," Ph.D. dissertation, Delft University of Technology, Delft, 1987.

CHAPTER 5

Amendola, A., "Systems Reliability Benchmark Exercises Parts I and II," EUR-10696, EN/I, 1986.

Apostolakis, G., "The Broadening of Failure Rate Distributions in Risk Analysis: How Good Are the Experts?" *Risk Analysis*, vol. 5, no. 2, 89–95, 1985.

Apostolakis, G. (ed.), *Reliability Engineering and System Safety*, vol. 23, no. 4, 1988.

Bernreuter, D. L., Savy, J. B., Mensing, R. W., and Chung, D. H., "Seismic Hazard Characterization of the Eastern United States: Methodology and Interim Results for Ten Sites," NUREG/CR-3756, 1984.

Brockhoff, K., "The Performance of Forecasting Groups in Computer Dialogue and Face to Face Discussion," in H. A. Linstone and M. Turoff (eds.), *The Delphi Method, Techniques and Applications*, Addison Wesley, Reading, Mass., 1975, pp. 291–321.

Dalkey, N., Brown, B., and Cochran, S., "Use of Self-Ratings to Improve Group Estimates," *Technological Forecasting*, vol. 1, no. 3, pp. 283–291, 1970.

Hofer, E., Javeri, V., and Loffler, H., "A Survey fof Expert Opinion and Its Probabilistic Evaluation for Specific Aspects of the SNR-300 Risk Study," *Nuclear Technology*, vol. 68, pp. 180–225, 1985.

Lewis, H. W. et al., *Risk Assessment Review Group Report to the U.S. Nuclear Regulatory Commission*, NUREG/CR-0400, 1979.

Linstone, H. A., and Turoff, M., *The Delphi Method, Techniques and Applications*, Addison Wesley, Reading, Mass., 1975.

Morris, J. M., and D'Amore, R. Y., "Aggregating and Communicating Uncertainty," Pattern Analysis and Recognition Corp., 228 Liberty Plaza, Rome, New York, 1980.

Office of Nuclear Regulatory Research, "Reactor Risk Reference Document," NUREG-1150, 1987.

Poucet, A., Amendola, A., and Cacciabue, P. C., "CCF-RBE Common Cause Failure Reliability Benchmark Exercises," EUR-11054, EN, 1987.

Sackman, H., *Delphi Critique, Expert Opinion, Forecasting and Group Processes*, Lexington Books, Lexington, Mass., 1975.

U.S. NRC, "Reactor Safety Study," U.S. Nuclear Regulatory Commision, WASH-1400, NUREG-751014, 1975.

CHAPTER 6

Allais, M., "The So-Called Allais Paradox and Rational Decisions Under Uncertainty," in M. Allais and O. Hagen (eds.), *The Expected Utility Hypothesis and the Allais Paradox*, Reidel, Dordrecht, 1979, pp. 437–683.

Allais, M., "Le comportement de l'homme rationeel devant le Risque," *Econometrica*, vol. 21, pp. 503–546, 1953.

Blach, M., and Fishburn, P., "Subjective Expected Utility for Conditional Primitives," in M. Blach, D. McFadden, and S. Wu (eds.), *Essays on Economic Behavior Under Uncertainty*, North-Holland, Amsterdam, 1974, pp. 57–69.

Blach, M., McFadden, D., and Wu, S., *Essays on Economic Behavior Under Uncertainty*, North-Holland, Amsterdam, 1974, pp. 57–69.

Cooke, R., "Conceptual Fallacies in Subjective Probability," *Topoi*, vol. 5, pp. 21–27, 1986.

Cooke, R., "A Result in Renyi's Theory of Conditional Probability with Application to Subjective Probability," *Journal of Philosophical Logic*, vol. 12, 1983.

Ellsberg, D., "Risk, Ambiguity and the Savage Axioms," *Quarterly Journal of Economics*, vol. 75, pp. 643–699, 1961.

Hogarth, R., *Judgement and Choice*, Wiley, New York, 1987.

Jeffrey, R., *The Logic of Decision*, McGraw-Hill, New York, 1966.

Kahneman, D., and Tversky, A., "Prospect Theory," *Econometrica*, vol. 47, no. 2, 1979.

Kraft, C. H., Pratt, J. W., and Seidenberg, A., "Intuitive Probability on Finite Sets," *Annals of Mathematical Statistics*, vol. 30, pp. 408–419, 1959.

Krantz, D., Luce, R., Suppes, P., and Tversky, A., *Foundations of Measurement*, vol. 1, Academic Press, New York, 1971.

Luce, R., and Krantz, D., "Conditional Expected Utility," *Econometrica*, vol. 39, no. 2, pp. 253–271, 1971.

MacCrimmon, K. R., "Descriptive and Normative Implications of the Decision-Theory Postulates," in K. Borch and J. Mossin (eds.), *Risk and Uncertainty*, MacMillan, London, 1968, pp. 3–24.

Machina, M., "'Rational' Decision Making versus 'Rational Decision Modelling'," *Journal of Mathematical Psychology*, vol. 24, pp. 163–175, 1981.

Pfanzagl, J., *Theory of Measurement*, Physica Verlag, Würzburg-Wien, 1968.

Ramsey, F., "Truth and Probability," in R. B. Braithwaite (ed.), *The Foundations of Mathematics*, Keegan Paul, London, 1931, pp. 156–198.

Savage, L., *The Foundations of Statistics*, Dover, New York, 1972, first published by John Wiley & Sons, 1954.

Shafer, G., "Savage Revisited," *Statistical Science*, vol. 1, no. 4, pp. 463–501, 1986.

Tversky, A., "Intransitivity of Preference," *Psychological Review*, vol. 76, no. 1, pp. 31–48, 1969.

Villegas, C., "On Qualitative Probability σ-Algebras," *Annals of Math. Stat.*, vol. 35, no. 4, pp. 1787–1796, 1964.

CHAPTER 7

Aldous, D., "Exchangeability and Related Topics," in *Lecture Notes in Mathematics*, vol. 117, Springer-Verlag, Berlin, 1985.

Berman, S. M., "Stationarity, Isotropy and Sphericity in I_p," *Z. Wahr. ver. Geb.*, vol. 54, pp. 21–23, 1980.

Box, G., and Tiao, G., *Bayesian Inference in Statistical Analysis*, Addison-Wesley, Reading, Mass., 1973.

Cooke, R., and Misiewicz, J., "l_p-Invariant Probability Measures," Delft University of Technology, Department of Mathematics, report 88-91, 1988.

Cooke, R., Misiewicz, J., and Mendel, M. "Applications of l_p-Symmetric Measures to Bayesian Inference," in W. Kasprzak and A. Weron (eds.), *Stochastic Methods in Experimental Sciences*, World Scientific, Singapire, 1990.

De Finetti, B., *Theory of Probability*, Wiley, New York, 1974.

De Finetti, B., "La Prevision; ses lois logique, ses source subjectives," *Annales de L'Instut Henri Poincare*, vol. 7, pp. 1–68, 1937. English translation in H. Kyburg and H. Smokler (eds.), *Studies in Subjective Probability*, Wiley, New York, 1964.

Heath, D., and Sudderth, W., "De Finetti's Theorem on Exchangeable Variables," *The Amer. Statis.*, vol. 30, no. 4, pp. 188–189, 1975.

Hewitt, E., and Savage, J., "Symmetric Measures on Cartesian Products," *Trans. Am. Math. Soc.*, vol. 80, pp. 470–501, 1955.

Schoenberg, J., "Metric Spaces and Completely Monotonic Functions," *Ann. Math.*, vol. 38, pp. 811–841, 1938.

Tucker, H., *A Graduate Course in Probability*, Academic Press, New York, 1967.

CHAPTER 8

Alpert, M., and Raiffa, H., "A Progress Report on the Training of Probability Assessors," in D. Kahneman, P. Slovic, and A. Tversky (eds.), *Judgment under Uncertainty, Heuristics and Biases*, Cambridge University Press, Cambridge, 1982, pp. 294–306.

Brier, G. "Verification of Forecasts Expressed in Terms of Probability," *Mon. Weath. Rev.*, vol. 75, pp. 1–3, 1950.

Cooke, R., Mendel, M., and Thys, W., "Calibration and Information in Expert Resolution; a Classical Approach," *Automatica*, vol. 24, pp. 87–94, 1988.

De Finetti, B., "La prevision: ses lois logique, ses source subjectives," *Annales de L'Instut Henri Poincare*, vol. 7, pp. 1–68, 1937. English translation in H. Kyburg and H. Smokler (eds.), *Studies in Subjective Probability*, Wiley, New York, 1964.

De Groot, M. H., *Optimal Statistical Decisions*, McGraw-Hill, New York, 1970.

ESRRDA, "Expert Judgment in Risk and Reliability Analysis; Experiences and Perspective," ESRRDA project group "Expert Judgment," (draft report), 1989.

Galanter, E., "The Direct Measurement of Utility and Subjective Probability," *Amer. J. of Psych.*, vol. 75, pp. 208–220, 1962.

Hoel, P., *Introduction to Mathematical Statistics*, Wiley, New York, 1971.

Kullback, S., *Information Theory and Statistics*, Wiley, New York, 1959.

Lichtenstein, S., and Fischhoff, B., "Do Those Who Know More Also Know More About How Much They Know?" *Orgl. Behavior Human Perform.*, vol. 20, pp. 159–183, 1977.

Lichtenstein, S., Fischhoff, B., and Phillips, D., "Calibration of Probabilities: The State of the Art to 1980," in D. Kahneman, P. Slovic, and A. Tversky (eds.), *Judgment under Uncertainty, Heuristics and Biases*, Cambridge University Press, Cambridge, 1982, pp. 306–335.

Lindley, D., *Introduction to Probability and Statistics from a Bayesian Viewpoint*, Cambridge University Press, Cambridge, 1970.

Murphy, A., "A New Vector Partition of the Probability Score," *J. Appl. Met.*, vol. 12, pp. 595–600, 1973.

Preyssl, C., and Cooke, R., "Expert Judgment: Subjective and Objective Data for Risk Analysis of Spaceflight Systems," Proceedings PSA '89 International Topical Meeting Probability, Reliability and Safety Assessment, Pittsburg, April 2–7, 1989.

Ramsey, F., "Truth and Probability," in Braithwaite (ed.), *The Foundations of Mathematics*, Kegan Paul, London, 1931, pp. 156–198.

von Winterfeld, D., "Eliciting and Communicating Expert Judgments: Methodology and Application to Nuclear Safety," Joint Research Centre, Commission of the Europen Communities, 1989.

Wheeler T., Hora, S., Cramond, W., and Unwin, S., "Analysis of Core Damage Frequency from Internal Events: Expert Judgment Solicitation," NUREG-CR-4550, vol. 2, U.S. Nuclear Regulatory Commission, 1989.

Winkler, R., "Scoring Rules and the Evaluation of Probability Assessors," *J. Amer. Statist. Assoc.*, vol. 64, pp. 1073–1078, 1969.

CHAPTER 9

Bayarri M., and De Groot, M., "Gaining Weight: A Bayesian Approach," Dept. of Stat., Carnegie Mellon University Tech. Report 388, January 1987.

De Groot, M., and Fienberg, S., "Comparing Probability Forecasters: Basic Binary Concepts and Multivariate Extensions," in P. Goel and A. Zellner (eds.), *Bayesian Inference and Decision Techniques*, Elsevier, New York, 1986.

De Groot, M., and Fienberg, S., (1983) "The Comparison and Evaluation of Forecasters," *The Statistician*, vol. 32, pp. 12–22, 1983.

Friedman, D., "Effective Scoring Rules for Probabilistic Forecasts," *Management Science*, vol. 29, no. 4, pp. 447–454, 1983.

Honglin, W., and Duo, D., "Reliability Calculation of Prestressing Cable System of the PCPV Model," *Trans. of the the 8th Intern. Conf. on Structural Mech. in Reactor Techn.*, Brussels, Aug. 19–23, 1985, pp. 41–44.

Lichtenstein, S., Fischoff, B., and Phillips, D., "Calibration of Probabilities: The State of the Art to 1980," in D. Kahneman, P. Slovic, and A. Tversky (eds.), *Judgment under Uncertainty: Heuristics and Biases*, Cambridge University Press, Cambridge, 1982, pp. 306–335.

Matheson, J., and Winkler, R., "Scoring Rules for Continuous Probability Distributions," *Management Science*, vol. 22, no. 10, pp. 1087–1096, 1976.

McCarthy, J., "Measures of the Value of Information," *Proc. of the National Academy of Sciences*, 1956, pp. 654–655.

Murphy, A., "A New Vector Partition of the Probability Score," *J. of Applied Meteorology*, vol. 12, pp. 595–600, 1973.

Roberts, H., "Probabilistic Prediction," *J. Amer. Statist. Assoc.*, vol. 60, pp. 50–62, 1965.

Savage, L., "Elicitation of Personal Probabilities and Expectations," *J. Amer. Statis. Assoc.*, vol. 66, no. 336, pp. 783–801, 1971.

Shuford, E., Albert, A., and Massengil, H., "Admissible Probability Measurement Procedures," *Psychometrika*, vol. 31, pp. 125–145, 1966.

Stael von Holstein, C., "Measurement of Subjective Probability," *Acta Psychologica*, vol. 34, pp. 146–159, 1970.

Tucker, H., *A Graduate Course in Probability*, Wiley, New York, 1967.

Wagner, C., and Lehrer, K., *Rational Consensus in Science and Society*, Reidel, Dordrecht, 1981.

Winkler, R., "On Good Probability Appraisers," in P. Goel and A. Zellner (eds.), *Bayesian Inference and Decision Techniques*, Elsevier, New York, 1986.

Winkler, R., (1969) "Scoring Rules and the Evaluation of Probability Assessors," *J. Amer. Statist. Assoc.*, vol. 64, pp. 1073–1078, 1969.

Winkler, R., and Murphy, A., "Good Probability Assessors," *J. of Applied Meteorology*, vol. 7, pp. 751–758, 1968.

CHAPTER 10

Adams, J. K., and Adams, P. A., "Realism of Confidence Judgment," *Psychol. Rev.*, vol. 68, pp. 33–45, 1961.

Alpert, M., and Raiffa, H., "A Progress Report on the Training of Probability Assessors," in D. Kahneman, P. Slovic, and A. Tversky (eds.), *Judgment under Uncertainty, Heuristics and Biases*, Cambridge University Press, Cambridge, 1982, pp. 294–306.

Bhola, B., Blaauw, H., Cooke, R., and Kok, M., "Expert Opinion in Project Management," appearing in *European Journal of Operations Research*, 1991.

Cooke, R., Mendel, M., and Thys, W., "Calibration and Information in Expert Resolution; a Classical Approach," *Automatica*, vol. 24, pp. 87–94, 1988.

Hodges, J. L., and Lehmann, E. L., *Basic Concepts of Probability and Statistics*, Holden-Day, San Francisco, 1970.

Lichtenstein, S., and Fischhoff, B., "Do Those Who Know More Also Know More About How Much They Know?" *Orgl. Behavior Human Perform*, vol. 20, pp. 159–183, 1977.

Lichtenstein, S., Fischhoff, B., and Phillips, D., "Calibration of Probabilities: The State of the Art to 1980," in D. Kahneman, P. Slovic, and A. Tversky (eds.), *Judgment under Uncertainty, Heuristics and Biases*, Cambridge University Press, Cambridge, 1982, pp. 306–335.

Sieber, J., "Effects of Decision Importance on Ability to Generate Warranted Subjective Uncertainty," *J. Personality Social Psychol.*, vol. 30, pp. 688–694, 1974.

Siegel, S., *Nonparametric Statistics*, McGraw-Hill, New York, 1956.

CHAPTER 11

Blanchard, P., Mitchell, M., and Smith, R., "Likelihood-of-Accomplishment Scale of Man-Machine Activities," Dunlop and Associates, Inc. Santa Monica, CA., 1966.

Bernreuter, D. L., Savy, J. B., Mensing, R. W., and Chung, D. H., "Seismic Hazard Characterization of the Eastern United States: Methodology and Interim Results for Ten Sites," NUREG/CR 3756, 1984.

Bradley, R., (1953) "Some Statistical Methods in Taste Testing and Quality Evaluation," *Biometrics*, vol. 9, pp. 22–38, 1953.

Clemen, R. T., and Winkler, R. L., (1987) "Calibrating and Combining Precipitation Probability Forecasts," in R. Viertl (ed.), *Probability and Bayesian Statistics*, Plenum Press, New York, 1987, pp. 97–110.

Comer, K., Seaver, D., Stillwell, W., and Gaddy, C., "Generating Human Reliability Estimates Using Expert Judgment, vols. I and II," NUREG/CR-3688, 1984.

Cooke, R., "Problems with Empirical Bayes," *Risk Analysis*, vol. 6, no. 3, pp. 269–272, 1986.

David, H. A., *The Method of Paired Comparisons*, Charles Griffin, London, 1963.

De Groot, M., "Reaching a Consensus," *J. Amer. Statist. Assoc.*, vol. 69, pp. 118–121, 1974.

Embrey, D., and Kirwan, B., "A Comparative Evaluation of Three Subjective Human Reliability Quantification Techniques," *Proceedings of the Ergonomics Society's Converence*, K. Coombes (ed.), Taylor and Francis, London, 1983.

French, S., "Group Consensus Probability Distributions: A Critical Survey," in J. M. Bernardo, M. H. De Groot, D. V. Lindley, and A. F. M. Smith (eds.), *Bayesian Statistics*, Elsevier, North Holland, 1985, pp. 183–201.

Genest, C., and Zidek, J., (1986) "Combining Probability Distributions: A Critique and an Annotated Bibliography," *Statistical Science*, vol. 1, no. 1, pp. 114–148, 1986.

Gokhale, D., and Press, S., "Assessment of a Prior Distribution for the Correlation Coefficient in a Bivariate Normal Distribution," *J. R. Statist. Soc. A*, vol. 145, P. 2, pp. 237–249, 1982.

Hardy, G. H., Littlewood, J. E., and Polya, G. *Inequalities*, Cambridge University Press, Cambridge, 1983, (first edition 1934).

Hofer, E., Javeri, V., and Loffler, H., "A Survey of Expert Opinion and Its Probabilistic Evaluation for Specific Aspects of the SNR-300 Risk Study," *Nuclear Technology*, vol. 68, pp. 180–225, 1985.

Humphreys, P., *Human Reliability Assessor's Guide*, United Kingdom Atomic Energy Authority, Warrington WA3 4NE, 1988.

Hunns, D., "Discussions Around a Human Factors Data-Base. An Interim Solution: The Method of Paired Comparisons," in A. E. Green (ed.), *High Risk Safety Technology*, Wiley, New York, 1982.

Hunns, D., and Daniels, B., "The Method of Paired Comparisons and the Results of the Paired Comparisons Consensus Exercise," *Proceedings of the 6th Advances in Reliability Technology Symposium*, vol. 1, NCSR R23, Culcheth, Warrington, 1980, pp. 31–71.

Huseby, A. B., "Combining Opinions in a Predictive case," presented at Third Valencia International Meeting on Bayesian Statistics, Altea, Spain, June 1–5, 1987.

IEEE, *IEEE Guide to the Collection and Presentation of Electrical, Electronic, and Sensing Component Reliability Data for Nuclear Power Generation Stations*, IEEE st-500, 1977.

Kadane, J., Dickey, J., Winkler, R., Smith, W., and Peters, S., "Interactive Elicitation of Opinion for a Normal Linear Model," *J. Amer. Statist. Assoc.*, vol. 75, pp. 845–854, 1980.

Kirwan, B., *Human Reliability Assessor's Guide*, vols. 1 and 2, (DRAFT) Human Reliability Associates, Dalton, 1987.

Laddaga, R., "Lehrer and the Consensus Proposal," *Synthese*, vol. 36, pp. 473–477, 1977.

Lehrer, K., and Wagner, C. G., *Rational Consensus in Science and Society*, Reidel, Dordrecht, 1981.

Lindley, D., "Reconciliation of Discrete Probability Distributions," in J. Bernardo, M. De Groot, D. Lindley, and A. Smith (eds.), *Bayesian Statistics 2*, North Holland, Amsterdam, 1985, pp. 375–390.

Lindley, D., "Reconciliation of Probability Distributions," *Operations Research*, vol. 31, no. 5, pp. 866–880, 1983.

Lindley, D., and Singpurwalla, N., "Reliability (and Fault Tree) Analysis Using Expert Opinions," *J. Amer. Statist. Assoc.*, vol. 81, no. 393, pp. 87–90, 1986.

Martz, H., "Reaction to 'Problems with Empirical Bayes'," *Risk Analysis*, vol. 6, no. 3, pp. 272–273, 1986.

McConway, K. J., "Marginalization and Linear Opinion Pools," *J. Amer. Statist. Assoc.*, vol. 76, pp. 410–414, 1981.

Mosleh, A., and Apostolakis, G., "The Assessment of Probability Distributions from Expert Opinions with an Application to Seismic Fragility Curves," *Risk Analysis*, vol. 6, no. 4, pp. 447–461, 1986.

Mosleh, A., and Apostolakis, G., "Models for the Use of Expert Opinions," presented at the workshop on low-probability high-consequence risk analysis, Society for Risk Analysis, Arlington, Va., June 1982.

Morris, P., "An Axiomatic Approach to Expert Resolution," *Management Science*, vol. 29, pp. 24–32, 1983.

Morris, P., "Combining Expert Judgments: A Bayesian Approach," *Management Science*, vol. 23, pp. 679–693, 1977.

Morris, P., "Decision Analysis Expert Use," *Management Science*, vol. 20, pp. 1233–1241, 1974.

Mosleh, A., and Apostolakis, G., "The Assessment of Probability Distributions from Expert Opinions with an Application to Seismic Fragility Curves," *Risk Analysis*, vol. 6, no. 4, pp. 447–461, 1986.

Mosleh, A., and Apostolakis, G., "Models for the Use of Expert Opinions," presented at the workshop on low-probability high-consequence risk analysis, Society for Risk Analysis, Arlington, Va., June 1982.

Pontecorvo, A., "A Method of Predicting Human Reliability," *Reliability and Maintenance*, vol. 4, 4th Annual Reliability and Maintainability Conference, pp. 337–342, 1965.

Raiffa, H., and Schlaifer, R., *Applied Statistical Decision Theory*, Harvard University, 1961.

Seaver, D., and Stillwell, W., "Procedures for Using Expert Judgment to Estimate Human Error Probabilities in Nuclear Power Plant Operations," NUREG/CR-2743, 1983.

Swain A., and Guttman, H., "Handbook of Human Reliability, Analysis with Emphasis on Nuclear Power Plant Applications," NUREG/CR-1278, 1983.

Thurstone, L., "A Law of Comparative Judgment," *Psychl. Rev.*, vol. 34, pp. 273–286, 1927.

Torgerson, W., *Theory and Methods of Scaling*, Wiley, New York, 1958.

Wagner, C. G., "Allocation, Lehrer Models, and the Consensus of Probabilities," *Theory and Decision*, vol. 14, pp. 207–220, 1982.

Williams, J. C., "Validation of Human Reliability Assessment Techniques," Proceedings of the 4th National Reliability Conference, Birmingham NEC, 6–8 July, 1983.

Winkler, R. L., "Combining Probability Distributions from Dependent Information Sources," *Management Science*, vol. 27, pp. 479–488, 1981.

Winkler, R. L., "The Consensus of Subjective Probability Distributions," *Management Science*, vol. 15, pp. B61–B75, 1968.

CHAPTER 13

Mendel, M., "Development of Bayesian Parametric Theory with Applications to Control," Ph.D Dissertation, Department of Mechanical Engineering, Massachusetts Institute of Technology, 1989.

Mendel, M., and Sheridan, T., "Filtering Information from Human Experts," *Trans. IEEE on Systems, Man and Cybernetics*, vol. 19, no. 1, pp. xxx, 1989.

Mendel, M., and Sheridan, T., (1986) "Optimal Combination of Information from Multiple Sources," Department of Mechanical Engineering, Massachusetts Institute of Technology, Contract n00014-83-K-0193, Work Unit no. 196–179, Office of Naval Research, Arlington, Virginia, 1986.

Morris, P., "Combining Expert Judgments: A Bayesian Approach," *Management Science*, vol. 23, no. 7, pp. 679–693, 1977.

CHAPTER 14

Bradley, R., "Some Statistical Methods in Taste Testing and Quality Evaluation," *Biometrica*, vol. 9, pp. 22–38, 1953.

Bradley, R., and Terry, M., "Rank Analysis of Incomplete Block Designs," *Biometrica*, vol. 39, pp. 324–345, 1952.

Comer, K., Seaver, D., Stillwell, W., and Gaddy, C., "Generating Human Reliability Estimates Using Expert Judgment, vols. I and II," NUREG/CR-3688, 1984.

David, H. A., *The Method of Paired Comparisons*, Charles Griffin, London, 1963.

Ford, L., "Solution of a Ranking Problem from Binary Comparisons," *Amer. Math. Monthly*, vol. 64, pp. 28–33, 1957.

Hanushek, E., and Jackson, J., *Statistical Methods for Social Scientists*, Academic Press, New York, 1977.

Hunns, D., "Discussions Around a Human Factors Data-Base. An Interim Solution: The Method of Paired Comparisons," in A. E. Green (ed.), *High Risk Safety Technology*, Wiley, New York, 1982.

Kendall, M., *Rank Correlation Methods*, Charles Griffin & Co. Limited, London, 1962.

Kirwan, B. et al., *Human Reliability Assessor's Guide*, vols. 1 and 2, Human Reliability Associates, 1987.

Mosteller, F., "Remarks on the Method of Paired Comparisons: I The Least Squares Solution Assuming Equal Standard Deviations and Equal Correlations," *Psychometrika*, vol. 16, no. 1, pp. 3–9, 1951a.

Mosteller, F., "Remarks on the Method of Paired Comparisons: II The Effect of an Aberrant Standard Deviation When Equal Standard Deviations and Equal Correlations Are Assumed," *Psychometrika*, vol. 16, no. 2, pp. 203–206, 1951b.

Mosteller, F., "Remarks on the Method of Paired Comparisons: III A Test of Significance for Paired Comparisons When Equal Standard Deviations and Equal Correlations Are Assumed," *Psychometrika*, vol. 16, no. 2, pp. 207–218, 1951c.

Siegel, S., *Nonparametric Statistics*, McGraw-Hill, New York, 1956.

Thurstone, L., "A Law of Comparative Judgment," *Psychl. Rev.*, vol. 34, pp. 273–286, 1927.

Torgerson, W., *Theory and Methods of Scaling*, Wiley, New York, 1958.

CHAPTER 15

Akkermans, D. E., "Crane Failure Estimates at DSM," presented at ESRRDA Conference on Expert Judgment in Risk and Reliability Analysis, Brussels, Oct. 11, 1989.

Claessens, M., *An Application of Expert Judgment in Ground Water Modeling* (in Dutch), Masters Thesis, Dept. of Mathematics, Delft. University of Technology, Delft, 1990.

Clemen, R. T., and Winkler, R. L., "Calibrating and Combining Precipitation Probability Forecasts," in R. Viertl (ed.), *Probability and Bayesian Statistics*, Plenum Press, New York, 1987, pp. 97–111.

Cooke, R., French, S., and van Steen, J., *The Use of Expert Judgment in Risk Assessment*, European Space Agency, Noordwÿk, the Netherlands, 1990.

Cooke, R. M., Goossens, L. H. J., Mendel, M., Oortman Gerlings, P. D., Stobbelaar, M. F., and van Steen, J. F. J., *Expert Opinions in Safety Studies*, vols. 1–5, Delft University of Technology/TNO, Delft/Apeldoorn, The Netherlands, 1989.

Daan, H., and Murphy, A., "Subjective Probability Forecasting in The Netherlands: Some Operational and Experimental Results," *Meteorol. Resch.*, vol. 35, pp. 99–112, 1982.

Meima, A., and Cooke, R. M., "Space Debris and Expert Judgment," Report to the European Space Agency, Delft, The Netherlands, 1989.

Murphy, A., and Daan, H., "Impacts of Feedback and Experience on the Quality of Subjective Probability Forecasts: Comparison of Results from the First and Second Years of the Zierikzee Experiment," *Monthly Weather Review*, vol. 112, pp. 413–423, 1984.

Offerman, J., "Safety Analysis of the Carbon Fibre Reinforced Composite Material of the Hermes Cold Structure," ESA ESTEC, Noordwijk, May 1990a.

Offerman, J., (1990b) "Study on Question Formats and Elicitation Procedures for the TU-Delft/ESA-ESTEC Expert Judgment Concept—Application of the Expert Judgment Concept Within the Safety Analysis of a LI/SOCL2 Battery for the ESA EVA/IVA Space suite," ESA ESTEC, Noordwijk, June 1990b.

Preyssl, C., and Cooke, R. M., "Expert Judgment; Subjective and Objective Data for Risk Analysis of Spaceflight System," in *Proceedings PSA '89 International Topical Meeting on Probability, Reliability and Safety Assessment*, Pittsburgh, April 2–7, 1989.

Roeleven, D., "Evaluatie en Combinatie van Expertmeningen," Masters thesis, Dept. of Mathematics, Delft University of Technology, Delft, The Netherlands, 1989.

APPENDIX A

Borel, E., "Valeur pratique et philosophie des probabilities," in E. Borel (ed.), *Traitè du Calcul des Probabilitès*, Gauthier-Villars, Paris, 1924.

Carnap, R., *Logical Foundations of Probability*, Univ. of Chicago Press, Chicago, 1950.

De Finetti, B., (1937) "La Prevision: ses lois logique, ses source subjectives," *Annales de L'Institut Henri Poincare*, vol. 7, pp. 1–68, 1937. English translation in H. Kyburg and H. Smokler (eds.). *Studies in Subjective Probability*, Wiley, New York, 1964.

De Finetti, B., *Theory of Probability*, vols. I and II, Wiley, New York, 1974.

Feller, W., *An Introduction to Probability Theory and its Applications*, vol. 2, Wiley, New York, 1971.

Keynes, J. M., *Treatise on Probability*, MacMillan, London, 1973 (first edition 1921).

Kullback, S., *Information Theory and Statistics*, Wiley, New York, 1959.

Laplace, P. S., "Probability and its Principles," in E. H. Madden (ed.), *The Structure of Scientific Theories*, Routledge Kegan Paul, London, 1960, pp. 250–255. Translated from sixth French edition of *A Philosophical Essay on Probabilities*.

Martin-Lof, P., "On the Notion of Randomness," in A. Kino and R. E. Vesley (eds.), *Intuitionism and Proof Theory*, North-Holland, 1970, pp. 73–78.

Ramsey, F., "Truth and Probability," in R. B. Braithwaite (ed.), *The Foundations of Mathematics*, Kegan Paul, London, 1931, pp. 156–198.

Reichenbach, H., "Axiomatik der Wahrscheinlichkeitsrechnung," *Math. Z.*, vol. 34, pp. 568–619, 1932.

Savage, L. J., *The Foundations of Statistics*, Wiley, New York, 1954. Second edition, Dover, New York, 1972.

Schnorr, C. P., "Zufäligkeit und Wahrscheinlichkeit," Lecture Notes in Mathematics, no. 218, Springer-Verlag, Berlin, 1970.

van Lambalgen, M., "Random Sequences," PhD. Dissertation, University of Amsterdam, 1987.

von Mises, R., *Probability Statistics and Truth*, Dover, New York, 1981. Translation of 3rd edition of *Wahrscheinlichkeit, Statistik und Wahrheit*, Springer, 1936.

von Mises, R., "Grundlagen der Wahrscheinlichkeitsrechnung," *Math Z.*, vol. 5, pp. 52–99, 1919.

von Neumann, J., and Morgenstern, O., *Theory of Games and Economic Behavior*, Wiley, New York, 1944.

Author Index

Subject Index

Printed in the United Kingdom
by Lightning Source UK Ltd.
2768